INDUSTRIAL LOCOMOTIVES

including preserved and minor
railway locomotives

HANDBOOK
9EL

Published by the INDUSTRIAL RAILWAY SOCIETY
at 47 Waverley Gardens, London, NW10 7EE.

© INDUSTRIAL RAILWAY SOCIETY 1991.

ISBN 0 901096 61 X (hardbound)
ISBN 0 901096 62 8 (softbound)

Compiled by G. Morton, Leeds, West Yorkshire.

Distributed by IRS Publications, 47 Waverley Gardens, London, NW10 7EE.

Printed in Great Britain by AB Printers Ltd, Leicester.

CONTENTS

The text of this book incorporates all amendments notified before
January 1991.

FOREWORD to the ninth edition

This is the ninth in a line of EL (existing locomotive information as opposed to historical) books which commenced in 1968. The lists now include all known locomotives of gauge 1'3" and above excluding only London Transport Capital Stock, British Rail Capital Stock and British Coal underground stock (formally N.C.B.) .

The lists are in accordance with our records for January 1st, 1991 and we would like to stress that the information derives from the observations and researches of members and others. We shall always rely on similar enthusiastic support to keep all the records up to date, so please send your observations (and your queries too) to the Hon. Records Officer and his Assistants, as listed below. All such information will be distributed to members through the bi-monthly bulletin.

Historical & Current Notes - British Coal (Formally N.C.B.)
           R.D. Darvill (Assistant Hon. Records Officer)
           119 Bath Street,
               Rugby,
               Warwickshire.
               CV21 3JA.

Current observations other owners.
           I.R. Bendall (Assistant Hon. Records Officer)
           25 Byron Way,
               Melton Mowbray,
               Leicestershire.
               LE13 1NY.

Historical and current preservation notes.
           E.S. Tonks
           87 Sunnymead Road,
               South Yardley,
               Birmingham.
               B26 1LL.

Historical - other owners, rolling stock, etc.
           G.W. Green (Hon. Records Officer)
           197 St Lawrence Road,
               Sheffield.
               S9 1SF.

Current Tunnelling Contractors Notes.
           J.A. Foster (Assistant Hon. Records Officer)
           27 Charnock Crescent,
               Gleadless Town End,
               Sheffield.
               S12 3HB.

INTRODUCTION

The format is a little different from that used in previous editions, with the locomotives listed under owners arranged alphabetically within industrial and preservation location subgroups within each county. Locomotive entries which may be subject to ambiguity are noted as 'Pvd'.

Locomotives at scrapyards purely for scrapping are not generally listed, only if they have been 'resident' for a year or more. Those at dealers yards are however listed in order that members can keep a full picture of their movements.

Contractors locos can be found on sites in all parts of the country, working on sewer and tunnel schemes, etc. The locomotive details are shown in the usual way in the lists for the County in which the depot or main plant depot, if more than one is in use, is situated. A list of Contractors/Plant Hire specialists and the relevant counties are shown after the Locomotive Builders table.

When a works is closed, this is indicated after the title or subtitle, thus; TRACK SUPPLIES LTD, (Closed).

When a works is still in operation but no longer uses a certain type of rail traffic, this is indicated by a note after the gauge concerned, thus; Gauge : 4'8½".   (TQ 741694)   RTC.   RTC = Rail Traffic Ceased.

GAUGE.   The gauge of the railway system is given at the head of the locomotive list. At preservation sites and museums, etc, where several gauges differ by only a fraction they are usually all listed under the nominal gauge of the majority.

GRID REFERENCE.   An indexed six-figure grid reference is given in brackets after the gauge, and denotes the location of the loco shed or stabling point. Most references are known but details of any missing references are most welcome.

NUMBER and NAME.   The title of the locomotive - number, name or both - is given in the first two columns. A name unofficially bestowed and used by staff but not carried on the loco is indicated by inverted commas "   ". Locomotives under renovation at preservation sites, etc, may not carry their intended title but these are shown, unless it is definitely known that the name will not be retained. Ex B.R. locomotives are further identified by the inclusion of the B.R. number in brackets, if not carried now.

TYPE.   The type of locomotive is given in column three.

           The Whyte system of wheel classification is used in the main, but when driving wheels are connected not by outside rods but by chains or other means (as in various 'Sentinel' steam locos and diesel locos) they are shown as 4w (Four-wheeled), 6w (Six-wheeled), or if only one axle is motorised it is shown as 2w-2.

           Trapped Rail System adhesion locomotives are shown using wheel type ad.

For ex B.R. diesel and electric locomotives the usual development of the Continental notation is used.

The following abbreviations are used :-

T	Side tank or similar - a tank positioned externally and fastened to the frame.
CT	Crane tank - a T type loco fitted with load lifting apparatus.
PT	Pannier tank - side tanks not fastened to the frame.
ST	Saddle tank.
STT	Saddle tank with tender.
WT	Well tank - a tank located between the frames under the boiler.
VB	Vertical boilered locomotive.
F	Fireless steam locomotive.
D	Diesel locomotive - unknown transmission.
DC	Diesel locomotive - compressed air transmission.
DM	Diesel locomotive - mechanical transmission.
DE	Diesel locomotive - electrical transmission.
DH	Diesel locomotive - hydraulic transmission.
P	Petrol or Paraffin locomotive - unknown transmission.
PM	Petrol or Paraffin locomotive - mechanical transmission.
PE	Petrol or Paraffin locomotive - electrical transmission.
PH	Petrol or Paraffin locomotive - hydraulic transmission.
R	Railcar - a vehicle primarily designed to carry passengers.
BE	Battery powered electric locomotive.
CE	Conduit powered electric locomotive.
RE	Third rail powered electric locomotive.
WE	Overhead wire powered electric locomotive.
FE	Flywheel electric locomotive.

CYLINDER POSITION. Shown in column four for steam locomotives.

IC	Inside cylinders.
OC	Outside cylinders.
VC	Vertical cylinders.
G	Geared transmission - coupled to IC, OC or VC.

FLAMEPROOF. Flameproof battery locos are denoted by FLP in column four.

RACK ADHESION. Locomotives that employ rack adhesion are shown by R/A in column four.

ROAD/RAIL. Certain locomotives are capable of working on either rail or road, (most are of either Unilok or Unimog manufacture). These are indicated by R/R in column four.

STEAM OUTLINE. Diesel or Petrol locomotives with a steam locomotive appearance added are shown by S/O in column four.

MAKERS. The builder of the locomotive is shown in column five; the abbreviations used are listed on a later page.

MAKERS NUMBER and DATE. The sixth column shows the works number, the next column the date which appears on the plate, or the date the loco was built if none appears on the plate.

c.	denotes circa, i.e. about the time of the date quoted.
reb.	denotes Rebuilt.
Pvd.	denotes Preserved on site.

DOUBTFUL INFORMATION.   Information known to be doubtful is denoted as such by the wording, or printed in brackets with a question mark. Thus, Wkm (7573?) 1956 denotes that the loco is a Wickham of 1956 vintage and possibly of the works number 7573.

Locomotives which have ceased work are indicated by OOU (Out of Use), or as Dsm (Dismantled) if incomplete. These are only shown when the condition is permanent, not of a temporary nature.   Wkm motorised trollies which have been converted to engineless trailers are shown as DsmT.
N.B. Battery locos not carrying a battery are not regarded as Dsm.

There are cases where all locomotives are OOU but the tramway is still in use; with a road tractor or man-power for example. Also some owners use locos on their own internal system after the B.R. connection has been severed.

Electric locomotives at Steelworks and British Coal sites are usually to be found working at Coke Ovens, the grid reference for the Ovens being shown in the usual place, when known.

Hints on recording observations

As already explained, the Society is almost entirely reliant on the observations of enthusiasts to keep the records up to date, and we are always pleased to receive reports of visits, which should be sent to the Hon. Records Officer or his assistants. The following will, we hope, be of assistance to those sending in reports.

Report anything you see, on the locomotive side this means reporting locos present, even if there is no change to the published list. Someone else may visit in the next six months and find a change, the date of which can then be narrowed down to this six months. In addition to locos, details of rolling stock, layouts and items of historical information gleaned, are all welcome and will be filed for future reference. It is surprising how often these items are required later by other enthusiasts. There is nothing to beat a note made on the spot; a note now is far better than trying to get the information from the recollection of employees in five years time.

On the other hand, care should be taken to distinguish observation from inference. If you see a 4wDH with worksplate say S 10019 and numbered 19, all well and good; but if it does not carry a worksplate, that fact should be stated. You may infer the loco is S 10019 from the running number, or by reference to the pocket book. If you say 'number on bonnet', 'number on cabside' or 'loco assumed to be S 10019' as the case may be, we will know the position precisely, which is important when bonnets are exchanged or cabs rebuilt, for example. If a loco carries no identification, you will probably be able to guess its make from design features, and a note of livery may help. Also with diesel locomotives, a note should be made of engine make, type and its number, if easily obtained.

A thorough search of all premises is worthwhile, as locos (particularly if OOU) are frequently hidden away - and how often have surprises turned up in this way. Further, if you search diligently, and a loco is missing it may be assumed to have gone and enquiries can be made, as to where it has gone or if it has been scrapped. Please try to ascertain such information from the staff. Similarly, in the case of new arrivals, dates should be obtained if possible.

Firm's titles change from time to time, so please check from the office or board at the entrance. Subsidiary companies frequently display the title of a parent company, but we always use the name under which the company trades, i.e. the subsidiary where such exists.

The necessity for obtaining permission to see locomotives applies with equal force to 'preserved' locations other than public parks and museums. Established systems which operate trains or have 'Open Days' will usually permit inspection of locos by arrangement. Locos shown as at private homes or farms are very often in storage or in the course of restoration and as a rule members are not advised to in any way possible embarrass the owners by writing for permission but to wait until they announce that their machines are available for inspection.

Finally, please try to establish friendly relations with the firms visited, as we are only allowed access by their courtesy. Don't be a nuisance or hold up production, nor expose yourself to danger in any way, but show a healthy interest in the processes being carried out. In this way, not only you, but other enthusiasts will be welcome later.

LOCOMOTIVE BUILDERS

AB	Andrew Barclay, Sons & Co Ltd, Caledonia Works, Kilmarnock.
A.C.Cars	A.C. Cars Ltd, Thames Ditton, Surrey.
Acton	Acton Works, London Transport Executive, Acton, London, then London Underground Ltd from /1987.
AE	Avonside Engine Co Ltd, Bristol.
AEC	Associated Equipment Co Ltd, Southall.
AEG	Allgemeine Elektrizitats Gesellschaft, Berlin, Germany.
Afd	South Eastern & Chatham Railway, Ashford Works, Kent, later S.R. and B.R.
Ageve	AB Galve Vagnvekstad, P.O. Box 655, S-801 27 Gavle, Sweden.
AH	A. Horlock & Co, North Fleet Iron Works, Kent.
AK	Alan Keef Ltd, Cote Farm, Cote, near Bampton, Oxon, Lea Line, Ross-on-Wye, Hereford & Worcester from 11/1986 (Successor to SMH).
AL	American Locomotive Co, U.S.A./Canada.
A&O	Alldays & Onions, Birmingham.
AP	Aveling & Porter Ltd, Invicta Works, Canterbury, Kent.
Arrols	Arrols, Glasgow.
Art	Artisair Ltd, Moorwell Road, Yaddlethorpe, Scunthorpe, Humberside.
ASEA	Allmanna Svenska Elektriska A.B., Vasteras, Sweden.
AT	Les Ateliers Metallurgiques Nivelles, Tubize & La Sambre Works, Tubize, Belgium.
Atlas	Atlas Loco & Mfg Co Ltd, Cleveland, Ohio, U.S.A.
AtW	Atkinson Walker Wagons Ltd, Preston, Lancashire.
AW	Sir W.G. Armstrong, Whitworth & Co (Engineers) Ltd, Newcastle-Upon-Tyne.
Ayle	Ayle Colliery Co Ltd, Ayle East Drift Mine, Alston, Northumberland.
Barlow	H.N. Barlow, Southport, Lancashire.
A.Barnes	A. Barnes & Co Ltd, Albion Works, Rhyl, Clwyd.
S.Battison	Samuel Battison, c/o Tathams Ltd, Nottingham Road, Ilkeston, Derbyshire.
B(C)	Peter Brotherhood, Engineers, Chippenham, Wiltshire.
BD	Baguley-Drewry Ltd, Burton-on-Trent.
BE	Brush Electrical Engineering Co Ltd, Loughborough, Leicestershire.
Berwyn	Berwyn Engineering, Thickwood, Chippenham.
BEV	British Electric Vehicles Ltd, Chapeltown, Southport, Lancashire.
Bg	E.E. Baguley Ltd, Burton-on-Trent.
BGB	Becorit (Mining) Ltd, Grove Street, Mansfield Woodhouse, Nottingham, then Hallam Fields Road, Ilkeston, Derbyshire from /1984.
BgC	Baguley Cars Ltd, Burton-on-Trent.
Bg/DC	Built by Bg for DC; makers numbers identical.
BH	Black, Hawthorn & Co Ltd, Gateshead.
Bicester	War Department, Arncott Workshops, Bicester Depot, Oxfordshire.
BL	W.J. Bassett Lowke Ltd, Northampton.
BLW	Baldwin Locomotive Works, Philadelphia, Pennsylvania, U.S.A.
BM	Brown Marshall & Co.
BMR	Brecon Mountain Railway Co Ltd, Pant, Mid Glamorgan.

BNFL	British Nuclear Fuels Ltd, Windscale Factory, Sellafield, Cumbria.
BnM	Bord na Mona (Irish Turf Board) : Various of the larger sites have built thier locos & railcars.
E.Booth	E. Booth, Lappa Valley Railway, St Newlyn East, near Newquay, Cornwall.
Borsig	A. Borsig G.m.b.H., Berlin, Germany.
Boston Lodge	Boston Lodge Workshops, Festiniog Railway.
Bow	Bow Locomotive Works, North London Railway.
BP	Beyer, Peacock & Co Ltd, Gorton, Manchester.
BPH	Beyer, Peacock (Hymek) Ltd, Gorton, Manchester.
BRCW	Birmingham Railway Carriage & Wagon Co Ltd, Smethwick.
Bredbury	Bredbury & Romiley U.D.C., Cheshire.
BRE(D)	British Rail Engineering Ltd, Derby Locomotive Works, Derbyshire.
BRE(S)	British Rail Engineering Ltd, Shildon Works, Co. Durham.
Bressingham	Alan Bloom, Bressingham Hall, Diss, Norfolk.
Bruff	Bruff Rail Ltd, Suckley, Worcestershire.
B&S	Bellis & Seekings Ltd, Birmingham.
B.S.C.	British Steel Corporation, Hartlepool Works, Cleveland.
BT	Brush Electrical Machines Ltd, Traction Divison, Falcon Works, Loughborough.
BT/WB	Built jointly by BT & WB.
BTH	British Thomson-Houston Co Ltd, Rugby.
Bton	London, Brighton and South Coast Railway, Brighton Works East Sussex, later S.R. and B.R.
Bury	Bury, Curtis & Kennedy, Liverpool.
BV	Brook Victor Electric Vehicles Ltd, Burscough Bridge, near Ormskirk, Lancashire.
BVR	Bure Valley Railway Ltd, Aylsham, Norfolk.
Cannon	Cannon, Dudley, West Midlands.
Carland	Carland Engineering Ltd, Harold Wood, Essex.
D.Carter	D. Carter, Tucking Mill Tramway, Midford, Avon.
Carn	Steamtown Railway Museum, Warton Road, Carnforth, Lancashire.
Cdf	Cardiff West Yard Locomotive Works, Taff Vale Railway.
CE	NEI Mining Equipment Ltd, Clayton Equipment, Hatton, near Derby, formerley Clarke Chapman Ltd, Clayton Works.
Chance	Chance Manufacturing Co Inc., Wichita, Kansas, U.S.A.
Chaplin	Alexander Chaplin & Co Ltd, Cranstonhill Works, Glasgow.
CHR	Chrzanow, Poland.
Clarkson	H. Clarkson & Son, York.
Clay Cross	Clay Cross Co Ltd, Clay Cross Iron Works, Spun Pipe Plant, Derbyshire.
Cleveland	Dorman Long (Steel) Ltd, Cleveland Works, Yorks N.R.
Coalbrookdale	Coalbrookdale Co, Coalbrookdale Ironworks, Shropshire.
Cockerill	Societe pour L'Exploitation Des Etablissements John Cockerill, Seraing, Belgium.
Consett	Consett Iron Co Ltd, Consett Works, Co. Durham.
Corpet	Corpet Louvet & Cie, La Courneuve, Seine, France.
CoSi	Coleby-Simkins Engineering, Melton Mowbray, Leics.
Couillet	S.A. des Usines Metallurgiques du Hainaut, Couillet, Belgium.
Cowlairs	North British Railway, Cowlairs Works, Glasgow later B.R.
E.A.Craven	E.A. Craven, North Shields, Tyne & Wear.
Cravens	Cravens Ltd, Darnall, Sheffield.
Crewe	London and North Western Railway, Crewe Works, Cheshire, later L.M.S.R. and B.R.

Crewe G.R.	British Rail, Gresty Road Depot, Crewe, Cheshire.
Curwen	A. Curwen, All Cannings, near Devizes, Wiltshire.
D	Dubs & Co, Glasgow Locomotive Works, Glasgow.
Dar	North Eastern Railway, Darlington Works, later L.N.E.R. and B.R.
DB	Sir Arthur P. Heywood, Duffield Bank Works.
DC	Drewry Car Co Ltd, London. (Suppliers only)
AR.Deacon	A.R. Deacon, Somerset.
Decauville	Societe Nouvelle Des Establissements Decauville Aine, Petit Bourg, Corbeil, S & O, France.
Decon	Decon Engineering Co (Bridgwater) Ltd, Wylds Road, Bridgwater, Somerset.
De Dietrich	De Dietrich, Niederbrown, France.
Derby	Midland Railway, Derby Works, later L.M.S.R. and B.R.
Derby C&W	Litchurch Lane Carriage Works, Midland Railway, Derby, later L.M.S.R. and B.R.
DeW	DeWinton & Co, Caernarfon.
Diema	Diepholzer Maschinenfabrik (Fr.Schöttler G.m.b.H.), Diepholz, Germany.
DK	Dick Kerr & Co Ltd, Preston, Lancashire.
DL	Dormon Long (Steel) Ltd, Middlesbrough, Yorkshire N.R.
DM	Davies & Metcalfe Ltd, Romiley, Stockport, Cheshire.
Dodman	Alfred Dodman & Co, Highgate Works, Kings Lynn.
Don	Great Northern Railway, Doncaster Works, later L.N.E.R. and B.R.
DonM	British Rail, Eastern Region, Marshgate Permanent Way Depot, Doncaster, South Yorkshire.
Donelli	F.L. Donelli, Reggio, Italy.
Donelon	J.F. Donelon Ltd, Horwich, Greater Manchester.
Dowty	Dowty Group Ltd, Ashchurch, Gloucestershire.
DP	Davey, Paxman & Co Ltd, Colchester, Essex.
Dtz	Motorenfabrik Deutz, A.G., Cologne, Germany.
Dundalk	Dundalk Works, Great Northern Railway of Ireland.
DundalkE	Dundalk Engineering.
EB	E. Borrows & Sons, St. Helens.
EE	English Electric Co Ltd, Preston, Lancashire.
EES	English Electric Co Ltd, Stephenson Works, Darlington. (Successors to R.S.H.D.)
EEV	English Electric Co Ltd, Vulcan Works, Newton-le-Willows, Lancashire. (Successors to V.F.)
EK	Marstaverken, Eksjo, Sweden.
Elh	London and South Western Railway, Eastleigh Works, Hampshire, later S.R. and B.R.
ESCA	ESCA Engineering Ltd, Wigan, Greater Manchester.
EW	E.B. Wilson & Co., Railway Foundry, Leeds.
Fairbourne	Fairbourne Railway Co, Fairbourne, Gwynedd.
FB	Societe Franco-Belge de Materiel de Chemins de Fer, La Croyere, Belgium.
Festiniog Rly	Festiniog Railway Co, Boston Lodge Works, Gwynedd.
F.F.P.	Fire Fly Project, Great Western Preservations Ltd, Bristol, Avon (1987) & Didcot, Oxon.(1989).
FH	F.C. Hibberd & Co Ltd, Park Royal, London; later at Butterley Works, Derbyshire.
Fisons	Fisons Ltd, British Moss Works, Swinefleet, near Goole, Humberside; Constructed from parts supplied by Diema.

FJ	Fletcher Jennings & Co, Lowca Engine Works, Whitehaven.
Ford	Ford Motor Co Ltd, (? Dagenham, Essex).
Foster Rastrick	Foster, Rastrick & Co, Stourbridge, Worcestershire.
Freud	Stahlbahnwerke Freudenstein & Co, Tempelhof, Dortmund, Berlin, Germany.
Frichs	A/S Frichs Maskinfabrik & Kedelsmedie, Arhus, Denmark.
S.Frogley	S. Frogley, Nent Valley Railway, Nentsberry, near Nenthead, Cumbria.
B.J.Fry	B.J. Fry Ltd, Dorchester, Dorset.
FW	Fox, Walker & Co, Atlas Engine Works, Bristol.
GB	Greenwood & Batley Ltd, Leeds.
GE	George England & Co Ltd, Hatcham Ironworks, London.
GEC	General Electric Co Ltd, Witton, Birmingham.
GECT	G.E.C. Traction Ltd, Newton-le-Willows, Lancashire.
Geevor	Geevor Tin Mines Ltd, Pendeen, near St Just, Cornwall.
Geismar	
Geo.Stephenson	George Stephenson, Hetton, Durham.
GEU	General Electric Co, Erie, Pennsylvania, U.S.A.
GH	Gibb & Hogg, Airdrie.
Ghd	Gateshead Works, North Eastern Railway.
GKN	GKN Sankey Ltd, Castle Works, Hadley, Telford, Shropshire.
Gleismac	
GM	General Motors Ltd, Electro-Motive Division, La Grange, Illinois, U.S.A.
GMT	Gyro Mining Transport Ltd, Victoria Road, Barnsley, then Bramley Way, Hellaby Industrial Estate, Hellaby, near Rotherham, South Yorkshire from c/1987.
Goold	J.R. Goold Eng. Ltd, Camerton, Avon.
Gorton	Great Central Railway, Gorton Works, Manchester, later L.N.E.R. and B.R.
GR	Grant, Ritchie & Co, Kilmarnock.
GRC	Gloucester Railway Carriage & Wagon Co Ltd, Gloucester.
Greaves	J.W. Greaves & Sons Ltd, Llechwedd Quarry, Gynwedd.
Group 4	Festiniog Railway, Group 4, Birmingham.
Guest	Guest Engineering & Maintenance Co Ltd, Stourbridge.
NL.Guinness	Nigel L. Guinnness, Cobham, Surrey.
G&S	G. & S. Light Engineering Co Ltd, Stourbridge, Worcs.
H	James & Fredk. Howard Ltd, Britannia Ironworks, Bedford.
HAB	Hunslet-Barclay Ltd, Caledonia Works, Kilmarnock, Strathclyde.
Hackworth	Timothy Hackworth, Soho Works, Shildon, Co. Durham.
Hano	Hannoversche Maschinenbau AG (vormals Georg Egestorff), Hannover, Germany.
K.Hardy	K. Hardy, Brookhouse, Badgeworth, near Cheltenham, Gloucestershire.
Hartmann	Sachsische Maschinenfabrik, vormals Richard Hartmann AG, Chemnitz, Germany.
Hayne	N. Haynes, Sheppards Tea Rooms & Boat House, near Saltford, Somerset.
HB	Hudswell Badger Ltd, Hunslet, Leeds.
H&B	Hill & Bailey Ltd, Gilfach Ddu, Llanberis, Gwynedd.
HC	Hudswell, Clarke & Co Ltd, Railway Foundry, Leeds.
HE	Hunslet Engine Co Ltd, Hunslet, Leeds.
R.Heath	Robert Heath & Sons Ltd, Norton Ironworks, Stoke-on-Trent
Wm.Hedley	William Hedley, Wylam Colliery, Northumberland.
WVO.Heiden	W. Van der Heiden, Rotterdam, Holland.
Hen	Henschel & Sohn G.m.b.H., Kassel, Germany.

A.Herschell	Allan Herschell, North Tonawanda, New York, U.S.A.
HF	Haydock Foundry Co Ltd, Haydock, Lancashire.
H(L)	Hawthorn & Co, Leith, Edinburgh.
HL	R. & W. Hawthorn, Leslie & Co Ltd, Forth Bank Works, Newcastle-Upon-Tyne.
HLT	Hughes Locomotive & Tramway Engine Works Ltd, Loughborough, Leicestershire.
Hor	Lancashire and Yorkshire Railway, Horwich Works, Lancs, later L.M.S.R. and B.R.
HT	Hunslet Taylor Consolidate (Pty) Ltd, Germiston, Transvaal, South Africa.
HU	Robert Hudson Ltd, Leeds.
R.Hutchings	
HW	Head, Wrightson & Co Ltd.
Inchicore	Inchicore Works, Dublin; Great Southern Railways. (Previously G.S. & W.R., later C.I.E.)
IOMT	Manx Electric Railway, Derby Castle Works, Douglas, Isle of Man.
Iso	Iso Speedic Co Ltd, Fabrications & Electric Vehicles, Charles Street, Warwick.
Jaco	Jaco Engineering Co Ltd, Edwards Road, Birmingham 24, West Midlands.
Jaywick Rly	Jaywick Light Railway, near Clacton, Essex.
JF	John Fowler & Co (Leeds) Ltd, Hunslet, Leeds.
Jubilee Min Rly	J.M.R. (Sales) Ltd, 173 Liverpool Road South, Birkdale, Southport.
Jung	Arn. Jung Lokomotivfabrik G.m.b.H., Jungenthal, Germany.
K	Kitson & Co, Airedale Foundry, Leeds.
KC	Kent Construction & Engineering Co Ltd, Ashford, Kent.
Kearsley	Central Electricity Generating Board, Kearsley Power Station, Radcliffe, Greater Manchester.
T.Kennan	Thos Kennan & Son, Dublin.
Kierstead	Kierstead, Telford, Shropshire.
Kilmarnock	British Rail, Scottish Region, Regional Civil Engineers Workshops, Kilmarnock Works, Strathclyde.
Kilroe	T. Kilroe & Sons Ltd, Lomax Street, Radcliffe, Greater Manchester.
Kitching	A. Kitching, Hope Town Foundry, Darlington, Co.Durham.
Krauss	Lokomotivfabrik Krauss & Co, Munich, Germany & Linz, Austria.
Krupp	Friedrich Krupp, Maschinenfabriken Essen, Abt. Lokomotivbau, Essen, Germany.
KS	Kerr, Stuart & Co Ltd, California Works, Stoke-on-Trent.
KTH	Kitson, Thompson & Hewittson, Leeds.
L	R. & A. Lister & Co Ltd, Dursley, Gloucestershire.
La Loire	Ateliers et Chantiers de la Loire, Penhoet, (pres. Nantes), France.
La Meuse	Societe Anonyme Des Ateliers De Construction De La Meuse Sclessin-Liege, Belgium.
Lake & Elliot	Lake & Elliot Ltd, Braintree, Essex.
Lancing	Southern Railway, Lancing Carriage Works, Sussex, later B.R.
Lancs Tanning	Lancashire Tanning Co Ltd, Littleborough, Lancashire.
Lane	Charles lane, Liphook, Hampshire.
LB	Lister Blackstone Traction Ltd, Dursley, Glos.
LBNGRS	Leighton Buzzard Narrow Gauge Railway Society, Stonehenge Workshops, Leighton Buzzard, Beds.

J.Lemon-Burton	J. Lemon-Burton, Paynesfield, Aldbourne Green, West Sussex, and Shelmerdine & Mulley Ltd, Edgeware Road, Cricklewood, London, NW2.
Lewin	Stephen Lewin, Dorset Foundry, Poole, Dorset.
Leyland	Leyland Vehicles Ltd, Workington, Cumbria.
Lima	Lima Locomotive Works Inc., Lima, Ohio, U.S.A.
Llanwern	British Steel Corporation, Welsh Division, Llanwern Works, Newport, Gwent.
LMM	Logan Mining & Machinery Co Ltd, Dundee.
Loco Ent	Locomotion Enterprises (1975) Ltd, Bowes Railway, Springwell, Gateshead, Tyne & Wear.
Locospoor	Holland ?
Longfleet Eng.	Longfleet Motor & Engineering Works Ltd, 46 Fernside Road, Poole, Dorset.
Longhedge	South Eastern & Chatham Rly, Longhedge Works, London.
Longleat	Longleat Light Railway, Longleat, Warminster, Wiltshire.
Ludlay Brick	Ludlay Brick Co, Berwick, near Eastbourne, Sussex.
M	Metropolitan Carriage & Wagon Co Ltd, Birmingham.
C.Mace	C. Mace, The Woodland Railway, Kent. (1 Maple Close, Larkfield, Maidstone, Kent)
MAK	Maschinenbau Kiel G.m.b.H., Kiel-Friedrichsort, Germany.
J.Marshall	J. Marshall, Spring Lane, Hockley Heath, Warwickshire.
Massey	G.D. Massey, 57 Silver Street, Thorverton, Exeter, Devon.
Massey Ferguson	
Matisa	
Maxi	Maxitrack, "Rothiemay", Offham Road, West Malling, Kent.
MC	Metropolitan Cammell Carriage & Wagon Co Ltd, Saltley, Birmingham.
P.McGarigle	P. McGarigle, Niagra Falls, near Buffalo, New York State, U.S.A.
Mercury	The Mercury Truck & Tractor Co, Gloucester.
Met.Amal.	Metropolitan Amalgamated Railway Carriage & Wagon Co Ltd
MH	Muir Hill Engineering Ltd, Trafford Park, Manchester.
Minilok	allrad-Rangiertecknik G.m.b.H., D-5628, Heiligenhaus Bez, Dusseldorf, Germany.
Minirail	Minirail Ltd, Frampton Cotterell, Bristol.
Mkm	Markham & Co Ltd, Chesterfield, Derbyshire.
MM	Meridian (Motioneering) Ltd, Bradley Way, Hellaby Industrial Estate, Hellaby, near Rotherham, South Yorkshire.
RP.Morris	R.P. Morris, 193 Main Road, Longfield, Kent.
RH.Morse	R.H. Morse, Potter Heigham, Norfolk.
Motala	A.B. Motala Verkstad, Motala, Sweden.
Moyse	Locotracteurs Gaston Moyse, La Courneuve, Seine, France.
MP	Mather & Platt Ltd, Park Works, Manchester.
MR	Motor Rail Ltd, Simplex Works, Bedford.
MV	Metropolitan-Vickers Electrical Co Ltd, Trafford Park, Manchester.
MW	Manning, Wardle & Co Ltd, Boyne Engine Works, Hunslet, Leeds.
N	Neilson & Co, Springburn Works, Glasgow.
NB	North British Locomotive Co Ltd, Glasgow.
NBH	North British Loco, Hyde Park Works, Glasgow.
NBQ	North British Loco, Queens Park Works, Glasgow.
NCC	L.M.S. (Northern Counties Committee), York Road Works, Belfast, Ireland.
Nea	Metropolitan Railway, Neasden Works, London.

NLP	North London Polytechnic.
NMW	National Museum of Wales, Industrial & Maritime Museum, Butetown, Cardiff, South Glamorgan.
NNM	Noord Nederlandsche Machinefabriek B.V., Winschoten, Holland.
Nohab	Nydquist & Holm A.B., Trollhattan, Sweden.
NR	Neilson Reid & Co, Glasgow.
NW	Nasmyth, Wilson & Co Ltd, Bridgewater Foundry, Patricroft, Manchester.
Oakeley	Oakeley Slate Quarries Co Ltd, Blaenau Ffestiniog, Merionydd.
Oerlikon M.C.	Oerlikon, Zurich, Switzerland.
OK	Orenstein & Koppel A.G., Berlin, Germany.
Oldbury	Oldbury Carriage & Wagon Co Ltd, Birmingham.
Omam	G. Mameli 65, 20058 Villasanta, Milan, Italy.
P	Peckett & Sons Ltd, Atlas Locomotive Works, St George, Bristol.
Parry	J.P.M. Parry & Associates Ltd, Corngreaves Trading Estate, Overend Road, Cradley Heath, West Midlands.
J.Peat	John Peat, Chicken Farm, Shaftsbury, Dorset.
Pendre	Talyllyn Railway Co, Pendre Works, Tywyn, Gwynedd.
Permaquip	The Permanent Way Equipment Co Ltd, Pweco Works, Lillington Road North, Bulwell, Nottingham, Later at 1 Giltway, Giltbrook, Nottingham.
Plasmor	Plasmor Ltd, Womersley Road, Knottingley, West Yorkshire.
Plasser	Plasser Railway Machinery (GB) Ltd, Drayton Green Road, West Ealing, London.
Potter	D.C. Potter, Yaxham Park, Yaxham, near Dereham, Norfolk.
PR	Park Royal Vehicles, Park Royal, London.
Pressed Steel	Pressed Steel Ltd, Swindon, Wiltshire.
PVRA	Plym Valley Railway Association, Marsh Mills, Plympton, Devon.
Ravenglass	Ravenglass & Eskdale Railway Co Ltd, Ravenglass, Cumbria
Red(F)	Redland Bricks Ltd, Funton Works, near Sittingbourne, Kent.
Red(W)	Redland Bricks Ltd, Warnham Works, West Sussex.
Redstone	Redstone, Penmaenmawr.
Regent St	Regent Street Polytechnic, London.
Resco	Resco (Railways) Ltd, Bilton Road, Manor Road Industrial Estate, Erith, Greater London.
RFS	R.F.S. Engineering Ltd, Doncaster Works, Hexthorpe Road, Doncaster, South Yorkshire.
RFSK	R.F.S. Engineering Ltd, Vanguard Works, Hooton Road, Kilnhurst, South Yorkshire. (Successors to TH).
RH	Ruston & Hornsby Ltd, Lincoln.
RHDR	Romney Hythe & Dymchurch Railway, New Romney, Kent.
Rhiwbach	Rhiwbach Quarries Ltd, Rhiwbach Slate Quarry, Merionethshire.
Richardsons	Richardsons Moss Litter Co Ltd, Letham Moss Works, near Airth Station, Central.
Riordan	Riordan Engineering Ltd, Surbiton, Surrey.
Robel	Robel & Co, Maschinenfabrik, Munchen, 25, Germany.
K.Rosewall	K. Rosewall, Cross Elms Nursery, Bristol, Avon.
RP	Ruston, Proctor & Co Ltd, Lincoln.
RR	Rolls Royce Ltd, Sentinel Works, Shrewsbury. (Successors to Sentinel).

R&R	Ransomes & Rapier Ltd, Riverside Works, Ipswich, Suffolk.
RRS	Rapido Rail Systems, Dudley, West Midlands.
RS	Robert Stephenson & Co Ltd, Forth Street, Newcastle-Upon-Tyne and Darlington.
RSC	Ransomes Sims & Co, Orwell Works, Ipswich, Suffolk.
RSH	Robert Stephenson & Hawthorns Ltd.
RSHD	Robert Stephenson & Hawthorns, Darlington Works.
RSHD/WB	Built by RSHD but ordered by WB.
RSHN	Robert Stephenson & Hawthorns, Newcastle-Upon-Tyne Works. (Successors to HL).
RSM	Royal Scottish Museum, Chambers Street, Edinburgh, Midlothian.
Ruhr	Ruhrthaler Maschinenfabrik Schwarz & Dyckerhoff, Mulheim, Germany.
RWH	R.& W. Hawthorn & Co, Newcastle-Upon-Tyne. (Later HL).
RYP	R.Y. Pickering & Co Ltd, Wishaw, Scotland.
S	Sentinel (Shrewsbury) Ltd, Battlefield, Shrewsbury.
Sabero	Hulleras de Sabero Y Anexas S.A., Sabero, Spain.
Sara	Sara & Burgess, Penryn, Cornwall.
Scammell	Scammell Trucks Ltd, Watford, Hertfordshire. (Road vehicle conversions.)
DJ.Scarrott	D.J. Scarrott, Kingsteignton, Newton Abbot, Devon.
Sch	Berliner Maschinenbau - A.G., vormals L.Schwartzkopff, Berlin, Germany.
Schalker	Schalker Eisenhette Maschinenfabrik G.m.b.H., 465, Gelsenkirken, Magdeburger Strasse 37, West Germany.
Schichau	F. Schichau, Maschinen-und Lokomotivfabrik, Elbing, Germany. (Now Elbtag, Poland).
Schöma	Christoph Schöttler Maschinenfabrik, G.m.b.H., Diepholz, Germany.
Science Mus.	Science Museum, South Kensington, London.
SCW	I.C.I. Ltd, South Central Workshops, Tunstead, Derbys.
Sdn	Great Western Railway, Swindon Works, Wiltshire. later B.R.
Sdn C.	Swindon College, Department of Engineering, North Star Avenue, Swindon, Wiltshire.
SE	Sharon Engineering Ltd, Leek, Staffordshire.
Selhurst	British Rail, Southern Region, Selhurst Maintenance Depot, near Croydon, Greater London.
S&H	Strachan & Henshaw Ltd, Ashton, Bristol.
Shackerstone	Market Bosworth Light Railway, Shackerstone Station, Market Bosworth, Leicestershire.
FG.Shepherd	F.G. Shepherd, Flow Edge Colliery, Middle Fell, Alston, Cumbria.
Siemens	Siemens Bros. Ltd.
SIG	Schweizerische Industriegesellschaft, Neuhausen am Rheinfall, Switzerland.
D.Skinner	D. Skinner, 660 Streetsbrook Road, Solihull, West Midlands.
SL	Severn-Lamb Ltd, Western Road, Stratford-Upon-Avon.
SLM	Schweizerische Lokomotiv-and Maschinenfabrik, Winterthur, Switzerland.
SMH	Simplex Mechanical Handling Ltd, Elstow Road, Bedford. (Successors to MR).
N.Smith	N. Smith, Heatherslaw Light Railway, Heatherslaw Mill, near Coldstream, Northumberland.
South Crofty	South Crofty Ltd, Pool, near Camborne, Cornwall.
Spence	Wm Spence, Cork Street Foundry, Dublin.

Spondon	Derbyshire & Notts. Electric Power Co Ltd, Spondon Power Station, Derbyshire.
SS	Sharp, Stewart & Co Ltd, Atlas Works, Manchester and Atlas Works, Glasgow. (Latter from 1888).
St.Rollox	Caledonian Railway, St Rollox Works, Glasgow, later L.M.S.R. and B.R.
T.Stanhope	T. Stanhope, Arthington Station, near Leeds, West Yorkshire.
WP.Stewart	W.P. Stewart, Washington Sheet Metal Works, Industrial Road, Hertburn Industrial Estate, Washington, Tyne & Wear.
Stewarts Lane	British Rail, Southern Region, Stewarts Lane Maintenance Depot, London.
Stockton	South Durham Steel & Iron Co Ltd, Stockton Works, Co. Durham.
Stoke	North Staffordshire Railway, Stoke Works, Staffordshire.
M.Stokes	M. Stokes, Little Garden West Railway, Southerndown, Mid Glamorgan.
Str	Great Eastern Railway, Stratford Works, London.
SSt	Swing Stage, Canada.
Syl	Sylvester Steel Co, Lindsay, Ontario, Canada.
B.Taylor	B. Taylor, 7 Abbey Road, Shepley, Huddersfield, West Yorkshire.
J.Taylor	J. Taylor, The Ford, Woolhope, Hereford & Worcester.
TG	T. Green & Son Ltd, Leeds.
TH	Thomas Hill (Rotherham) Ltd, Vanguard Works, Hooton Road, Kilnhurst, South Yorkshire.
TH/S	Built by TH, utilising frame of Sentinel steam loco.
Thakeham	Thakeham Tiles Ltd, Thakeham, Sussex.
R.Thomas	Richard, Thomas & Co Ltd, Crowle Brickworks, Lincs. Constructed from parts supplied by FH.
THub	Taylor & Hubbard Ltd, Kent Street, Leicester.
Thwaites	Thwaites Engineering Co Ltd, Leamington Spa, Warwicks.
TMA	TMA Automation Ltd, Feeds Automated Systems, Jubilee Works, Tyburn Road, Erdington, Birmingham 24.
TMS	Tramway Museum Society, Cliff Quarry, Crich, near Matlock, Derbyshire.
Todd Kitson & Laird	Todd, Kitson & Laird, Leeds.
Towyn	Talyllyn Railway Co, Pendre Works, Tywyn, Gwynedd.
Track Supplies	Track Supplies & Services Ltd, Old Wolverton Road, Old Wolverton, Milton Keynes, Buckinghamshire.
TU	Task Undertakings Ltd, Birmingham, West Midlands.
Tunnequip	Tunnequip Ltd, Nowhurst Lane, Broadbridge Heath, Horsham, West Sussex.
Tyseley	Birmingham Railway Museum, The Steam Depot, Warwick Road, Tyseley, Birmingham, West Midlands.
UC	Union Construction.
UMM	Underground Mining Machinery Ltd, Aycliffe, Co. Durham.
Unilok	Hugo Aeckerle & Co, Hamburg, Germany.
Unimog	Mercedes Benz Ltd, (Unimog), West Germany.
D.Vanstone	D. Vanstone, Pixieland Mini-Zoo, Kilkhampton, near Bude, Cornwall.
VE	Victor Electrics Ltd, Burscough Bridge, Lancashire.
VF	Vulcan Foundry Ltd, Newton-le-Willows, Lancashire.
VIW	Vulcan Iron Works, Wilkes-Barre, Pennsylvania, U.S.A.
VL	Vickers Ltd, Barrow-in-Furness.

Vollert	Hermann Vollert K.G., Maschinenfabrik, 7102 Weinsberg/Wurtt, West Germany.
G.Walker	G. Walker, Lakeside Miniature Railway, Marine Lake, Southport, Merseyside.
RJ.Washington	R.J. Washington, Holly Lodge, 404 Gloucester Road, Cheltenham, Gloucestershire.
RA.Watson	R.A. Watson Ltd, White Moss Peat Works, Bogsbank, West Linton, Borders.
WB	W.G. Bagnall Ltd, Castle Engine Works, Stafford.
Wcb	Whitcomb Locomotive Co, Wilkes-Barre, Rochelle, Illinois, U.S.A.
WCI	Wigan Coal & Iron Co Ltd, Kirklees, Lancashire.
P.Weaver	P. Weaver, New Farm, Lacock, near Corsham, Wiltshire.
Wentscher	
WhC	Whiting Corporation, Harvey, Illinois, U.S.A.
WHR	Welsh Highland Light Railway (1966) Ltd, Gelert Farm, Porthmadog, Gwynedd.
Wilson	A.J. Wilson, 6 Trentdale Road, Carlton, Nottinghamshire.
Wilton	I.C.I. Ltd, Wilton Works, Middlesbrough, Cleveland.
WkB	Walker Bros (Wigan) Ltd, Wigan, Lancashire.
Wkm	D. Wickham & Co Ltd, Ware, Hertfordshire.
WkmR	Wickham Rail, Bush Bank, Suckley, Hereford & Worcester, and Ware, Hertfordshire. (Successors to Bruff & Wkm).
WMD	Waggon & Maschinenbau G.m.b.H., Donauworth, Germany.
Wolf	R. Wolf A.G., Abteilung Lokomotivfabrik Hagans, Erfurt, Germany.
Wolverton	British Rail Engineering Ltd, Wolverton Works, Buckinghamshire.
Woolwich	Woolwich Arsenal, London.
WR	Wingrove & Rogers Ltd, Kirkby, Liverpool.
WSO	Wellman, Smith, Owen Engineering Corporation Ltd, Darlaston, Staffordshire.
YE	Yorkshire Engine Co Ltd, Meadow Hall Works, Sheffield.
York	British Rail Engineering Ltd, York Works, North Yorkshire.
Zweiweg	Zweiweg-Fehrzeug, Allrad-Ranglerttechnik G. m.b.H., D-5628, Heiligenhaus, Bez, Dusseldorf, Germany.
9E	London & South Western Railway, Nine Elms Works, London.

Listed below are the Civil Engineering Contractors / Plant Hire specialists who own locos for use on tunnelling and sewer contracts, etc. The locos are to be found in all parts of the Country but the details of the loco fleets are listed under the firms main depot in the County shown below.

TITLE OF FIRM	COUNTY
Cementation Mining Ltd.	South Yorkshire.
Centriline Ltd.	Lancashire.
Costain Ltd.	North Yorkshire.
Delta Civil Engineering Company Ltd.	Somerset.
Edmund Nuttall Ltd.	Strathclyde.
Fairclough Civil Engineering Ltd.	Staffordshire.
G.L. Plant Ltd.	Greater London.
Grant Lyon Eagre Ltd.	Humberside.
J.C. Gillespie Civil Engineers Ltd.	Greater Manchester.
J.F. Donelan & Co Ltd.	Greater Manchester.
J.J. Gallagher & Co Ltd.	West Midlands.
J. Murphy & Sons Ltd.	Greater London.
Johnston Construction Ltd.	Warwickshire.
Kilroe Civil Engineering Ltd.	Greater Manchester.
Lilley Plant Ltd.	Warwickshire.
May Gurney & Co Ltd.	Norfolk.
McNicholas Construction Co Ltd.	Hertfordshire.
Miller Construction Ltd.	Warwickshire.
Norwest Holst Plant Ltd.	Merseyside.
Raynesway Plant Ltd.	Derbyshire.
Rivertower Construction Ltd.	Greater Manchester.
Sir Robert McAlpine & Sons Ltd.	Northamptonshire.
South Western Mining & Tunnelling Ltd.	Cornwall.
Specialist Plant Associates Ltd.	Bedfordshire.
Stepney Tunnelling Ltd.	Humberside.
Tarmac Construction Ltd.	West Midlands.
Taylor Woodrow Plant Ltd.	Greater London.
Thyssen (Great Britain) Ltd.	West Yorkshire.
Tickhill Plant Ltd.	South Yorkshire.
Transmanche-Link.	Kent.

SECTION 1 ENGLAND

AVON

A.E. MURPHY LTD, CANADA WAREHOUSE, CHITTENING ESTATE, AVONMOUTH
Gauge : 4'8½".   (ST 529815)

-		0-4-0DH	EEV	D1124	1966
-		0-4-0DH	JF	4220001	1959

IMPERIAL CHEMICAL INDUSTRIES LTD, AGRICULTURAL DIVISION,
SEVERNSIDE WORKS, HALLEN, BRISTOL
Gauge : 4'8½".   (ST 536831, 539828)

IBURNDALE	0-6-0DE	YE	2725	1958
KILDALE	0-6-0DE	YE	2741	1959

PASMINCO EUROPE SMELTING DIVISION, KINGSWESTON LANE, AVONMOUTH
Gauge : 4'8½".   (ST 522797)

6		4wDH	S	10048	1960
7	FLYING VICAR	4wDH	S	10023	1960

WESSEX WATER PLC/J.F. DONELON & CO LTD
NORTHERN FOUL WATER INTERCEPTOR, PHASE 1, PORTWAY, BRISTOL
Gauge : 75cm.   (ST 563745)

1		4wDH	Ageve			+
-		4wDH	Ageve	898	1980	+
3909		4wDH	Ruhr	3909	1969	+
3920		4wDH	Ruhr	3920	1969	+

> Locos in use underground - sometimes to be found on surface.
> Tunnelling Contract Januray 1990 to November 1992.
> + Loco owned by J.F. Donelon & Co Ltd.

WESTERN FUEL CO LTD, FILTON COAL CONCENTRATION DEPOT
Gauge : 4'8½".   (ST 611788)

-		0-6-0DM	HC	D1171	1959
DOUGAL		0-4-0DM	(VF	D77	1947
			(DC	2251	1947

PRESERVATION SITES

AVON VALLEY RAILWAY, BITTON STEAM CENTRE, BITTON STATION
Gauge : 4'8½".   (ST 670705)

34058	SIR FREDERICK PILE	4-6-2	3C	Bton		1947	
44123		0-6-0	IC	Crewe	5658	1925	
45379		4-6-0	OC	AW	1434	1937	
47324		0-6-0T	IC	NB	23403	1926	
48173		2-8-0	OC	Crewe		1943	
	EDWIN HULSE	0-6-0ST	OC	AE	1798	1918	
	LITTLETON No 5	0-6-0ST	IC	MW	2018	1922	
	FONMON	0-6-0ST	OC	P	1636	1924	
No.9		0-6-0T	OC	RSHN	7151	1944	
2		0-4-0ST	OC	WB	2842	1946	
	KINGSWOOD	0-4-0DM		AB	446	1959	
	-	0-4-0DM		Bg/DC	2158	1941	
	STAR OF INDIA	4wDM		RH	210481	1942	
	-	4wDM		RH	235519	1945	
252823		4wDM		RH	252823	1947	Dsm
128		0-4-0DH		S	10128	1963	
D2	ARMY 610	0-8-0DH		S	10143	1963	
	PWM 3769	2w-2PMR		Wkm	6648	1953	
(TP 57P)	ENGINEER'S No.1	2w-2PMR		Wkm	8267	1959	

BLAISE CASTLE MINIATURE RAILWAY, BLAISE CASTLE, BRISTOL
Gauge : 1'3".   (ST 559786)

2	FOXGLOVE	2w-2-4BER	Hayne	1977	
	-	2w-2BER	Hayne	1983	

BRISTOL INDUSTRIAL MUSEUM, BRISTOL HARBOUR RAILWAY, PRINCES WHARF, CITY DOCK, BRISTOL
Gauge : 4'8½".   (ST 585722)

	PORTBURY	0-6-0ST	OC	AE	1764	1917
3		0-6-0ST	OC	FW	242	1874
6	HENBURY	0-6-0ST	OC	P	1940	1937
	-	0-4-0DM		RH	418792	1959

B. CLARKE, 11 PENN GARDENS, BATH
Gauge : 2'0".   (   )

	ADAM	4wDM	MR	9978	1954
	SYLVIA	4wDM	RH	213834	1942

Locos are not on public display.

J.S. CRITCHLEY, "WEST WINDS", GROVESEND ROAD, THORNBURY, BRISTOL
Gauge : 4'8½".   (ST 653891)

ROSEDALE	4wDH	S	10070	1961

I.GUNN, BRISTOL
Gauge : 1'10".   (   )

-	4wBE	WR	2489	1943

Loco in storage.

ANDREW JOHNSON, BATH
Gauge : 2'0".   (   )

TOBY	4wPM	D.Carter		1986

MONKTON FARLEIGH MINE MUSEUM, near BATH
Gauge : 2'0".   (   )

-	4wDM	FH	2525	1941

BEDFORDSHIRE

INDUSTRIAL SITES

AMPTHILL SCRAP METAL PROCESSING CO LTD, STATION ROAD, AMPTHILL
Gauge : 4'8½".   (TL 022372)

-	0-4-0DE	RH	425477	1959	OOU

Yard with locos for scrap or resale occasionally present.

C.A.E.C. HOWARD LTD, ST JOHNS WORKS, ST JOHNS STREET, BEDFORD
Gauge : 4'8½".   (   )

-	4wDM	MR	9921	1959	OOU

P.J. MACKINNON, WALNUT LODGE SAWMILLS, 36 LUTON ROAD, WILSTEAD
Gauge : 1'8".   (TL 063433)

-	4wDM	OK	6703	Pvd

SPECIALIST PLANT ASSOCIATES LTD, PLANT DEPOT,
23 PODINGTON AIRFIELD, HINWICK, WELLINGBOROUGH
Gauge : 2'0"/1'6". (SP 948608)

JM 82	4wBE	CE	5806	1970
JM 83	4wBE	CE	5942A	1972
JM 84	4wBE	CE	5942B	1972
JM 85	4wBE	CE	5942C	1972
JM 88	4wBE	CE	B0402A	1974
JM 90	4wBE	CE	B0402C	1974

Gauge : 1'6".

428001		4wBE	CE	B0156	1973
428002		4wBE	CE	B0176A	1974
428005		4wBE	CE	B0182C	1974
	-	2w-2BE	Iso	T15	1972
	-	2w-2BE	Iso	T40	1973
	-	2w-2BE	Iso	T41	1973
	-	2w-2BE	Iso	T46	1974
	-	2w-2BE	Iso	T49	1974
	-	2w-2BE	Iso	T57	1974
	-	2w-2BE	Iso	T66	1974
JM 100		2w-2BE	WR	L800	1983
JM 101		2w-2BE	WR	L801	1983
	-	2w-2BE	WR	544901	1984
	-	2w-2BE	WR	546001	1987
JM 103		2w-2BE	WR	546601	1987

Locos present in yard between contracts.

PRESERVATION SITES

LEIGHTON BUZZARD NARROW GAUGE RAILWAY SOCIETY
Gauge : 2'0".   Locos are kept at :-

Pages Park Shed   (SP 929242)
Stonehenge Workshops (SP 941275)

1	CHALONER		0-4-0VBT	VC	DeW		1877
No.2	PIXIE		0-4-0ST	OC	KS	4260	1922
No.3	RISHRA		0-4-0T	OC	BgC	2007	1921
4	DOLL		0-6-0T	OC	AB	1641	1919
5	ELF		0-6-0WT	OC	OK	12740	1936
11	P.C.ALLEN		0-4-0WT	OC	OK	5834	1912
6	CARAVAN		4wDM		MR	7129	1936
No.6			4wDM		MR	5875	1935 b
No.7	6619	9303/507	0-4-0DM		HE	6619	1966
7	8986	FALCON	4wDM		OK	8986	
8	GOLLUM		4wDM		RH	217999	1942
9	MADGE		4wDM		OK	7600	
No.10	HAYDN TAYLOR		4wDM		MR	7956	1945
No.11			4wDM		MR		
12	CARBON		4wPM		MR	6012	1930
No.13	ARKLE		4wDM		MR	7108	1936
14			4wDM		HE	3646	1946
15			4wDM		FH	2514	1941
16	THORIN OAKENSHIELD		4wDM		L	11221	1939
No.17	DAMREDUB		4wDM		MR	7036	1936

18	FëANOR	4wDM	MR	11003	1956	
20		4wDM	MR	60S317	1966	
No.21	FESTOON	4wPM	MR	4570	1929	
22		4wDM	LBNGRS	1	1989	c
23		4wDM	MR	11298	1965	.
24		4wDM	MR	4805	1934	Dsm
24		4wDM	MR	11297	1965	
25		4wDM	MR	7214	1938	
26	YIMKIN	4wDM	RH	203026	1942	
No.28	R.A.F. STANBRIDGE	4wDM	RH	200516	1940	
29	CREEPY					
	YARD No. P 19774	4wDM	HE	6008	1963	
30		4wDM	MR	8695	1941	
43		4wDM	MR	10409	1954	
No.44		4wDM	MR	7933	1941	
	-	4wDM	FH	3582	1954	
LOD/758054		4wDM	HE	2536	1941	
	-	4wDM	L	4228	1931	
	-	4wDM	MR	5603	1931	Dsm
"No.5"		4wDM	MR	5608	1931	b
5612	R8	4wDM	MR	5612	1931	Dsm
R7	No.131	4wDM	MR	5613	1931	Dsm a
164346		4wDM	RH	164346	1932	
	-	4wDM	RH	172892	1934	
	-	4wDM	RH	218016	1943	
	POPPY	4wDH	RH	408430	1957	
LM 39	T.W.LEWIS	4wDM	RH		1954	+

```
+   Either 375315 or 375316
a   In use as a work bench
b   Converted into a brake van
c   Built from parts of RH 425798/1958 & RH 444207/1961.
```

THE LIGHT RAILWAY ASSOCIATION
STEVINGTON & TURVEY LIGHT RAILWAY, TURVEY
Gauge : 2'0".   (TL 970524)

	-	4wPM	FH	1767	1931	
No.15	OLDE	4wDM	HE	2176	1940	
	-	4wDM	MR	7128	1936	
	-	4wDM	MR	9655	1951	
	-	4wDM	(MR?)			Dsm
	-	4wDM	OK	6504	1936	
5	COLLINGWOOD	4wDM	RH	373359	1958	

PLEASURE-RAIL LTD, WHIPSNADE & UMFOLOZI RAILWAY, WHIPSNADE ZOO
Gauge : 3'6".   (TL 004172)

390		4-8-0	OC	SS	4150	1896

Gauge : 2'6".

No.1	CHEVALLIER	0-6-2T	OC	MW	1877	1915
No.2	EXCELSIOR	0-4-2ST	OC	KS	1049	1908
No.3	CONQUEROR	0-6-2T	OC	WB	2192	1922
No.4	SUPERIOR	0-6-2T	OC	KS	4034	1920
	VICTOR	0-6-0DM		JF	4160004	1951
9	HEATHER	0-6-0DM		JF	4160005	1951

WOBURN ABBEY (NARROW GAUGE) RAILWAY, WOBURN PARK
Gauge : 2'0".   (SP 968328)

	FLYING SCOTSMAN	4wDM	S/O	RH	223749	1944

BERKSHIRE

PRESERVATION SITES

T. BUCK, DEEP MEADOWS, LEDGER LANE, FIFIELD, near WINDSOR
Gauge : 4'8½".   (   )

	-	0-6-0ST	OC	HL	3138	1915
	HORNPIPE	0-4-0ST	OC	P	1756	1928
No.43	22	0-4-0DH		JF	4220031	1964

F. STAPLETON, near NEWBURY
Gauge : 2'0".   (   )

-	4wBE		BE	16306	c1917 Dsm
-	4wDM		HE	2024	1940 Dsm

N. WILLIAMS, READING
Gauge : 2'0".   (   )

-	4wDM		MR	11264	1964
-	4wDM		RH	296091	1949

WINDSOR ROYALTY EXHIBITION, WINDSOR CENTRAL STATION.
Gauge : 4'8½".   (   )

3041	THE QUEEN	4-2-2		Carn	1983

Non-working replica of Sdn 1401/1894.

BUCKINGHAMSHIRE

INDUSTRIAL SITES

CASTLE CEMENT (PITSTONE) LTD, PITSTONE WORKS, near IVINGHOE
Gauge : 4'8½".   (SP 932153)

No.1		0-6-0DM	WB	3160	1959
2		4wDH	S	10159	1963
3		4wDH	RR	10264	1966

PRESERVATION SITES

W.H. McALPINE, FAWLEY HILL, FAWLEY GREEN, near HENLEY-on-THAMES
Gauge : 4'8½".   (SU 755861)

No.31		0-6-0ST	IC HC	1026	1913
(D2120)	03120	0-6-0DM	Sdn		1959
No.5	FLYING FLEA	4wDM	RH	294266	1951

QUAINTON RAILWAY SOCIETY LTD, BUCKINGHAMSHIRE RAILWAY CENTRE,
QUAINTON ROAD STATION, near AYLESBURY
Gauge : 4'8½".   (SP 736189, 739190)

6989	WIGHTWICK HALL	4-6-0	OC	Sdn		1948
7200		2-8-2T	OC	Sdn		1934
7715		0-6-0PT	IC	KS	4450	1930
9466		0-6-0PT	IC	RSH	7617	1952
(30585)	E0314	2-4-0WT	OC	BP	1414	1874
41298		2-6-2T	OC	Crewe		1951
41313		2-6-2T	OC	Crewe		1952
46447		2-6-0	OC	Crewe		1950
	SWANSCOMBE	0-4-0ST	OC	AB	699	1891
	-	0-4-0F	OC	AB	1477	1916
	TOM PARRY	0-4-0ST	OC	AB	2015	1935
	LAPORTE	0-4-0F	OC	AB	2243	1948
	SYDENHAM	4wWT	G	AP	3567	1895
No.1	SIR THOMAS	0-6-0T	OC	HC	1334	1918
	-	0-4-0ST	OC	HC	1742	1946
	ARTHUR	0-6-0ST	IC	HE	3782	1953
	JUNO	0-6-0ST	IC	HE	3850	1958
	-	0-6-0ST	IC	HE	3890	1964
3		0-4-0ST	OC	HL	3717	1928
No.1	COVENTRY No.1	0-6-0T	IC	NB	24564	1939
1		0-4-4T	IC	Nea	3	1898
	-	0-4-0T	OC	P	1900	1936
2087		0-4-0ST	OC	P	2087	1948
	-	0-4-0ST	OC	P	2105	1950 +
L.N.E.R. 49		4wVBT	VCG	S	6515	1926
11		4wVBT	VCG	S	9366	1945
7		4wVBT	VCG	S	9376	1947
5208		2w-2-2-2w-4-4	12CGR	S	9418	1950
7	SUSAN	4wVBT	VCG	S	9537	1952
	-	0-4-0ST	OC	WB	2469	1932
	-	0-6-0ST	OC	YE	2498	1951

D2298	LORD WENLOCK	0-6-0DM	(RSH	8157	1960	
			(DC	2679	1960	
(D5207)	25057	Bo-BoDE	Derby		1963	
53028		2w-2-2-2wRER	BRCW		1938	
54233		2w-2-2-2wRER	GRC		1939/40	
		Rebuilt	Acton		1941	
	-	0-4-0DM	Bg/DC	2161	1941	
T1		4wDM	FH	2102	1937	
	WALRUS	0-4-0DM	FH	3271	1949	
	-	4wDM	FH	3765	1955	
	-	0-4-0DM	HE	2067	1940	
	-	0-4-0DM	JF	20067	1933	
	REDLAND	4wDM	KS	4428	1929	
FLEET No.1139	HILSEA	4wDM	RH	463153	1961	
No.24		4wDH	TH/S	188C	1967	
9040		2w-2PMR	Wkm	6963	1955	
9037		2w-2PMR	Wkm	8197	1958	
	-	2w-2PMR	Wkm	8263	1959	

+ Actually built in 1948 but plates dated as shown

Gauge : 2'0".

803		2w-2-2-2wRE	EE	803	1931 +

+ Built 1931 but originally carried plates dated 1930.

CAMBRIDGESHIRE

<u>INDUSTRIAL SITES</u>

<u>BRITISH SUGAR CORPORATION LTD, WOODSTON FACTORY, PETERBOROUGH</u>
Gauge : 4'8½".   (TL 175976)

(D2089)	03089	0-6-0DM	Don		1960	
(D2112	03112)	0-6-0DM	Don		1960	
	-	0-4-0DH	JF	4220033	1965	
	SIR ALFRED WOOD	0-6-0DM	RH	319294	1953	OOU

<u>CIBA-GEIGY PLASTICS LTD, DUXFORD</u>
Gauge : 4'8½".   (TL 486454)

03030		4wDM	R/R	Unilok	2109	1980
03038		4wDM	R/R	Unimog	12/961	1982

<u>G.G. PAPWORTH LTD, RAIL DISTRIBUTION CENTRE, QUEEN ADELAIDE, ELY</u>
(Member of Potter Group)
Gauge : 4'8½".   (TL 563810)

(D2302)		0-6-0DM	(RSH	8161	1960
			(DC	2683	1960
(D3272)	08202	0-6-0DE	Derby		1956
	-	0-6-0DH	TH/S	150C	1965

J. & K. HARRIS, HARRIS'S SCRAP METAL MERCHANTS,
NORWOOD ROAD INDUSTRIAL ESTATE, MARCH
Gauge : 3'0".   (TL 415980)

06/22/6/2		4wDM	RH	224337	1945 OOU

MAYER NEWMAN & CO LTD, KNAPPETTS SCRAPYARD, SNAILWELL, near NEWMARKET
Gauge : 4'8½".   (TL 638678)

(D2012)	03012	F 135 L	0-6-0DM	Sdn		1958	Dsm
(D2020)	03020	F 134 L	0-6-0DM	Sdn		1958	
(D2180)	03180		0-6-0DM	Sdn		1962	Dsm
-			0-4-0DM	JF	4210080	1953	OOU
ARMY 410			0-4-0DH	NB	27645	1958	

SEADYKE FREIGHT SYSTEMS LTD, ALUMINIUM PANEL DEALERS, NENE PARADE, WISBECH
Gauge : 4'8½".   (TF 426099)

-	0-4-0DM	HE	2068	1940 Dsm +

+ Retained for use as an emergency generator

THE RUGBY GROUP PLC, BARRINGTON CEMENT WORKS
Gauge : 4'8½".   (TL 396504)

7	0-4-0DE	RH	499435	1963
8	0-4-0DE	RH	499436	1963
15	4wDH	TH	127V	1963
	Rebuilt	TH	240V	1972
17	4wDH	S	10040	1960
18	4wDH	S	10035	1960
19	4wDH	RR	10260	1966
-	4wDH	TH	164V	1966

PRESERVATION SITES

Mr. DRAGE, NEW BUILDINGS FARM, HEYDON, near ROYSTON
Gauge : 4'8½".   (TL 419409)

-		0-4-0ST	OC	AB	1219	1910
44	CONWAY	0-6-0ST	IC	K	5469	1933
11 13	NEWCASTLE	0-6-0ST	IC	MW	1532	1901
960236		2w-2PMR		Wkm	1519	1934

IMPERIAL WAR MUSEUM, DUXFORD AERODROME
Gauge : 4'8½".   (TL 461462)

-	4wPM	MR	1364	1918

Gauge : 2'0".

	TIGER	4-6-0	OC	BLW	44699	1917 +
	-	4wDM		MR	22211	1964 +

+ Currently under restoration at John Appleton Engineering, Leiston, Suffolk.

C.J. & A.M. PEARMAN, 4 TANGLEWOOD, ALCONBURY WESTON, HUNTINGDON
Gauge : 1'11½". (   )

	-	4wPM		MR	2059	1920

PETERBOROUGH RAILWAY SOCIETY LTD, NENE VALLEY INTERNATIONAL STEAM RAILWAY
Gauge : 4'8½". Locos are kept at :-

Peterborough Nene Valley	(TL 188982)
Wansford Steam Centre	(TL 093979)
Woodston Sugar Factory, Peterborough	(TL 184979)

No.	Name	Type	Cyl	Builder	Works No.	Year	Notes
34081	92 SQUADRON	4-6-2	3C	Bton		1948	
(45231)	3RD (VOLUNTEER) BATTALION,						
5231	THE WORCESTERSHIRE AND SHERWOOD FORESTERS REGIMENT						
		4-6-0	OC	AW	1286	1936	
73050	CITY OF PETERBOROUGH						
		4-6-0	OC	Derby		1954	
90432		0-4-0ST	OC	AB	2248	1948	
	-	0-6-0ST	OC	AE	1945	1926	
	-	0-4-0VBT	OC	Cockerill	1626	1890	
Nr.740	D.F.D.S. DANISH SEAWAYS						
62015		2-6-4T	3C	Frichs	86	1928	
Nr.656		0-6-0T	OC	Frichs	360	1949	
	RYAN	0-6-0ST	IC	HC	1539	1924	
1	THOMAS	0-6-0T	OC	HC	1800	1947	
	JACKS GREEN	0-6-0ST	IC	HE	1953	1939	
68081		0-6-0ST	IC	HE	2855	1943	
3.628		4-6-0	4CC	Hen	10745	1911	
64-305		2-6-2T	OC	Krupp	1308	1934	
1178		2-6-2T	OC	Motala	516	1914	
No.35	RHIWNANT	0-6-0ST	IC	MW	1317	1895	
SVBJ 101		4-6-0	OC	Nohab	2082	1944	
80-014		0-6-0T	OC	Wolf	1228	1927	
D306	(40106)	1Co-Co1DE		(EE	2726	1960	
	ATLANTIC CONVEYOR		—	(RSH	8136	1960	
D9516		0-6-0DH		Sdn		1964	
D9523		0-6-0DH		Sdn		1964	
(D9529)	14029	0-6-0DH		Sdn		1965	
11187		4w-4wRER				1937	
804		Bo-BoDE		AL	77778	1950	
	-	0-4-0DH		EEV	D1123	1966	
	-	4wPM		FH	2894	1944	Dsm
	-	4wDM		FH	2896	1944	OOU
	-	4wDM		RH	294268	1951	
	-	0-4-0DM		RH	304469	1951	
	-	4wDM		RH	321734	1952	
	HORSA	0-4-0DH		RSHD/WB	8368	1962	
	DONCASTER	0-4-0DE		YE	2654	1957	
VR 001		2w-2PMR		Wkm		1944	

RAILWORLD, (MUSEUM OF WORLD RAILWAYS), WOODSTON, PETERBOROUGH
Gauge : 4'8½".   (   )

996		4-6-2	4C	Frichs	415	1950

Gauge : 2'0".

740	MATHERN	0-6-0T	OC	OK	2343	1907

Mr. THORNHILL, BARNWELL JUNCTION STATION, near CAMBRIDGE
Gauge : 4'8½".   (TL 472596)

960225	2w-2PMR	Wkm	1308	1933

UPWELL FEN LIGHT RAILWAY, (C.CROSS), 8 THURLANDS DROVE, UPWELL, WISBECH
Gauge : 4'8½".   (   )

-	2w-2PMR	Wkm	1548	1934	DsmT

Gauge : 2'0".

No.1	CRAIG	4wDM	HE	4556	1954
13	MAC	4wDM	MR	8994	1946

CHANNEL ISLANDS

INDUSTRIAL SITES

STATES OF GUERNSEY, TECHNICAL SERVICES, GROSNEZ FORT, ALDERNEY
Gauge : 4'8½".

ALD 7	MOLLY 2	4wDM	RH	425481	1958	OOU

PRESERVATION SITES

ALDERNEY RAILWAY SOCIETY, MANNEZ QUARRY, ALDERNEY
Gauge : 4'8½".

6	ALD 40	4wVBT	VCG	S	6909	1927	Dsm
3	J.T.DALY	0-4-0ST	OC	WB	2450	1931	
D100	ELIZABETH	0-4-0DM		(VF	D100	1949	
				(DC	2271	1949	
(10177)		2w-2-2-2wRER		MC		1938	
(11177)		2w-2-2-2wRER		MC		1938	
PWM 3776	CADENZA	2w-2PMR		Wkm	6655	1953	+
PWM 3954	MARY LOU	2w-2PMR		Wkm	6939	1955	
1	GEORGE RLC/009025	2w-2PMR		Wkm	7091	1955	
9028		2w-2PMR		Wkm	7094	1955	a
2	SHIRLEY RLC/009029	2w-2PMR		Wkm	7095	1955	
7		2w-2PMR		Wkm	8086	1958	

+ Currently under renovation by R. Warren, Little Street,
  St Anne, Alderney.
a Currently stored at Mount Hale Ltd, Fort Albert.

THE PALLOT WORKING STEAM MUSEUM, RUE DE BECHET, TRINITY, JERSEY
Gauge : 4'8½".

	LA MEUSE	0-6-0T	OC	La Meuse	3442	1931
	-	0-4-0ST	OC	P	2085	1948
60130	KESTREL	0-4-0ST	OC	P	2129	1952

Gauge : 2'0".

| 2 | | 4wDM | S/O | MR | 11143 | 1960 |
| | - | 4wDM | | MR | 60S383 | 1969 Dsm + |

+ In use as a brake van.

CHESHIRE

INDUSTRIAL SITES

ASSOCIATED OCTEL CO LTD, OIL SITES ROAD, ELLESMERE PORT
Gauge : 4'8½".   (SJ 415767)

2	0-4-0DM	RH	313394	1952
3	0-4-0DM	RH	313396	1952 OOU
(4)	0-4-0DM	RH	319284	1952

BRITISH OXYGEN CO LTD, MARSH WORKS, WIDNES
Gauge : 4'8½".   (SJ 503846)

| | - | 0-6-0DH | | HE | 7189 | 1970 |
| D23 | | 4wDH | | RR | 10194 | 1964 |

BRITISH RAIL ENGINEERING (1988) LTD, CREWE WORKS
Gauge : 4'8½".   (SJ 691561)

(D3585)	08470	0-6-0DE		Crewe		1958
7158		4wDH	R/R	NNM	77501	1980
	-	8wDH	R/R	Minilok	130	1986

BRITISH SALT LTD, CLEDFORD LANE, MIDDLEWICH
Gauge : 4'8½".   (SJ 718645)

| (D2150) | 0-6-0DM | Sdn | | 1960 |
| | - | 4wDH | RR | 10194 | 1964 |

HAYS CHEMICALS, MURGATROYD DIVISION, ELWORTH WORKS, SANDBACH
Gauge : 4'8½".   (SJ 729633)

-	0-4-0DE	RH	544997	1969

IMPERIAL CHEMICAL INDUSTRIES LTD, MOND DIVISION,
Winnington, Wallerscote & Lostock Works, Northwich
Gauge : 4'8½".   (SJ 641746, 642737, - Workshops 649747, 682744)

JOHN BRUNNER	0-6-0DE	EE	1901	1951
PERKIN	0-6-0DE	EE	1904	1951
LUDWIG MOND	6wDE	GECT	5578	1980
FARADAY	0-4-0DE	RH	402803	1957
KELVIN	0-4-0DE	RH	402807	1957
RUTHERFORD	0-4-0DE	RH	412710	1957
-	4wBE	Vollert		1980
-	4wBE	Vollert		1980

Winnington Works, Crystal Plant
Gauge : 2'6".   (SJ 643749)

-	4wBE	WR		c1948
-	4wBE	WR	K7070	1970

KEMIRA (UK) LTD, FERTILISER FACTORY, INCE MARSHES
Gauge : 4'8½".   (SJ 472765)

12.082	0-6-0DE	Derby		1950	
PLANT No.Q7202	0-6-0DH	EEV	D1233	1968	Dsm
-	0-6-0DH	EEV	D3986	1970	

MACCLESFIELD CORPORATION, ENGINEERS DEPARTMENT STORE, MACCLESFIELD
Gauge : 2'0".   (SJ 922738)

-	4wPM	MR	7033	1936	OOU

MANCHESTER SHIP CANAL CO LTD, ELLESMERE PORT & STANLOW
Gauge : 4'8½".   (SJ 399777, 419763)

3001	0-6-0DH	S	10144	1963
3003	0-6-0DH	S	10146	1963
3004	0-6-0DH	S	10147	1963
3005	0-6-0DH	S	10162	1963
CE 9604	4wDM	Robel		

Locos return to Mode Wheel Workshops, Greater Manchester
for repairs, etc.

MIDDLEBROOK MUSHROOMS LTD, PEAT FARM, LINDOW MOSS, MOOR LANE, WILMSLOW
Gauge : 2'0".   (SJ 823803)

	-	4wDM	AK	No.4	1979	
	-	4wDH	LB	50888	1959	
No.2		4wDM	LB	52528	1961	OOU

NORTH WEST WATER AUTHORITY, MOULDSWORTH DEPOT, near CHESTER
Gauge : 2'0".   (   )

81.04	4wDM	HE	6298	1964	OOU
81.02	4wDM	RH	260712	1948	OOU
-	4wDM	RH	452294	1960	Dsm

ROYAL ORDNANCE PLC, SMALL ARMS DIVISION, RADWAY GREEN, ALSAGER
         See Section Five for loco details.

SHELL U.K. OIL LTD, STANLOW & THORNTON-LE-MOORS
Gauge : 4'8½".   (SJ 425766, 432761, 440758)

2	0-4-0DH	S	10065	1961
3	4wDH	TH	235V	1971
5	4wDH	TH	220V	1970
7	4wDH	TH	236V	1971
8	4wDH	TH	288V	1980
9	4wDH	TH	287V	1980
10	4wDH	TH	234V	1971

PRESERVATION SITES

CREWE HERITAGE CENTRE, CREWE
Gauge : 4'8½".   (ST 708553)

3020	CORNWALL	2-2-2	OC	Crewe		1858 +
No.7	ROBERT	0-6-0ST	IC	HC	1752	1943
	JOSEPH	0-6-0ST	IC	HE	3163	1944
		Rebuilt		HE	3885	1964
	-	0-4-0ST	OC	KS	4388	1926
(D120)	45108	1Co-Co1DE		Crewe		1961
D1842	(47192)	Co-CoDE		Crewe		1965
(D5233)	25083	Bo-BoDE		Derby		1963
D7523	(25173)					
	JOHN F. KENNEDY	Bo-BoDE		Derby		1965
M 49002		Bo-BoWE		Derby		
M 49006		Bo-BoWE		Derby		

    + Currently in store at BREL Crewe Works.

DERBYSHIRE CAVING CLUB, HOUGH LEVEL, ALDERLEY EDGE
Gauge : 2'0".   (SJ 862778)

    P396   81A03   4wDM   RH   497542 1963

GULLIVERS WORLD, LOST WORLD RAIL ROAD, SHACKLETON CLOSE, WARRINGTON
Gauge : 1'3".   (SJ 590900)

            -                   6+2w-2-2wDE   S/O   MM   1988

CLEVELAND

<u>INDUSTRIAL SITES</u>

A.V. DAWSON LTD, DEPOT ROAD, MIDDLESBROUGH WHARF, MIDDLESBROUGH,
& AYRTON STORE AND RAILHEAD, FORTY FOOT ROAD
Gauge : 4'8½".   (NZ 488213, 493215)

    (D3942   08774)
    ARTHUR VERNON DAWSON   0-6-0DE   Derby   1960
            -                   0-4-0DM   Bg/DC   2725   1963
    7900   0-4-0DM   RSH   7900   1958

BASF CHEMICALS LTD, SEAL SANDS, near GREATHAM
Gauge : 4'8½".   (NZ 535241)

    25-1   0-6-0DH   EEV   3870   1969

BLACKETT, HUTTON & CO LTD, STEELFOUNDERS, RECTORY LANE, GUISBOROUGH
Gauge : 4'8¼".   (NZ 612155)

            -                   4wDM   RH   265617   1948

BRITISH STEEL PLC, GENERAL STEELS DIVISION, TEESSIDE WORKS
Apprentice Training Centre, Middlesbrough
Gauge : 4'8½".   (NZ 546211)

    42   0-4-0DH   JF   4220020 1961 OOU

Hartlepool Works, Hartlepool
Gauge : 4'8½".   (NZ 511286, 507288)

    450   4wDH   TH   231V   1971
    451   4wDH   TH   232V   1971
    452   4wDH   TH   233V   1971 OOU
    453   4wDH   TH   221V   1970 OOU

Redcar Coke Ovens, Redcar
Gauge : 4'8½".   (NZ 562257)

1		4wWE	GB	420355/1	1976	
2		4wWE	GB	420355/2	1976	
3		4wWE	GB	420408	1977	

Skinningrove Works, Carlin How, Saltburn-By-The-Sea
Gauge : 4'8½".   (NZ 708194)

1		0-4-0DH	S	10125	1963	OOU
2		0-4-0DH	S	10126	1963	
3		0-4-0DH	S	10127	1963	

South Bank Coke Ovens
Gauge : 4'8½".   (NZ 536214)

B.S.C. No.1		4wWE	B.S.C.		1986
B.S.C. No.2		4wWE	B.S.C.		1986

Teeside Works, Middlesbrough
Gauge : 4'8½".   (NZ 563223)

T 35 85	BARRIER WAGON No.2	4wDH	Cleveland		1972	Dsm b
T 36 40	BARRIER WAGON No.1	4wDH	Cleveland		1972	Dsm b
54		0-6-0DH	S	10054	1961	
55		0-6-0DH	TH/S	109C	1961	
56		0-6-0DH	S	10056	1961	
64		0-4-0DH	S	10064	1961	OOU
68		0-4-0DH	S	10068	1961	OOU
78		0-6-0DH	S	10078	1961	
101		0-6-0DH	S	10101	1962	
105		0-4-0DH	S	10105	1962	OOU
167		0-6-0DH	S	10167	1964	
168		0-6-0DH	S	10168	1964	
210		0-6-0DH	RR	10210	1964	
211		0-6-0DH	RR	10211	1964	
224		0-6-0DH	RR	10224	1965	
225		0-6-0DH	RR	10225	1965	
234		0-6-0DH	RR	10234	1965	
251	WALTER URWIN	6wDE	GECT	5414	1976	
252	BOULBY	6wDE	GECT	5415	1976	
253	ESTON	6wDE	GECT	5416	1976	
254	BROTTON	6wDE	GECT	5417	1976	
255	LIVERTON	6wDE	GECT	5418	1976	
256	LINGDALE	6wDE	GECT	5425	1977	
257	NORTH SKELTON	6wDE	GECT	5426	1977	
258	GRINKLE	6wDE	GECT	5427	1977	
259	CARLIN HOW	6wDE	GECT	5428	1977	OOU
260	ROSEDALE	6wDE	GECT	5429	1977	
261	STAITHES	6wDE	GECT	5430	1977	
262	LOFTUS	6wDE	GECT	5431	1977	
263	LUMPSEY	6wDE	GECT	5432	1977	
264	PORT MULGRAVE	6wDE	GECT	5461	1977	
265	ROSEBERRY	6wDE	GECT	5462	1977	
266	SHERRIFFS	6wDE	GECT	5463	1977	
267	SLAPEWATH	6wDE	GECT	5464	1977	
268	KIRKLEATHAM	6wDE	GECT	5465	1977	

```
269      LONGACRES         6wDE      GECT      5466    1977
270      CHALONER          6wDE      GECT      5467    1977
271      GLAISDALE         6wDE      GECT      5469    1978
272      GROSMONT          6wDE      GECT      5470    1978
273      KILTON            6wDE      GECT      5471    1978
274      ESKDALESIDE       6wDE      GECT      5472    1978
275      RAITHWAITE        6wDE      GECT      5473    1978
276      SPAWOOD           6wDE      GECT      5474    1978
277      WATERFALL         6wDE      GECT      5475    1978
LM 20    LACKENBY 1        4wDM      Robel 54.12-107 AD184  1980
LM 21    REDCAR 2          4wDM      Robel 54.12-107 AD183  1980
```

   b   Rebuilds of 4wDH S 10027/1960 & S 10080/1960 respectively.

CASTLE CEMENT (RIBBLESDALE) LTD, FORTY FOOT ROAD, MIDDLESBROUGH
Gauge : 4'8½".   (NZ 487207)

```
         -                0-4-0DE      RH         312989  1952
```

CLEVELAND POTASH LTD, BOULBY MINE, LOFTUS
Gauge : 4'8½".   (NZ 763183)

```
No.1                      0-6-0DE      YE         2723    1958
No.2                      0-6-0DE      YE         2713    1958
```

Gauge : 1'6".   (NZ 774189)   (Underground)

```
         -                0-4-0BE      WR         7654    1974
```

COBRA RAILFREIGHT LTD, NORTH ROAD, MIDDLESBROUGH
Gauge : 4'8½".   (NZ 488209)

```
 (D3984    08816)         0-6-0DE      Derby              1960
```

DOUGLAS ENGINEERING SERVICES, MIDDLESBROUGH
Gauge : 4'8½".   (NZ 555225)

```
No.1                      0-6-0DH      AB         608     1976
No.3                      0-6-0DH      AB         614     1977
```

IMPERIAL CHEMICAL INDUSTRIES LTD, AGRICULTURAL DIVISION, BILLINGHAM
Gauge : 4'8½".   (NZ 475228)

```
1                         6wDH         TH         296V    1981
D 2                       6wDH         TH         297V    1981
D 3                       0-6-0DH      TH         167V    1966
T1                        4wDH   R/R   NNM        82503   1983
T 2                       4wDH   R/R   NNM        83501   1983
T 3                       4wDH   R/R   NNM        83502   1984
T 4      05/273           4wDH   R/R   NNM        83503   1984
T 5                       4wDH   R/R   NNM        83504   1984
```

I.C.I. CHEMICALS & POLYMERS LTD, WILTON WORKS, MIDDLESBROUGH
Gauge : 4'8½".   (NZ 564218)

(60532)	532	BLUE PETER	4-6-2	3C	Don	2023	1948	Pvd
(D2989)	07005	LANGBAURGH	0-6-0DE		RH	480690	1962	OOU
(D2995)	07011	CLEVELAND	0-6-0DE		RH	480696	1962	
(D3657)	08502		0-6-0DE		Don		1958	
(D3658)	08503		0-6-0DE		Don		1958	
		WILTONIA	2w-2DMR		Wkm	7591	1957	
	-		2w-2PMR		Wkm	7603	1957	DsmT

MINISTRY OF DEFENCE, NAVY DEPARTMENT, EAGLESCLIFFE SPARE PARTS CENTRE
Gauge : 4'8½".   (NZ 410148)

-		4wDM	R/R	Unimog	2660	1978

NATIONAL POWER, HARTLEPOOL POWER STATION, SEATON CAREW
(A Division of C.E.G.B.)
Gauge : 4'8½".   (NZ 532270)

-	0-6-0DH	JF	4240015	1962
-	4wDH	RH	544996	1968

STEETLEY QUARRY PRODUCTS LTD, SPECIAL PRODUCTS DIVISION, HARTLEPOOL WORKS, HARTLEPOOL
Gauge : 4'8½".   (NZ 509352)

DL2		0-4-0DH	HC	D1346	1965
	-	0-4-0DH	HE	7425	1981

STOCKTON HAULAGE LTD, SCOTS ROAD, MIDDLESBROUGH
Gauge : 4'8½".   (NZ 510207)

1		BESSEMER BOY	4wDE	Moyse	1464	1979	
2	M43	AUTUMN GOLD	4wDE	Moyse	1364	1976	
3			4wDE	Moyse	1365	1976	OOU

TEES & HARTLEPOOL PORT AUTHORITY, TEES DOCK, GRANGETOWN
Gauge : 4'8½".   (NZ 546232)

No.1	0-4-0DH	S	10137	1962	
No.2	0-4-0DH	S	10170	1964	
No.6	0-6-0DH	RR	10215	1965	
No.7	0-6-0DH	S	10095	1962	
	Rebuilt	AB	6007	1982	
-	2w-2DMR	Wkm	(6607	1953?)	

TEESBULK HANDLING LTD, TEES DOCK, GRANGETOWN
(Subsidiary of Cleveland Potash Ltd)
Gauge : 4'8½". (NZ 549235)

1		0-6-0DH	S	10102	1962
2		0-6-0DH	AB	487	1964
-		0-6-0DH	RR	10257	1966

TEES STORAGE CO LTD, CLEVELAND DOCKYARD, MIDDLESBROUGH
Gauge : 4'8½". (NZ 508207)

-		4wDM	FH	3958	1961
-		0-4-0DH	NB	27644	1959

T.J. THOMSON & SON LTD, MILLFIELD SCRAP WORKS, STOCKTON
Gauge : 4'8½". (NZ 438193)

B.G.		0-4-0DH	AB	559	1970	
	-	0-6-0DM	RH	395303	1956	OOU
No.198	ELIZABETH	0-4-0DE	RH	421436	1958	
	HELEN	4wDH	TH	264V	1976	

Also other locos for scrap occasionally present.

PRESERVATION SITES

SALTBURN MINIATURE RAILWAY, VALLEY GARDENS, SALTBURN
Gauge : 1'3". (NZ 667216)

PRINCE CHARLES	4-6-2DE	S/O	Barlow		1953

STOCKTON-ON-TEES BOROUGH COUNCIL, PRESTON HALL MUSEUM
PRESTON PARK, EAGLESCLIFFE
Gauge : 4'8½". (NZ 430158)

-	0-4-0VBT	VCG	HW	21	1870

CORNWALL

INDUSTRIAL SITES

A. & P. APPLEDORE (FALMOUTH) LTD, FALMOUTH DOCKS
Gauge : 4'8½". (SW 822324)

129	0-4-0DH	S	10129	1963

CONCORD TIN MINES LTD, WHEAL CONCORD MINE, SKINNERS BOTTOM,
BLACKWATER, near TRURO   (Closed)
Gauge : 2'0".   (SW 728460)

	-	0-4-0BE	WR	3867	1948	OOU
	-	0-4-0BE	WR	G7179	1967	OOU
	-	0-4-0BE	WR	S7950	1978	OOU

DELABOLE SLATE (1977) LTD, OLD DELABOLE QUARRY, LOWER PENGELLY,
DELABOLE, CAMELFORD
Gauge : 1'11".   (SX 075835)

No.2	4wDM	MR	3739	1925	Pvd

ENGLISH CHINA CLAYS LTD.
Blackpool Dries, Burngullow
Gauge : 4'8½".   (   )

(D3390	08320)				
P400D	SUSAN	0-6-0DE	Derby		1957

Moorswater Works, near Liskeard
Gauge : 4'8½".   (   )

401D	SHARON	0-4-0DH	EEV	3987	1970

Treviscoe Dries
Gauge : 4'8½".   (   )

P404D	G924 ACV ELAINE	4wDM	R/R		1989

ENGLISH CLAYS LOVERIN POCHIN & CO LTD, ROCKS WORKS, near BUGLE
Gauge : 4'8½".   (SX 025586)

P403D	DENISE	4wDH	S	10029	1960

GEEVOR TIN MINES LTD, PENDEEN, near ST.JUST
(On care and maintenance basis only, from 16/02/90)
Gauge : 1'6".   (SW 375346)   (Locos are used mainly underground)

1		4wBE	CE	5514	1968
2		4wBE	CE	5764	1970
3		4wBE	CE	5739	1970
4		4wBE	CE	5712	1969
5		4wBE	CE	5623	1969
6		4wBE	CE	B0485.A	1975
7		4wBE	CE	B0485.B	1975
-		4wBE	CE	B1501	1977
-		4wBE	CE	B1558A	1977
-		4wBE	CE	B1558B	1977
-		4wBE	CE	B01592A	1978

-	4wBE	CE	BO1592B	1978
-	4wBE	CE	B1851A	1978
-	4wBE	CE	B1851B	1978
-	4wBE	CE	B3132A	1984
-	4wBE	CE	B3132B	1984
-	4wBE	CE	B3218	1985
-	4wBE	CE	B3606A	1989
-	4wBE	CE	B3606B	1989
-	4wBE	CE	B3606C	1989
-	4wBE	CE	B3606D	1989
No.1	0-4-0BE	Geevor		1973
2	0-4-0BE	Geevor		1950
3	0-4-0BE	Geevor		1950
4	0-4-0BE	Geevor		1952
5	0-4-0BE	Geevor		
6	0-4-0BE	Geevor		
7	0-4-0BE	Geevor		1954
8	0-4-0BE	Geevor		
9	0-4-0BE	Geevor		
10	0-4-0BE	Geevor		
11	0-4-0BE	Geevor		
No.12	0-4-0BE	Geevor		
13	0-4-0BE	Geevor		1973
14	0-4-0BE	Geevor		
15	0-4-0BE	Geevor		1973
16	0-4-0BE	Geevor		
17	0-4-0BE	Geevor		
18	0-4-0BE	Geevor		
19	0-4-0BE	Geevor		
20	0-4-0BE	Geevor		
21/1	4wBE	WR	H6583	1968
22/2	4wBE	WR	K6916	1970
23/3	4wBE	WR	K6915	1970
24/4	4wBE	WR	L7496	1971
25/5	4wBE	WR	L7495	1971

The Geevor locos are based on, and use parts of WR locos;
some of which are : 1386/1939; 2556/1943; 3555 & 3556/1946;
4206/1950; 4537/1950; 4884/1952; 5932/1958; 6130/1959;
6135/1960; 6402/1961.

M.P.S. METAL PROTECTION SERVICES, ST. AGNES STATION YARD, near TRURO
Gauge : 4'8½". (SW 721492)

YARD No.5200	4wDM	FH	3776	1956	Pvd

SOUTH CROFTY LTD, POOL near CAMBORNE
Gauge : 1'10". (SW 664409, 668413)

6	4wBE	CE	5876	1971
9	4wBE	CE	5932	1972
10	4wBE	CE	BO163B	1973
11	4wBE	CE	BO163A	1973
01	4wBE	CE	BO960	1976
02	4wBE	CE	B1524	1977
03	4wBE	CE	B1524/2	1977

04	4wBE	CE	B1524	1977
05	4wBE	CE	B1557A	1977
06	4wBE	CE	B1557C	1977
07	4wBE	CE	B1557B	1977
08	4wBE	CE	B1810A	1978
09	4wBE	CE	B1810B	1978
010	4wBE	CE	B1827	1978
011	4wBE	CE	B1851A	1978 +
012	4wBE	CE	B1851B	1978 +
013	4wBE	CE	B1899A	1979 +
014	4wBE	CE	B1899B	1979 +
015	4wBE	CE	B2247A	1980 +
016	4wBE	CE	B2247B	1980 +
017	4wBE	CE	B2247C	1980
018	4wBE	CE	B2930A	1981 +
019	4wBE	CE	B2930B	1981 +
020	4wBE	CE	B2930C	1981 +
021	4wBE	CE	B2930D	1981 +
022	4wBE	CE	B2930E	1981 +
023	4wBE	CE	B2930F	1981 +
024	4wBE	CE	B2944B	1981
025	4wBE	CE	B2944A	1981
026	4wBE	CE	B2944C	1981
027	4wBE	CE	B2944I	1982
028	4wBE	CE	B2944E	1982
029	4wBE	CE	B2944D	1982 +
030	4wBE	CE	B2944F	1982 +
031	4wBE	CE	B2944G	1982 +
032	4wBE	CE	B2944J	1982
033	4wBE	CE	B2944H	1982 +
034	4wBE	CE	B2944K	1982 +
035	4wBE	CE	B2247D	1980
036	4wBE	CE	B2944N	1982
037	4wBE	CE	B2944O	1982
038	4wBE	CE	B2944L	1982 +
039	· 4wBE	CE	B2944Q	1982
040	4wBE	CE	B2944P	1982 +
041	4wBE	CE	B2944M	1982
042	4wBE	CE	B2944R	1982
043	4wBE	CE	B2944T	1982
044	4wBE	CE	B2944W	1982
045	4wBE	CE	B2944V	1982
046	4wBE	CE	B2944S	1982
047	4wBE	CE	B2944X	1982
048	4wBE	CE	B2944U	1982
049	4wBE	CE	B3077C	1983
050	4wBE	CE	B3077A	1983 +
051	4wBE	CE	B3077B	1983
052	4wBE	CE	B3063A	1983 +
053	4wBE	CE	B3063B	1983 +
054	4wBE	CE	B3149A	1984
055	4wBE	CE	B3149B	1984
056	4wBE	CE	B3149C	1984
057	4wBE	CE	B3346C	1987
058	4wBE	CE	B3346A	1987
059	4wBE	CE	B3346B	1987
-	4wBE	CE	B2247E	1980
-	4wBE	CE	B2938	1981
-	4wBE	CE	B3149D	1984
-	4wBE	CE	B3149E	1984
-	4wBE	CE	B3149F	1984

-	4wBE	CE		B3149G	1984
-	4wBE	CE		B3150	1984
-	4wBE	CE		B3358A	1987
-	4wBE	CE		B3358B	1987
-	4wBE	CE		B3358C	1987
-	4wBE	WR		L4000	1981
1	0-4-0BE	WR Rebuilt South Crofty			OOU
2	0-4-0BE	WR Rebuilt South Crofty			
3	0-4-0BE	WR Rebuilt South Crofty			
4	0-4-0BE	WR Rebuilt South Crofty			OOU
5	0-4-0BE	WR Rebuilt South Crofty			
6	0-4-0BE	WR Rebuilt South Crofty			OOU
7	0-4-0BE	WR Rebuilt South Crofty			OOU
8	0-4-0BE	WR Rebuilt South Crofty			
9	0-4-0BE	WR Rebuilt South Crofty			OOU
10	0-4-0BE	WR Rebuilt South Crofty			OOU
11	0-4-0BE	WR Rebuilt South Crofty			OOU
12	0-4-0BE	WR Rebuilt South Crofty			OOU
13	0-4-0BE	WR Rebuilt South Crofty			OOU
14	0-4-0BE	WR Rebuilt South Crofty			OOU
-	4wDM	HE		6342	1970
-	4wDM	HE		7083	1971
-	4wDM	HE		7084	1972
-	4wDM	HE		7087	1972
-	4wDM	HE		7273	1972
-	4wDM	HE		7320	1973
-	4wDM	HE		7516	1977

+ Identity assumed.

Principally underground; battery locos are brought to the surface at around 1500 hours daily for battery charging.

The WR locos are continually being rebuilt with both original and newly fabricated parts. The original WR locos were :-

1706/1940; 4309-10/1950; 4820/1952; 4879/1952; 5314/1955; 5843/1956; 5927/1957; 5928-30/1957; 6303/1960; 6393/1961; C6712/1963; 6880/1964; F7029-30/1966; F7113-15/1966; J7293/1970; J7373-76/1969; L7526-31/1971; L7532/1972; L7533/1971; 7719-21/1974; N7833-39/1974.

SOUTH WESTERN MINING & TUNNELLING LTD, PLANT DEPOT,
COTTONWOOD, NANSTALLON, BODMIN
Gauge : 2'0".   (   )

-	0-4-0BE	WR	G7174	1967

T. WARE, SCRAP DEALER, CARHARRACK, near ST. DAY
Gauge : 4'8½".   (SW 741417)

TO 9362	4wDM	RH	349041	1953 OOU

WHEAL JANE LTD, CLEMO'S SHAFT, BALDHU, near TRURO
Gauge : 2'0".   (SW 772427)   (Locos used mainly underground)

2		4wBE	CE	B0134B	1973	
3		4wBE	CE	5554/3	1968	
4		4wBE	CE	BI599B	1978	
5		4wBE	CE	5728	1969	
7		4wBE	CE	5876	1971	
8		4wBE	CE	5923	1972	
10		4wBE	CE	5839D	1971	Dsm
11		4wBE	CE	5918	1972	Dsm
17		4wBE	CE	B0139A	1973	
18		4wBE	CE	B0139B	1973	
20		4wBE	CE	B0466A	1976	
21		4wBE	CE	B0466B	1976	
	-	4wBE	CE			+
	-	4wBE	CE	5839A	1971	
	-	4wBE	CE	5839B	1971	
	-	4wBE	CE	5946/1	1972	
	-	4wBE	CE	5946/2	1972	
	-	4wBE	CE	5957/1	1972	
	-	4wBE	CE	5957/2	1972	
	-	4wBE	CE	B0174	1974	
	-	4wBE	CE	B0466C	1976	
	-	4wBE	CE	B2289A	1980	
	-	4wBE	CE	B2289B	1980	

+   One of 5512/1, 5512/2 of 1968 or 5766 of 1970

WHEAL PENDARVES LTD, PENDARVES MINE, LITTLE PENDARVES, CAMBORNE   (Closed)
Gauge : 60cm.   (SW 647385)

1	4wBE	CE	5554/1	1968 OOU
2	4wBE	CE	5554/2	1968 OOU

PRESERVATION SITES

BODMIN & WENFORD RAILWAY SOCIETY.
BODMIN GENERAL STATION & BODMIN PARKWAY STATION
Gauge : 4'8½".   (SX 108641)

3802			2-8-0	OC	Sdn		1939
5552			2-6-2T	OC	Sdn		1928
	SWIFTSURE		0-6-0ST	IC	HE	2857	1943
	-		0-4-0ST	OC	P	1611	1923
62	UGLY		0-6-0ST	IC	RSHN	7673	1950
19			0-4-0ST	OC	WB	2962	1950
	ALFRED		0-4-0ST	OC	WB	3058	1953
	-		0-4-0F	OC	WB	3121	1957
D1048	WESTERN LADY		C-CDH		Crewe		1962
D3452	73601		0-6-0DE		Dar		1957
3559	(08444)		0-6-0DE		Derby		1958
W79976			4wDMR		A.C.Cars		1958
No.13	AD705 V3 LUCY		0-4-0DM		HE	3133	1944
No.1	PETER		0-4-0DM		JF	22928	1940
No.2	PROGRESS		0-4-0DH		JF	4000001	1945
1135			4wDM		RH	242867	1946 +

No.3	LEE		4wDM	RH	443642	1960	
	-		2w-2PMR	Wkm	509	1932	DsmT
4162	TR 10	PWM 2185	2w-2PMR	Wkm	4162	1948	DsmT
4168	TR 41	PWM 2191	2w-2PMR	Wkm	4168	1948	

+   In storage at Boscarne.

G.J.A. EVANS, 'THE INNY VALLEY RAILWAY', TRECARRELL MILL, TREBULLETT, LAUNCESTON
Gauge : 1'10¾".   (SX 320772)

No.3		4wDM	MR	9546	1950

LAPPA VALLEY RAILWAY, BENNY HALT, ST. NEWLYN EAST, near NEWQUAY
Gauge : 1'3".   (SW 838574, 839564)

No.4	MUFFIN	0-6-0	OC	Berwyn		1967
No.1	ZEBEDEE	0-6-4T	OC	SL	34	1974
4	POOH	4wDM		L	20698	1942
	LAPPA LADY	4w-4DE		Minirail		c1960
	DOUGAL	4w-4PM		E.Booth		1975

LAUNCESTON STEAM RAILWAY CO, LAUNCESTON
Gauge : 60cm.   (SX 328850)

	LILIAN	0-4-0ST	OC	HE	317	1883
	VELINHELI	0-4-0ST	OC	HE	409	1886
	COVERTCOAT	0-4-0ST	OC	HE	679	1898
	-	0-4-0ST	OC	HE	763	1901
	SYBIL	0-4-0ST	OC	WB	1760	1906
	-	4wDM		FH	1896	1935
2		4wDM		MR	5646	1933
	-	2w-2BER				

PARADISE RAILWAY, BIRD PARADISE, PARADISE PARK, HAYLE
Gauge : 1'3".   (SW 555365)

No.3	ZEBEDEE	4wDM	L	10180	1938

PIXIELAND SPECIAL TRAIN, PIXIELAND MINI FARM AND PETS CORNER, WEST STREET, KIRKHAMPTON, near BUDE
Gauge : 1'3".   (   )

1	PIONEER	2-2wPM	K.Rosewell	1947
		Rebuilt	D.Vanstone	1980

POLDARK MINING CO LTD, WENDRON FORGE MUSEUM, WENDRON near HELSTON
Gauge : 4'8½".   (SW 683315)

POLDARK MINING CO LTD 6   0-4-0ST   OC P   1530 1919

D. PREECE, PENGALLY FARM, TAVISTOCK ROAD, CALLINGTON
Gauge : 2'0".   (SX 364698)

                    -           0-4-0DM   s/o Bg   3235 1947
No.1   SAMSON   4wDM   FH   1887 1934
                    -           4wDM   RH   186318 1937

ROSEVALE MINING HISTORICAL SOCIETY, ZENNOR
Gauge : 2'0".   (SW 458380)

              -                 0-4-0BE   WR Rebuilt South Crofty

ST. AUSTELL CHINA CLAY MUSEUM LTD, WHEAL MARTYN MUSEUM,
CARTHEW, near ST. AUSTELL
Gauge : 4'8½".   (SX 004555)

          JUDY   0-4-0ST   OC WB   2572 1937

Gauge : 4'6".

LEE MOOR No.1   0-4-0ST   OC P   783 1899

Gauge : 2'6".

              -                 4wDM   RH   244558 1946

CUMBRIA

INDUSTRIAL SITES

BRITISH FUEL CO, CARLISLE COAL CONCENTRATION DEPOT, LONDON ROAD, CARLISLE
Gauge : 4'8½".   (NY 417550)

   24   D2004   LANCELOT   0-6-0DM   HC   D851 1955

BRITISH GYPSUM LTD, THISTLE PLASTER WORKS, KIRKBY THORE
Gauge : 4'8½".   (NY 646268)

          J.W.H.   0-4-0DH   AB   558 1970

BRITISH STEEL CORPORATION, SCOTTISH DIVISION,
SHAPFELL LIMESTONE QUARRIES, SHAP, PENRITH
Gauge : 4'8½".   (NY 571134)

No.10	72/21/44	0-4-0DH	AB	601	1975 OOU
63.000.443	WESTERN PROGRESS	6wDE	GECT	5468	1978
	7222/70/01	0-4-0DE	RH	323599	1953 OOU
No.2	7222/70/02	0-4-0DE	RH	323605	1954
	-	0-4-0DE	RH	423659	1958 OOU

BRITISH STEEL PLC, GENERAL STEELS DIVISION,
WORKINGTON WORKS, MOSS BAY, WORKINGTON
Gauge : 4'8½".   (NX 988269)

No.305		0-4-0DH	YE	2952	1965
No.309	21	0-6-0DH	YE	2825	1961
No.311		0-6-0DH	YE	2827	1961
No.314	153/1508	0-6-0DH	YE	2832	1962
No.402		0-6-0DH	HE	7409	1976
No.403		0-6-0DH	HE	7543	1978
No.404		0-6-0DH	HE	8978	1979

COPELAND DISTRICT COUNCIL, WELLINGTON COLLIERY SITE, WHITEHAVEN HARBOUR
Gauge : 2'6".   (NX 968183)

| - | 0-4-0BE | WR | (5931 | 1958?) Pvd |

CUMBRIA COUNTY COUNCIL, THE PORT OF WORKINGTON
Gauge : 4'8½".   (NX 993294)   RTC.

| No.211 | 0-4-0DE | YE | 2628 | 1956 |
| No.212 | 0-4-0DE | YE | 2684 | 1958 |

DEPARTMENT OF THE ENVIRONMENT, BROUGHTON MOOR, near MARYPORT
Gauge : 2'6".   (NY   )

YARD No. 10448	4wDM	HE	6007	1963 OOU
-	4wDM	MR	7485	1940
-	4wDM	MR	8774	1942

EGREMONT MINING CO LTD, FLORENCE IRON ORE MINE, EGREMONT
Gauge : 2'6".   (NY 018103)   (Underground)

| - | 4wBE | WR | 6218 | 1961 |
| - | 4wBE | WR | C6694 | 1963 |

F.G. SHEPHERD, FLOW EDGE COLLIERY, MIDDLE FELL, ALSTON
Gauge : 2'0".   (NY 734442)

-		0-4-0BE	WR		
-		4wBE	WR		

FISONS LTD, CUMBRIA WORKS, KIRKBRIDE, WIGTON
Gauge : 2'0".   (NY 238539)

K 11128		4wDM	MR	7463	1939
	GLASSON	4wDM	MR	8627	1941
	WEDHOLME	4wDM	MR	8885	1944
11140		4wDM	MR	8905	1944
	-	4wDM	MR	9231	1947

GLAXOCHEM LTD, ULVERSTON
Gauge : 4'8½".   (SD 306777)

	GLAXO	0-4-0F	OC	AB	2268	1949
(D3845	08678)					
678	ULVERSTONIAN	0-6-0DE		Hor		1959

JONATHAN T. SCOTT, DEALER, CLIBURN
Gauge : 1'8".   (NY 585256)

-		4wDM	HE	3595	1948	OOU

L. & P. PEAT LTD, SOLWAY MOSS WORKS, LONGTOWN, near GRETNA
Gauge : 2'6".   (NY 338682)

S30	FANNY	4wDM	MR	21619	1957	
S31		4wDM	MR	26014	1967	OOU
S32		4wDM	MR	9710	1952	
S33		4wDM	MR	40S371	1970	
S34	L2	4wDM	MR	5879	1935	
S35		4wDM	MR	9709	1952	
S36		4wDM	AK	3	1978	OOU
S37		4wDM	AK	6	1981	OOU
	-	4wBE	WR	1393	1939	

MINISTRY OF DEFENCE, ARMY RAILWAY ORGANISATION
Eskmeals Establishment
See Section Five for full details.

Longtown Depot, near Carlisle
See Section Five for full details.

MINISTRY OF DEFENCE, N.A.T.O. AMMUNITION DEPOT, BROUGHTON MOOR, near MARYPORT
Gauge : 4'8½".   (NY 062318)

Name/No.	ND No.	Other	Type	Builder	Works No.	Date	Status
YARD No.9970			4wDM	FH	3995	1963	OOU
	ND 3815		0-4-0DM	HE	2389	1941	
YARD No.54			0-4-0DM	HE	2642	1941	
YARD No.343	ND 3644	56	0-4-0DM	HE	3282	1945	
YARD No.DP35			0-4-0DM	RH	313390	1953	

Gauge : 2'6".   (NY 055320)

Name/No.	ND No.	Other	Type	Builder	Works No.	Date	Status
YARD No.3			0-4-0DM	HE	2011	1939	OOU
No.4			0-4-0DM	HE	2017	1939	OOU
YARD No.25			0-4-0DM	HE	2018	1939	
YARD No.26			0-4-0DM	HE	2019	1939	
YARD No.B1	ND 6441	B1	0-4-0DM	HE	2021	1939	OOU
B7			0-4-0DM	HE	2022	1939	
13			0-4-0DM	HE	2242	1941	Dsm
4	ND 10394		0-4-0DM	HE	2243	1941	
YARD No.B3			0-4-0DM	HE	2263	1940	
No.27			0-4-0DM	HE	2264	1940	
YARD No.B5	ND 3054		0-4-0DM	HE	2265	1940	
YARD No.B6	ND 3055		0-4-0DM	HE	2266	1940	
YARD No.B7	ND 3056		0-4-0DM	HE	2267	1940	
YARD No.B10	ND 3059		0-4-0DM	HE	2270	1940	
YARD No.B19	ND 3060		0-4-0DM	HE	2398	1941	
YARD No.B20	ND 3061		0-4-0DM	HE	2399	1941	
YARD No.B21	ND 3062		0-4-0DM	HE	2400	1941	
No.23			0-4-0DM	HE	2402	1941	
YARD No.70		R4	4wDM	RH	221623	1942	
YARD No.72		R6	4wDM	RH	221626	1942	
YARD No.73	ND 6455	R7	4wDM	RH	221625	1942	
YARD No.83		R9	4wDM	RH	235727	1943	
YARD No.84		R10	4wDM	RH	235728	1943	
YARD No.86		R12	4wDM	RH	235730	1943	
YARD No.103		R8	4wDM	RH	242918	1947	

NEW COLEDALE MINING CO, FORCE CRAG BARYTES MINE, BRAITHWAITE, near KESWICK
Gauge : 2'0".   (NY 200216,202218)

Name/No.	Type	Builder	Works No.	Date
-	4wBE	WR	3557	1946
-	4wBE	WR	6092	1958

NUCLEAR ELECTRIC, WINDSCALE FACTORY, SELLAFIELD
(A Division of C.E.G.B.)
Gauge : 4'8½".   (NY 025034)

Name/No.	Other	Type		Builder	Works No.	Date
1	4300/B/0001	0-4-0ST	OC	P	2027	1942
B.N.F.L.1	PB 1056.860	0-4-0DH		HE	7426	1982
B.N.F.L.2	PB 1056.860	0-4-0DH		HE	7427	1982
B.N.F.L.4	PB 1056.860	0-6-0DH		HE	9000	1983
		Rebuilt		HE		1987
B.N.F.L.3	PB 1056.860	0-4-0DH		HE	9200	1983
B.N.F.L.5		0-6-0DH		S	10111	1963
		Rebuilt		TH		1987

SAMUEL MORGAN SERVICES LTD, HINDPOOL ROAD, BARROW-IN-FURNESS
(SD 192695)   Locos here occasionally for scrap or resale.

SCOTTISH AGRICULTURAL INDUSTRIES LTD, BOLTON FELL MILL, near HETHERSGILL
Gauge : 2'0".   (NY 487699)

	-	4wDM	L	37366	1951	
	-	4wDM	LB	52726	1961	
	-	4wDM	LB	55730	1968	
	-	4wDM	MR	7037	1936	
	-	4wDM	MR	7188	1937	
24		4wDM	MR	7498	1940	
	-	4wDM	MR	8638	1941	Dsm
No.1		4wDM	MR	8655	1941	
	-	4wDM	MR	8696	1941	
	-	4wDM	MR	8825	1943	
	-	4wDM	MR	20058	1949	

VICKERS SHIPBUILDING & ENGINEERING LTD,
BARROW SHIPBUILDING WORKS & BARROW ENGINEERING WORKS, BARROW-IN-FURNESS
(Member Company of British Shipbuilders)
Gauge : 4'8½".   (SD 192686)

6331	PLUTO	0-4-0DM	HC	D1089	1958
6614	MERCURY	0-4-0DM	HC	D1217	1960

PRESERVATION SITES

LAKESIDE & HAVERTHWAITE RAILWAY CO LTD, HAVERTHWAITE
Gauge : 4'8½".   (SD 349843)

5643							
5643		0-6-2T	IC	Sdn		1925	
(42073)	2073	2-6-4T	OC	Bton		1950	
(42085)	2085	2-6-4T	OC	Bton		1951	
12	ALEXANDRA	0-4-0ST	OC	AB	929	1902	+
	-	0-6-0F	OC	AB	1550	1917	
1	DAVID	0-4-0ST	OC	AB	2333	1953	
	ASKHAM HALL	0-4-0ST	OC	AE	1772	1917	
11	REPULSE	0-6-0ST	IC	HE	3698	1950	
	CUMBRIA	0-6-0ST	IC	HE	3794	1953	
	PRINCESS	0-6-0ST	OC	WB	2682	1942	
(D2072)	03072	0-6-0DM		Don		1959	
(D2117)	No.8	0-6-0DM		Sdn		1959	
7120		0-6-0DE		Derby		1945	
No.2	FLUFF	0-4-0DM		JF	21999	1937	
21		0-4-0DH		JF	4220045	1967	
	RACHEL	4wPM		MR	2098	1924	
	-	0-4-0DH		TH	108C	1961	
		Rebuild of 0-4-0DM		JF	22919	1940	
	-	0-4-0D					Dsm a
52071		2-2w-2w-2DMR		BRCW		1961	
52077		2-2w-2w-2DMR		BRCW		1961	
	L.H.R. P.T.1	2w-2PMR		Wkm	691	1932	
	-	2w-2PMR		Wkm			DsmT

+ Carries plate AB 1054/1906.
a Either 0-4-0DM AB 344/1941 or 0-4-0DH JF 4220044/1967.

NENT VALLEY RAILWAY, HORSE & WAGON INN, NENTSBERRY, near NENTHEAD
Gauge : 1'10".   (NY 764452)

No.1		4wBE S/O	FG Shepherd		c1973
		Rebuilt	S Frogley		1988

RAVENGLASS & ESKDALE RAILWAY CO LTD, RAVENGLASS
Gauge : 3'2¼".   (SD 086967)

	-	4wDM		RH	320573	1951 Dsm

Gauge : 1'3½"

	RIVER MITE	2-8-2	OC	Clarkson	4669	1966
No.3	RIVER IRT	0-8-2	OC	DB	3	1894
		Rebuilt		Ravenglass		1927
	RIVER ESK	2-8-2	OC	DP	21104	1923
11	BONNIE DUNDEE	0-4-2WT	OC	KS	720	1901
		Rebuilt		Ravenglass		1981
No.10	NORTHERN ROCK	2-6-2	OC	Ravenglass	10	1976
	-	4wBE		GB	2782	1957
I.C.L.9	CYRIL	4wDM		L	4404	1932
	-	4wDM		L	40009	1954 DsmT
	QUARRYMAN	4wPM		MH	2	1926
	PERKINS	4w-4DH		Ravenglass		1933
	Rebuild of 4wPM			MH	NG39A	1929
	SILVER JUBILEE	4-4w+4w-4DMR		Ravenglass		1976
I.C.L.8	LADY WAKEFIELD	4w-4wDM		Ravenglass		1980
	SHELAGH OF ESKDALE	4-6-4DH		SL		1969

THE ESKDALE (CUMBRIA) TRUST, RAILWAY MUSEUM, RAVENGLASS
Gauge : 1'3".   (SD 086967)

	SYNOLDA	4-4-2	OC	BL	30	1912
I.C.L.No.1	BUNNY	4w-4wPM	reb	Ravenglass		1925 +
	SCOOTER 1	2-2wPMR		Ravenglass		1970

+ Used as a tool wagon.

Also here are the side frames of ELLA 4-6-2 OC DB 2/189x,
            and the side frames of KATIE 0-4-0T OC DB 4/1896.

THE LAKELAND MINES & QUARRIES TRUST, MINING & QUARRYING MUSEUM,
THRELKELD QUARRY, near KESWICK
Gauge : 2'6".   (   )

-		0-4-0DM		HE	2248	1940	

Gauge : 2'0".

-		4wDM		RH	217993	1943	Dsm
-		4wDM		RH	223744	1944	Dsm

SOUTH TYNEDALE RAILWAY PRESERVATION SOCIETY, ALSTON STATION
Gauge : 2'6".   (NY 717467)

ND 10363	MAINT.No.17.200 722	4wDM		HE	6646	1967

Gauge : 2'0".

No.10	NAKLO	0-6-0WTT	OC	CHR	3459	1957	
No.3	SAO DOMINGOS	0-6-0WT	OC	OK	11784	c1925	
No.6	CHAKA'S KRAAL	0-4-2T	OC	HE	2075	1940	
6	THOMAS EDMONDSON	0-4-0WTT	OC	Hen	16047	1918	
No.1	THE PHOENIX	4wDM		FH	2325	1941	
No.4	NAWORTH	0-6-0DM		HC	DM819	1952	
2	AYLE	4wDM		HE	2607	1942	
No.9	2103/50	0-4-0DM		HE	4109	1952	
	-	0-4-0DM		HE	4110	1953	
	-	0-4-0DM		HE	5222	1958	
20		4wDM		MR	5880	1935	Dsm +
	-	4wDM		RH	222101	1943	
P.W. No.1	DB 965082	2w-2DMR		Wkm	7597	1957	

+   Converted into a brake-van.

DERBYSHIRE

INDUSTRIAL SITES

BALFOUR BEATTY RAILWAY ENGINEERING,
MIDLAND FOUNDRY, OSMASTON STREET, SANDIACRE
Gauge : 4'8½".   (SK 482362)

-		4wDM	FH	3884	1958

BIWATER PIPES & CASTINGS, THE CLAY CROSS CO LTD, CLAY CROSS, CHESTERFIELD
Gauge : 3'0½".   (SK 398644, 401644)

-		2w-2WE	Clay Cross	c1969
-		2w-2WE	Clay Cross	c1971
26004		2w-2WE	Clay Cross	c1973

Gauge : 2'10½".

-		2w-2CE	Clay Cross		c1975
27408		2w-2CE	Clay Cross		1977

Gauge : 2'0".

-		4wDM	L	41803	1955
-		4wDM	LB	54684	1965

BLUE CIRCLE INDUSTRIES LTD, HOPE CEMENT WORKS, HOPE
Gauge : 4'8½".   (SK 167823)

	BLUE JOHN	B-BDH	HAB	773	1990
PM 99/01	PEVERIL	0-6-0DH	S	10087	1963
		Rebuilt	AB	6140	1989
	-	0-6-0DH	S	10107	1963
	DERWENT	0-6-0DH	S	10156	1963
		Rebuilt	AB	6004	1988

BRITISH COAL
See Section Four for full details.

BRITISH RAIL ENGINEERING (1988) LTD
Derby Carriage Works, Litchurch Lane, Derby
Gauge : 4'8½".   (SK 364345)

(D4014 08846)	D4414	0-6-0DE	Hor		1961
1656		0-4-0WE	Derby C&W		1913 OOU
No.4		0-4-0WE	Derby C&W		(1913?) OOU
8		0-4-0WE	Derby C&W		1935
3335		0-4-0WE	Derby C&W		1935
(4953)		0-4-0WE	Derby C&W		1956

Derby Locomotive Works, Derby
Gauge : 4'8½".   (SK 364353)

480		2w-2BE	BRE(D)		c1980
No.600	463	2w-2BE	Derby C&W		c1960
602		2w-2BE	Derby C&W		c1960 OOU
1710	(DRT 81110)	4w-2DM	THub	DM1490	1940 +

+ Converted from a crane at Derby C & W c1988.

BUTTERLEY ENGINEERING LTD, ENGINEERING WORKS, RIPLEY
Gauge : 4'8½".   (SK 405518)

(D2858)		0-4-0DH	YE	2817	1960
	TEUCER	0-4-0DM	_(VF	D294	1955
			‾(DC	2567	1955

CHESTERFIELD CYLINDERS LTD, CHESTERFIELD
Gauge : 4'8½".   (SK 383702)

850		4wDH	TH	166V	1966

COALITE FUELS & CHEMICALS LTD, BOLSOVER COALITE WORKS,
BUTTERMILK LANE, BOLSOVER   (Part of Coalite Group Plc)
Gauge : 4'8½".   (SK 457715)

7		0-6-0DH	RR	10279	1968
8		0-6-0DH	RR	10291	1970
9		0-6-0DH	TH	237V	1971
	-	0-6-0DH	TH	238V	1971
11		0-6-0DH	YE	2839	1962

COURTAULDS ACETATE LTD
Spondon Power Station, Thulston Road, Borrowash, Derby
Gauge : 4'8½".   (SK 405344)

	-	0-4-0DH	RH	518190	1965
1		4wDH	RR	10251	1966

Spondon Works
Gauge : 4'8½".   (SK 405347)

DL3		4wDH	RR	10280	1968
DL2		4wDH	S	10177	1964

DEPARTMENT OF THE ENVIRONMENT, PROPERTY SERVICES AGENCY,
HEALTH & SAFETY EXECUTIVE, RESEARCH & LABORATORY, SERVICES DIVISION,
HARPUR HILL, BUXTON
Gauge : 3'0".   (SK 057706)

BB 307		2w-2DE	GB	6099	1964

FLOYD OIL & GAS (UK) LTD, DOE LEA COLLIERY, DOE LEA, near CHESTERFIELD
Gauge : 2'0".   (SK 454667)   (Underground)

-		4wDM	RH	252809	1948	OOU
-		4wDM	RH	296047	1950	OOU

GARDNER MACHINERY & METALS, SCRAP MERCHANTS,
DOVE HOLES STATION, near BUXTON
Gauge : 2'0".   (SK 074780)

-		4wDM	RH	244487	1946	OOU

I.C.I. CHEMICALS & POLYMERS LTD, TUNSTEAD LIMEWORKS & QUARRIES, TUNSTEAD
(Including Lime Group Workshops & Stores)
Gauge : 4'8½".   (SK 101743, 097755)

SCW/1/29	DOVEDALE	0-6-0DH	RR	10284	1969
		Rebuilt	TH		1974
SCW/1/10	NIDDERDALE	0-6-0DH	S	10149	1964
		Rebuilt	TH		1987
SCW/1/04	CHEEDALE	4wDH	TH	284V	1979
SCW/1/05	HARRY TOWNLEY	4wDH	TH	289V	1980

LAPORTE INDUSTRIES LTD, SALLET HOLE MINE, STONEY MIDDLETON
Gauge : 2'0".   (SK 223743)   RTC.

3		4wBE	CE	5180	1966 OOU
7		4wBE	CE	B1560	1977 OOU
9		4wBE	WR	H7205	1968 OOU
10		4wBE	CE	5950	1972 OOU

MARKHAM & CO LTD, BROADOAKS WORKS, CHESTERFIELD
Gauge : 4'8½".   (SK 389710)

—	4wDM	RH	476141	1963

McDONALD ENGINEERING LTD, STORES ROAD, DERBY
Gauge : 3'0".   (SK 359374)

2	812	4wBE	GB	1729	1941

NATIONAL POWER, WILLINGTON POWER STATION, near BURTON-ON-TRENT
(A Division of C.E.G.B.)
Gauge : 4'8½".   (SK 306292)

1		0-6-0DM	(RSHN	7859	1956 OOU
			(DC	2573	1956
2		0-6-0DM	(RSHN	7860	1956 OOU
			(DC	2574	1956

N.E.I. BECORIT LTD, HALLAM FIELDS ROAD, ILKESTON
New BGB locos under construction, and locos for repair, usually present.

N.E.I. MINING LTD, CLAYTON WORKS, HATTON
(SK 212298)
New CE locos under construction, and locos for repair, usually present.

POWER GEN, DRAKELOW POWER STATION, near BURTON-ON-TRENT
(A Division of C.E.G.B.)
Gauge : 4'8½".   (SK 235200, 241198)

No.5		0-6-0DH	AB	484	1963	OOU
No.6		0-6-0DH	AB	485	1963	
No.7		0-6-0DH	HE	7541	1976	

RAYNESWAY PLANT LTD, PLANT DEPOT, RAYNESWAY, DERBY
Gauge : 2'0".   (SK 384352)

17/2/9490	4wBE	SIG	705-709	1975
PLANT 8203	4wBE	WR	7296	1975
PLANT 8204	4wBE	WR	7297	1975

Locos present in yard between contracts.

R.M.C. INDUSTRIAL MINERALS LTD. HOLDERNESS LIMEWORKS, PEAK DALE
Gauge : 4'8½".   (SK 088773)

-		0-6-0DH	S	10108	1963	+
-		0-6-0DH	S	10186	1964	
	Rebuilt		AB	6459	1989	

+   Loco carries plate "UNIT No.10216".

SHAWS METALS LTD, HAYDOCK PARK ROAD, ASCOT DRIVE, DERBY
Gauge : 4'8½".   (SK 373336)

34101	HARTLAND	4-6-2	3C	Elh		1950	Pvd

STANTON PLC, STANTON WORKS, ILKESTON
Gauge : 4'8½".   (SK 468387, 469389, 471390, 476388)

52	689/162		0-6-0DH	S	10139	1962
53	689/163		0-6-0DH	S	10140	1962
60	689/170	LBH 47	0-4-0DH	RR	10253	1967
61	689/171	LBH 44				
	GRAHAM		0-4-0DH	RR	10207	1965
62	689/172		0-4-0DH	TH	227V	1970
63	689/173		0-4-0DH	TH	228V	1970

STEETLEY MINERALS LTD, MINERALS DIVISION, DOWLOW WORKS,
STERNDALE MOOR, BUXTON
Gauge : 4'8½".   (SK 102678)

	-	0-4-0DH	RH	418793	1957	OOU
QMP 154	HEATHCOTE	0-4-0DE	RH	461959	1961	

STEETLEY QUARRY PRODUCTS LTD, CHADDESDEN QUARRY, DERBY
Gauge : 4'8½".   (SK 378358)

		0-6-0DE	YE	2895	1964 OOU
-					

TARMAC ROADSTONE HOLDINGS LTD, BUXTON QUARRY
Gauge : 4'8½".   (SK 083694)   RTC.

550/6/0357		4wDM	RH	252842	1948 OOU

TARMAC ROADSTONE (NORTHERN) LTD
Middle Peak Quarry, Wirksworth
Gauge : 4'8½".   (SK 287548)

-		0-4-0DM	Bg	3357	1952
	DERBYSHIRE STONE	4wDM	FH	1891	1934 OOU
-		6wDH	RR	10274	1968

Topley Pike Quarry, Buxton
Gauge : 4'8½".   (SK 102724)

-		0-4-0DE	RH	424839	1959

W.H. DAVIS & SONS (1984) LTD, LANGWITH JUNCTION
Gauge : 4'8½".   (SK 529683)

23373	ARMY 808	4wDM	RH	224347	1945 OOU
-		4wDM	RH	321732	1952

WILMOT BROS (PLANT SERVICES) LTD, DIESEL ENGINE AND LOCOMOTIVE SPECIALISTS,
MANNERS INDUSTRIAL ESTATE, ILKESTON
Gauge : 4'8½".   (SK 458425)

BADDERSLEY No.3		0-6-0DH	TH	257V	1975
DL 12		0-6-0DH	YE	2911	1963

> Locos under repair occasionally present;
> locos also hired out to various concerns.

PRESERVATION SITES

DERBY CITY COUNCIL, MARKEATON PARK, DERBY
Gauge : 1'3".   (   )

	INVICTA	4wDH	Maxi		1989

DERBY INDUSTRIAL MUSEUM, THE SILK MILL, off FULL STREET, DERBY
Gauge : 4'8½".   (SK 354366)

VICTORY	0-4-0ST	OC	P	1547	1919
-	4wPM		MR	1930	1919

DINTING RAILWAY CENTRE LTD, DINTING, near GLOSSOP
Gauge : 4'8½".   (SK 021946)

JOAN		0-6-0ST	OC	AE	1883	1922
-		0-4-0VBTram		BP	2734	1886
SOUTHWICK		0-4-0CT	OC	RSHN	7069	1942
JAMES		0-4-0DE		RH	431763	1959
"RS 8"		0-4-0DH		SCW		1960
	Rebuild of	0-4-0ST	OC	AE	1913	1923
950021		2w-2PMR		Wkm	590	1932
1307		2w-2PMR		Wkm	(1307	1933?) Dsm

B. EWART, THE HOMESTEAD, YELDERSLEY LANE, BRADLEY, near ASHBOURNE
Gauge : 4'8½".   (SK 223446)

GEORGE	0-4-0ST	OC	RSHN	7214	1945

GULLIVERS KINGDOM LTD, MATLOCK BATH
Gauge : 1'9".   (SK 289578)

HILLARY	4w-4RE	S/O	SE		1988

M.A.G. JACOB, 53 RIBBLESDALE ROAD, SAWLEY, LONG EATON
Gauge : 2'0".   (   )

3916	4wPM		L	3916	1931

LITTLE MILL INN, ROWARTH, STOCKPORT
Gauge : 4'8½".   (SK 011890)

3051 CAR No.289 DERBYSHIRE BELLE	4w-4wRER		MC		1932

MIDLAND RAILWAY CENTRE, (MIDLAND RAILWAY TRUST LTD)
Gauge : 4'8½".   Locomotives are kept at :-

```
            Butterley        (SK 403520)
            Butterley Park   (SK 412519)
            Hammersmith      (SK 397519)


    158A                      2-4-0     IC  Derby               1866
    4247                      2-8-0T    OC  Sdn                 1916
   44027                      0-6-0     IC  Derby               1924
   44932                      4-6-0     OC  Hor                 1945
   46203    PRINCESS MARGARET ROSE
                              4-6-2     4C  Crewe         253   1935
  (47327)                     0-6-0T    IC  NBH         23406   1926
   47357                      0-6-0T    IC  NBQ         23436   1926
                                Rebuilt     Derby               1973
  (47445)                     0-6-0T    IC  HE           1529   1927
  (47564)                     0-6-0T    IC  HE           1580   1928 Dsm
  (49395)   9395              0-8-0     IC  Crewe        5662   1921
   53809    BEAUMONT          2-8-0     OC  RS           3895   1925
   73129                      4-6-0     OC  Derby               1956
   80080                      2-6-4T    OC  Bton                1954
   80098                      2-6-4T    OC  Bton                1954
   92214                      2-10-0    OC  Sdn                 1959
            STANTON No.24     0-4-0CT   OC  AB           1875   1925
  No.2      PN 8292           0-4-0F    OC  AB           2008   1935
            GLADYS            0-4-0ST   OC  Mkm           109   1894
  No.4                        0-4-0ST   OC  NW            454   1894
            KILMERSDON        0-4-0ST   OC  P            1788   1929
  No.1      LYTHAM No.1       0-4-0ST   OC  P            2111   1949
            -                 4wVBT     VCG S            9370   1947
   D4       (44004)
            GREAT GABLE       1Co-Co1DE     Derby               1959
  (D40)     45133             1Co-Co1DE     Derby               1961
   212      40012             1Co-Co1DE     (EE          2668   1959
            AUREOL                          (VF          D429   1959
   D2138                      0-6-0DM       Sdn                 1960
   D7671    (25321)           Bo-BoDE       Derby               1967
  (D9015)   55015             Co-CoDE       (EE          2920   1961
            TULYAR                          (VF          D572   1961
   12077                      0-6-0DE       Derby               1950
  M 59404                     4w-4wWE       MC                  1948
   1311-G                     2-2wDM        Mercury      5337   1927
            ALFRED FIELDS     0-6-0DM       HC          D1114   1958
   77       ANDY              0-4-0DM       JF          16038   1923
   RS 12                      4wDM          MR            460   1918
   RS 9                       4wDM          MR           2024   1920
   D2959                      0-4-0DE       RH         384139   1955
   PWM 3949                   2w-2PMR       Wkm          6934   1955
   DX 68062
  (TR34   DB 965566   TP52P)  2w-2PMR       Wkm          8272   1959
```

Gauge : 3'0".

```
            HANDYMAN          0-4-0ST   OC  HC            573   1900
```

Gauge : 2'6".

```
   1                          4wDM          RH         480678   1961 +
   3                          4wDM          RH         480680   1963
```

        +   Stored at a private location.

Gauge : 2'0".

No.1		4wVBT	G	J.Marshall		1970
E.D. No.1	FURIOUS	4wDM		Dtz	10249	1932
H85		4wDM		HE	7178	1971
	-	4wDM		MR	5906	1934
15		4wDM		MR	11246	1963
C.B. No.1	6111/34	4wDM		MR	60S364	1968
	-	4wPM		Thakeham		1950

PARK HALL LEISURE (DERBYSHIRE) LTD, THE AMERICAN ADVENTURE
SHIPLEY, near ILKESTON
Gauge : 1'3".   (SK 442444)

No.1		2-6-0DH	S/0	SL	RG 11-86	1986
	GENERAL GEORGE A. CUSTER	2-6-0DH	S/0	SL	76.3.88	1988

PEAK DISTRICT MINES HISTORICAL SOCIETY
Cliff Quarry, Crich
Gauge : 2'0".   (SK 342554)

4	4wBE	WR	3492	1946

Peak District Mining Museum, Temple Mine, South Parade, Matlock Bath
Gauge : 1'5".   (SK 292581)

-	4wBE	GB	1445	1936

PEAK RAIL LTD
Gauge : 4'8½".   Locos are kept at :-

        Midland Railway Station, Buxton   (SK 060738)
        Darley Dale Station   (SK 274626)
        Matlock Station   (SK 296603)

48624		2-8-0	OC	Afd		1943	
76084		2-6-0	OC	Hor		1957	a
92219		2-10-0	OC	Sdn		1960	
No.8	ROBERT	0-6-0ST	OC	AE	2068	1933	
No.1	"LUCY"	0-4-0VBT	OC	Cockerill	1625	1890	
BROOKES No.1		0-6-0ST	IC	HE	2387	1941	
W.D.150	WARRINGTON	0-6-0ST	IC	RSH	7136	1944	
		Rebuilt		HE	3892	1969	
No.31	METEOR	0-6-0T	IC	RSH	7609	1950	
No.47	MOORBARROW	0-6-0ST	OC	RSH	7849	1955	
	WILLIAM	4wVBT	VCG	S	9599	1956	
	VULCAN	0-4-0ST	OC	VF	1828	1902	
	THE DUKE	0-6-0ST	IC	WB	2746	1944	
D8	(44008) PENYGHENT	1Co-Co1DE		Derby		1959	
D99	45135						
	3RD CARABINIER	1Co-Co1DE		Crewe		1961	

D100	(45060)				
	SHERWOOD FORESTER 1Co-Co1DE		Crewe		1961
D3429	(08359)	0-6-0DE	Crewe		1958
D5705	TDB 968006	Co-BoDE	MV		1958
	HARLECH CASTLE/CASTELL HARLECH				
(D7615)	25265	Bo-BoDE	Derby		1966
CASTLEFIELD ENGINEER'S E1		0-6-0DM	HC	D1199	1960
	-	4wDM	MR	5765	1959
	-	4wDM	RH	412431	1957
No.6		0-6-0DE	YE	2748	1959
DB 6	TR6	2w-2PMR	Wkm	6901	1954

      a Currently under renovation at P Rollin,
         South Leverton, near Retford, Notts.

STEEPLE GRANGE LIGHT RAILWAY, STEEPLEHOUSE JUNCTION, WIRKSWORTH
Gauge : 1'6".   (SK 288554)

6		4wBE	GB	2493	1946 +
	-	4wBE	GB	6061	1961
	-	4wPM			

      + Currently in store at Tarmac Roadstone Ltd, Middleton Mine.

THE NATIONAL TRAMWAY MUSEUM
Clay Cross Store, Hepthorne Lane, Clay Cross
Gauge : Metre.   (SK 402652)

	-	4wDM	RH	373363	1954

Crich, near Matlock
Gauge : 4'8½".   (SK 345549)

47	JOHN BULL	0-4-0VBTram	BP	2464	1885
	-	4wWE	EE	717	1927
	RUPERT	4wDM	RH	223741	1944 +
G.M.J.		4wDM	RH	326058	1952 a

      + Rebuilt T.M.S. 1964 from 60cm gauge.
      a Rebuilt T.M.S. 1969 from 3'3" gauge.

UNICON MARINE LTD, c/o DOROTHEA RESTORATIONS, MEVRIL SPRING WORKS,
NEW ROAD, and CANAL STREET, WHALEY BRIDGE
Gauge : 4'8½".   (SK 013818 014804)

	-	0-4-0VBT	OC	Cockerill	3083	1924

DEVON

BRITISH COAL, EXETER FORWARD LANDSALE, WEST OF ENGLAND SALES REGION.
Gauge : 4'8½".   (SX 934938)

THOMAS	0-4-0DM	(VF	D98	1949
		(DC	2269	1949

ENGLISH CLAYS LOVERING POCHIN & CO LTD, MARSH MILLS DRYING WORKS, PLYMPTON
Gauge : 4'8½".   (SX 521574)

(D3513	08398)			
402D	ANNABEL	0-6-0DE	Derby	1958

MINISTRY OF DEFENCE, NAVY DEPARTMENT
Devonport Dockyard
Gauge : 4'8½".   (SX 449558)

YARD No.4858	4wDM	FH	3744	1955
YARD No.4860	4wDM	FH	3774	1955
YARD No.5199	4wDM	FH	3775	1955

Royal Naval Armament Depot, Ernesettle
Gauge : 4'8½".   (SX 45x60x)

52 RN 33	4wDM	R/R	Unimog
52 RN 35	4wDM	R/R	Unimog

**THE ESCOMBE GROUP LTD, BETHELL GWYN, SHIPOWNERS,
VICTORIA WHARVES, PLYMOUTH**
(Subsidiary of the P. & O. Group)
Gauge : 4'8½".   (SX 490537)

C.L.S.	4wDM	FH	3281	1948	00U

BICTON WOODLAND RAILWAY, BICTON PARK, EAST BUDLEIGH
Gauge : 1'6".   (SY 074862)

	WOOLWICH	0-4-0T	OC	AE	1748	1916	
	BICTON	4wDM		RH	213839	1942	
B.W.R. 2	CARNEGIE	0-4-4-0DM		HE	4524	1954	
LOD/758235		4wDM		RH	235624	1945	00U
	CLINTON	4wDM		HE	2290	1941	

'BYGONES' VICTORIAN EXHIBITION STREET AND RAILWAY MUSEUM,
FORE STREET, ST MARYCHURCH, TORQUAY
Gauge : 4'8½".   (SX 922658)

No.5   PATRICIA   0-4-0ST   OC HC   1632 1929

COMBE MARTIN WILDLIFE PARK, COMBE MARTIN, near ILFRACOMBE
Gauge : 1'3".   (   )

           -                  2-8-0PH S/O   SL   70.5.87 1987

COUNTRY LIFE MUSEUM, SANDY BAY HOLIDAY CENTRE, near EXMOUTH
Gauge : 1'6".   (ST 035808)

           -                  4-2-2   OC Regent St   1898
        SIR FRANCIS DRAKE   4-6-0   OC DJ. Scarrott   1987

G.W. GLOVER, UPTON PYNE, near EXETER
Gauge : 3'0".   (SX 906984)

           -                  4wDM   JF   3930048 1951

LYNTON & BARNSTAPLE RAILWAY ASSOCIATION,
c/o TANNERS GARAGE, LANDKEY
Gauge : 2'0".   (SS 596314)

        SYBIL MARY   0-4-0ST   OC HE   921 1906 Dsm
           -                  0-6-0T   OC KS   2451 1915
           -                  4wDM   RH   183773 1937

        The frame of 0-4-2ST OC KS 2395/1917 is here in two halves.
        Locos are not on public display.

G.D. MASSEY, 57 SILVER STREET, THORVERTON, EXETER
Gauge : 2'0".   (SS 932018)

     A.R.40   4wBER   Massey   1967

MUSEUM OF DARTMOOR LIFE, OKEHAMPTON
Gauge : 2'6".   (SX 592952)

     WD 767138   2w-2PM   Wkm   3284 1943

NATIONAL TRUST, SALTRAM HOUSE, PLYMPTON
Gauge : 4'6".   (SX 521556)

LEE MOOR No.2   0-4-0ST   OC P   784 1899

PAIGNTON & DARTMOUTH STEAM RAILWAY
Gauge : 4'8½".   Locos are kept at :-

	Churston Park Siding, Paignton			(SX 896564) (SX 889606)			

4555			2-6-2T	OC	Sdn		1924	
4588	WARRIOR		2-6-2T	OC	Sdn		1927	
7827	LYDHAM MANOR		4-6-0	OC	Sdn		1950	
D3014			0-6-0DE		Derby		1952	
D7535	(25185) MERCURY		Bo-BoDE		Derby		1965	
(900391)			2w-2PMR		Wkm	671	1932	Dsm
PWM 2210	4		2w-2PMR		Wkm	4127	1946	DsmT
PWM 3290			2w-2PMR		Wkm	4840	1948	DsmT
PWM 2802			2w-2PMR		Wkm	4980	1948	DsmT
PWM 3957	DART VALLEY No.4 ANGUS		2w-2PMR		Wkm	6942	1955	

Locos are returned to S.D.R. Buckfastleigh for overhaul as required.

PLYM VALLEY RAILWAY ASSOCIATION, MARSH MILLS, PLYMPTON
Gauge : 4'8½".   (SX 520571)

34007	WADEBRIDGE		4-6-2	3C	Bton		1945	
75079			4-6-0	OC	Sdn		1956	
No.3			0-4-0ST	OC	HL	3597	1926	
(D3002)	11	DULCOTE	0-6-0DE		Derby		1952	
	-		4wDH		TH	125V	1963	
	-		2w-2PMR		Wkm			
			Rebuilt		PVRA		1989	
	-		2w-2PMR		Wkm			
			Rebuilt		PVRA		1989	
	-		2w-2PMR		Wkm			DsmT
	-		2w-2PMR		Wkm			DsmT

Gauge : 3'6".

4112	SPRINGBOK		4-8-2+2-8-4T	4C	(BP	7827	1957	
					(NB	27770	1957	

W.L.A. PRYOR, LYNTON RAILWAY STATION, STATION HILL, LYNTON
Gauge : 2'0".   (SS 719488)

No.3	BRUNEL	42	4wDM	RH	179880	1936	

SEATON & DISTRICT ELECTRIC TRAMWAY CO, SEATON
Gauge : 2'9".   (SY 252904

-		4wDM	RH	435398	1959

SOUTH DEVON RAILWAY TRUST, THE PRIMROSE LINE.
Locos are kept at :-

Buckfastleigh   (SX 747663)
Staverton Bridge   (SX 785638)

Gauge : 7'0¼".

| 151 | TINY | | 0-4-0VBWT | VCG | Sara | | 1868 | Pvd |

Gauge : 4'8½".

1369			0-6-0PT	OC	Sdn		1934	
1420	BULLIVER		0-4-2T	IC	Sdn		1933	
1450	ASHBURTON		0-4-2T	IC	Sdn		1935	
1638	DARTINGTON		0-6-0PT	IC	Sdn		1951	
4920	DUMBLETON HALL		4-6-0	OC	Sdn		1929	
5239	GOLIATH		2-8-0T	OC	Sdn		1924	
6435			0-6-0PT	IC	Sdn		1937	
7027	THORNBURY CASTLE		4-6-0	4C	Sdn		1949	
(30587)	3298		2-4-0WT	OC	BP	1412	1874	
	BARBARA		0-6-0ST	IC	HE	2890	1943	
		Rebuilt			HE	3882	1962	
	GLENDOWER		0-6-0ST	IC	HE	3810	1954	
	LADY ANGELA		0-4-0ST	OC	P	1690	1926	
1	ASHLEY		0-4-0ST	OC	P	2031	1942	
	-		0-6-0ST	IC	WB	2766	1944	
D1023	WESTERN FUSILIER		Co-CoDH		Sdn		1963	
D2192	No.2 ARDENT		0-6-0DM		Sdn		1961	
M51592			2-2w-2w-2DHR		Derby C&W		1959	
M51604			2-2w-2w-2DHR		Derby C&W		1959	
	M.F.P. No.4		0-4-0DM		JF	4210141	1958	
	-		2w-2PMR		Wkm	946	1933	DsmT
	-		2w-2PMR		Wkm	4146	1947	
	-		2w-2PMR		Wkm	4149	1947	DsmT
PWM 3773			2w-2PMR		Wkm	6652	1953	
PWM 3944			2w-2PMR		Wkm	6929	1955	
	-		2w-2PMR		Wkm	8198	1958	
	-		2w-2PMR		Wkm	11717	1976	

THE MORWELLHAM RECREATION COMPANY LTD, (DARTINGTON HALL TRUST ENTERPRISE), MORWELLHAM QUAY, near TAVISTOCK
Gauge : 2'0".   (SX 448699)

No.1	S259	GEORGE	4wBE	WR	H7197	1968
No.2	B.E.A. 13	BERTHA	4wBE	WR	6298	1960
No.3	PLANT 5494	CHARLOTTE	4wBE	WR	G7124	1967
No.4		LUDO	4wBE	WR	6769	1964
No.5		WILLIAM	4wBE	WR	C6770	1964
No.6	JM 75	MARY	4wBE	WR	5665	1957
No.7	JM 77	HAREWOOD	4wBE	WR	D6800	1964
	-		0-4-0BE	WR		

TIVERTON MUSEUM SOCIETY, TIVERTON MUSEUM, ST. ANDREW STREET, TIVERTON
Gauge : 4'8½".   (SS 955124)

| 1442 | | | 0-4-2T | IC | Sdn | | 1935 |

DORSET

MINISTRY OF DEFENCE, ARMY RAILWAY ORGANISATION,
LULWORTH RANGES, EAST LULWORTH
        See Section Five for full details.

SYLVASPRINGS WATERCRESS LTD, WATERCRESS GROWERS, DODDINGS FARM, BERE REGIS
Gauge : 1'6".   (SY 847947)

| | WATERCRESS QUEEN | 4wPH | B.J.Fry | | 1948 |

PRESERVATION SITES

AVON CAUSEWAY HOTEL, HURN, near BOURNEMOUTH
Gauge : 4'8½".   (   )

| | - | 0-4-0DM | JF | 22871 | 1939 |

MICKY FINN RAILWAY, near POOLE
Gauge : 2'0".   (   )

| 8 | | 4wDM | L | 28039 | 1945 |
| | - | 4wDM | MR | 26007 | 1964 |

SWANAGE RAILWAY SOCIETY, SWANAGE STATION
Gauge : 4'8½".   (SZ 026789)

6695			0-6-2T	IC	AW	983	1928
35022	HOLLAND-AMERICA LINE		4-6-2	3C	Elh		1948
41708			0-6-0T	IC	Derby		1880
80078			2-6-4T	OC	Bton		1954
No.1704	NUNLOW		0-6-0T	OC	HC	1704	1938
	-		0-4-0T	OC	HE	1684	1931
47160	CUNARDER		0-6-0T	OC	HE	1690	1931
	LINDA		0-6-0ST	OC	HL	3931	1938
45160			2-8-0	OC	NB	24648	1941
MARDY No.1			0-6-0ST	OC	P	2150	1954
D3591	(08476)		0-6-0DE		Crewe		1958
7594	25244		Bo-BoDE		Dar		1964
1012	(S60016		4w-4wDER		Elh		1957
	(S60018		4w-4wDER		Elh		1957
3051	CAR No.88	S288S	4w-4wRER		MC		1932
2054	BERYL		4wPM		FH	2054	1938
7069			0-6-0DM		HL	3841	1935
M.St.C.7	MAY		0-4-0DM		JF	4210132	1957
655/29/38			4wDM		RH	518494	1967
7090			2w-2PMR		Wkm	7090	1955

DURHAM

BLUE CIRCLE INDUSTRIES LTD, WEARDALE CEMENT WORKS, near EASTGATE
Gauge : 4'8½".   (NY 949384)

-	4wDH	RR	10197	1965
ELIZABETH	4wDH	RR	10232	1965

BRITISH COAL
        See Section Four for full details.

CHEMICAL & INSULATING CO LTD, FAVERDALE, DARLINGTON
Gauge : 2'0".   (NZ 272166)

-	4wBE	GB	2848	1957

CLEVELAND BRIDGE & ENGINEERING CO LTD, YARM ROAD, DARLINGTON
Gauge : 4'8½".   (NZ   )

-	4wDH	TH/S	111C	1961

RAISBY QUARRIES LTD, GARMONDSWAY QUARRY, near COXHOE
Gauge : 4'8½".   (NZ 338353)

M 14	4wDH	S	10077	1961

SEAHAM HARBOUR DOCK CO LTD, SEAHAM HARBOUR
Gauge : 4'8½".   (NZ 432493)

D4	0-6-0DH	EEV	D1194	1967
D5	0-6-0DH	EEV	D1195	1967

WEARDALE MINERALS LTD
Cambokeels Mine, Eastgate, Stanhope   (Closed)
Gauge : 60cm.   (NY 934383)

2	4wBE	WR	S7968	1978
4	4wBE	WR	R7964	1977
5	4wBE	WR		(c1936?) OOU
6	4wBE	WR	S7967	1978
7	4wBE	WR		
7-8	4wBE	WR	P7789	1975
8	4wBE	WR	Q7796	1976
8	4wBE	WR	S7966	1978

Frasers Grove Mine Complex, near Groverake
Gauge : 2'0". (NY 889444)

3		4wBE	CE	B0475	1975
3		4wBE	CE	B01502	1977
4	1905/70/06	4wBE	CE	B0495	1975
2		4wBE	WR	7888R	1977
3		4wBE	WR	R7965	1977
-		0-4-0BE	WR		
-		0-4-0BE	WR		

Groverake Mine, Frasers Grove Mine Complex, near Rookhope
Gauge : 2'0". (NY 895441)

1		0-4-0BE	WR	T8009	1979	
1		0-4-0BE	WR	T8013	1979	
2		0-4-0BE	WR	T8005	1979	
3		0-4-0BE	WR	N7549	1973	
4	6	4wBE	CE	B1854	1979	
5		0-4-0BE	WR	T8008	1979	
-		4wBE	CE	B1599A	1978	Dsm
-		4wBE	WR	5299	1955	
-		4wBE	WR	D6805	1964	
-		0-4-0BE	WR	7481	1972	
-		0-4-0BE	WR	M7544	1972	
-		0-4-0BE	WR	N7644	1973	
-		0-4-0BE	WR	7728	1976	
-		0-4-0BE	WR	R7735	1977	
-		0-4-0BE	WR	P7846	1975	
-		0-4-0BE	WR	R7949	1977	
-		0-4-0BE	WR	T8007	1979	
-		0-4-0BE	WR	T8011	1979	
-		0-4-0BE	WR	T8012	1979	
-		0-4-0BE	WR			
-		0-4-0BE	WR			
-		0-4-0BE	WR			
-		0-4-0BE	WR			

WEARDALE STEEL (WOLSINGHAM) LTD, WOLSINGHAM STEELWORKS
Gauge : 4'8½". (NZ 081370) RTC.

-	4wDM	RH	432480	1959

PRESERVATION SITES

DARLINGTON (NORTH ROAD STATION) RAILWAY MUSEUM SOCIETY, DARLINGTON
Gauge : 4'8½". (NZ 289157)

No.1463		2-4-0	IC	Dar		1885
910		2-4-0	IC	Ghd		1875
17		0-4-0VBT	OC	HW	33	1873
25	DERWENT	0-6-0	OC	Kitching		1845
	MET	0-4-0ST	OC	HL	2800	1909
	LOCOMOTION	0-4-0	VC	RS	1	1825
	PATONS	0-4-0F	OC	WB	2898	1948
51203		2-2w-2w-2DMR		MC		1958

DARLINGTON RAILWAY PRESERVATION SOCIETY, STATION ROAD, HOPETOWN, DARLINGTON
Gauge : 4'8½".   (NZ 290157)

78018		2-6-0	OC	Dar		1954
N.G.B.No.1		0-4-0ST	OC	P	2142	1953
No.39	GREAT CENTRAL	0-6-0T	OC	RSH	6947	1938
No.1		4wWE		GEC		(1928?)
185	DAVID PAYNE	0-4-0DM		JF	4110006	1950
	-	0-4-0DM		JF	4200018	1947
	-	4wDM		RH	279591	1949
	DERWENT II	0-4-0DE		RH	312988	1952
2		0-4-0DM		(RSH	7925	1959
				(DC	2592	1959

Gauge : 1'8".

No.1		4wDM	RH	375360	1955
3		4wDM	RH	476124	1962
4	MOSELEY	4wDM	RH	354013	1953

NORTH OF ENGLAND OPEN AIR MUSEUM, BEAMISH HALL
Gauge : 4'8½".   (NZ 217547)

(65033) 876		0-6-0	IC	Ghd		1889
E No.1		2-4-0VBCT	OC	BH	897	1887
	-	0-4-0	VC	Geo Stephenson		1822
L.& H.C. 14		0-4-0ST	OC	HL	3056	1914
	-	0-4-0VBT	VC	HW		1871
		Rebuilt		Wilton		1984
	LOCOMOTION	0-4-0	VC	Loco Ent	No.1	1975
No.3.	2320/69 TWIZELL	0-6-0T	IC	RS	2730	1891
SOUTH DURHAM MALLEABLE No.5		0-4-0ST	OC	Stockton		1900
	JACOB	0-4-0PM		Bg	680	1916
	-	4wDM		RH	476140	1963

SEDGEFIELD DISTRICT COUNCIL, TIMOTHY HACKWORTH MUSEUM, SOHO WORKS, SHILDON
Gauge : 4'8½".   (NZ 233257)

	SANS PAREIL	0-4-0	VC	BRE(S)	1980
	(BRADYLL)	0-6-0	OC	Hackworth	1835

WHORLTON LIDO, near BARNARD CASTLE
Gauge : 1'3".   (NZ 106146)

No.102	JOHN	4-4-2	OC	A.Barnes	103	1924
	WENDY	4-4wDM		CoSi		1972

ESSEX

INDUSTRIAL SITES

BRITISH FERRIES LTD, HARWICH HARBOUR
Gauge : 4'8½.   (TM 234326)

PQ 364   4wDH R/R   NNM   80505 1980 OOU

BUTTERLEY BUILDING MATERIALS LTD
Gauge 2'0".   Locos are kept at :-

C = Milton Hall Rochford Brickworks, Cherry Orchard Lane,
    Hawkwell, Rochford.   (TQ 859899)
S = Milton Hall Star Lane Brickworks, Star Lane,
    Great Wakering.   (TQ 934873)

    -                 4wDH   AK   No.10   1983 C
    -                 4wDM   AK   40SD530   1987 S
    -                 4wDM   AK   26   1988 C
    -                 4wDM   AK   28   1989
    -                 4wDM   MR   8614   1941 C
    -                 4wDM   MR   21520   1955 S
    -                 4wDH   RH   283513   1949
                   Rebuilt   AK   20R   1986 C
    -                 4wDM   RH   441951   1960 S OOU

CARLESS SOLVENTS LTD, HARWICH REFINERY, REFINERY ROAD, PARKESTON, HARWICH
Gauge : 4'8½".   (TM 232323)

No.35   0-4-0F   OC RSHN   7803 1954 Pvd
    -              4wDH   R/R NNM   73511 1979
    BWC 687F FP 41 CO   4wDM   R/R S&H/Whc   4001

ESSO LUBRICANTS, LUBRICANTS PLANT, LONDON ROAD, PURFLEET
Gauge : 4'8½".   (TQ 563777)

    -              0-6-0DH   HC   D1373 1965

MINISTRY OF DEFENCE, ARMY RAILWAY ORGANISATION, SHOEBURYNESS ESTABLISHMENT
        See Section Five for full details.

MOBIL OIL CO LTD, CORYTON BULK TERMINAL, STANFORD-LE-HOPE
Gauge : 4'8½".   (TQ 746828)

506/1	0-4-0DH	AB	506/1	1969 +
506/2	0-4-0DH	AB	506/2	1969 +
-	0-6-0DH	TH	291V	1980

+ Rebuild of AB 506/1965.

PROCTOR & GAMBLE LTD, SOAP MANUFACTURERS, WEST THURROCK
Gauge : 4'8½".   (TQ 595773)

-	4wDH	TH	144V	1964

PURFLEET DEEP WHARF & STORAGE CO LTD, PURFLEET
Gauge : 4'8½".   (TQ 565776)

1	0-4-0DH	RH	437362	1960	
2	0-4-0DH	RH	457303	1963	
3	0-4-0DH	RH	512463	1965	
4	0-4-0DH	RH	512464	1965	
5	0-4-0DM	(RSH	7922	1957	
		(DC	2589	1957	
6	0-4-0DM	(VF	D297	1956	OOU
		(DC	2583	1956	

SHARPES AUTOS (LONDON) LTD, GABLES SERVICE STATION, RAWRETH, near RAYLEIGH
Gauge : 4'8½".   (TQ 784920)

139	BEATTY	0-4-0ST	OC	HL	3240	1917	Pvd

SHELL U.K., SHELL HAVEN REFINERIES, STANFORD-LE-HOPE
Gauge : 4'8½".   (TQ 729815, 729821, 740817)

No.25	4wDH	TH	279V	1978
No.26	4wDH	TH	280V	1978
No.27	4wDH	TH	281V	1978
No.28	4wDH	TH	282V	1979

THAMES MATEX LTD, OLIVER ROAD, WEST THURROCK
Gauge : 4'8½".   (TQ 578766)

-	0-6-0DH	RSHD/WB	8343	1962

PRESERVATION SITES

BRENT WALKER LEISURE LTD, SOUTHEND PIER RAILWAY, SOUTHEND-ON-SEA
Gauge : 3'0".   (TQ 884850)

A	SIR JOHN BETJEMAN	4w-4wDH	SL		SE4	1986
B	SIR WILLIAM HEYGATE	4w-4wDH	SL		SE4	1986

COLNE VALLEY RAILWAY, near CASTLE HEDINGHAM
Gauge : 4'8½".   (TL 774362)

2199	VICTORY	0-4-0ST	OC	AB	2199	1945
1875	BARRINGTON	0-4-0ST	OC	AE	1875	1921
WD 190	CASTLE HEDINGHAM	0-6-0ST	IC	HE	3790	1952
No.1		0-4-0ST	OC	HL	3715	1928
No.60	JUPITER	0-6-0ST	IC	RSH	7671	1950
No.40		0-6-0T	OC	RSH	7765	1954
72	2235/72	0-6-0ST	IC	VF	5309	1945
D2041		0-6-0DM		Sdn		1959
(D2063)	03063	0-6-0DM		Don		1959
D2184		0-6-0DM		Sdn		1962
E79978		4wDMR		A.C.Cars		1958
3211		0-4-0DM		AB	349	1941
	-	4wDM		FH	3147	1947
	-	4wPM		Lake & Elliot		c1924
	-	4wDM		RH	221639	1943
	-	2w-2PMR		Wkm	1946	1935

GLENDALE FORGE, MONK STREET, near THAXTED
Gauge : 2'0".   (TL 612287)

145	C.P.HUNTINGTON	4w-2-4wPH	S/O	Chance 76-50145-24	1976
	ROCKET	0-2-2+4wPH	S/O	Group 4, B'ham	1970

HARLOW DEVELOPMENT CORPORATION, LOWER MEADOW PLAY CENTRE, off PARINGDON ROAD, SOUTHERN WAY, HARLOW
Gauge : 4'8½".   (TL 452078)

-	4wDM	FH	3596	1953

MANGAPPS FARM RAILWAY MUSEUM, SOUTHMINSTER ROAD, BURNHAM-ON-CROUCH
Gauge : 4'8½".   (TQ 944980)

	DEMELZA	0-6-0ST	OC	WB	3061	1954
D2325		0-6-0DM		(RSH	8184	1961
				(DC	2706	1961
D2399	03399	0-6-0DM		Don		1961
D8233		Bo-BoDE		BTH	1131	1959
ELLAND No.1	AUSTIN WALKER	0-4-0DM		HC	D1153	1959

SAIL and STEAM ENGINEERING LTD, BRIGHTLINGSEA
Gauge : 4'8½".   (   )

4248		2-8-0T	OC	Sdn		1916
92134		2-10-0	OC	Crewe		1957

L.J. SMITH, THE BUNGALOW, RECTORY LANE, BATTLESBRIDGE
Gauge : 2'0".   (TQ 783965)

02	HAYLEY	0-4-0DM	S/O	Bg	3232	1947
	SMUDGE	4wDM		MR	8729	1941

STOUR VALLEY RAILWAY PRESERVATION SOCIETY, EAST ANGLIAN RAILWAY MUSEUM, CHAPPEL & WAKES COLNE STATION
Gauge : 4'8½".   (TL 898289)

69621		0-6-2T	IC	Str		1924	
80151		2-6-4T	OC	Bton		1957	
No.11	STOREFIELD	0-4-0ST	OC	AB	1047	1905	
2350	BELVOIR	0-6-0ST	OC	AB	2350	1954	
68067	GUNBY	0-6-0ST	IC	HE	2413	1941	
	-	0-4-0ST	OC	P	1438	1916	
	JEFFERY	0-4-0ST	OC	P	2039	1943	
54	PENN GREEN	0-6-0ST	IC	RSH	7031	1941	
	JUBILEE	0-4-0ST	OC	WB	2542	1936	
D2279		0-6-0DM		(RSH	8097	1960	
				(DC	2656	1960	
A.M.W.No.144	PAXMAN	0-4-0DM		AB	333	1938	
22624		2w-2-2-2wRER		GRC		1938	
		Rebuilt		GRC		1950	
	-	0-4-0DH		JF	4220039	1965	
	-	4wWE		KS	1269	1912	
	-	4wPM		MR	2029	1920	
	NITWERK RIDGWAY	4wDM		Thwaites			
	-	2w-2PMR		Wkm	1583	1934	DsmT
	(PWM 2830)	2w-2PMR		Wkm	5008	1949	
(TR 37	PWM 2797)	2w-2PMR		Wkm	6896	1954	
	(RT 960232)	2w-2PMR		Wkm			

WICKFORD NARROW GAUGE RAILWAY GROUP, 2A VISTA ROAD, WICKFORD
Gauge : 2'0".   (   )

	DOE 3983	4wDM	FH	3983	1962	
No.29	AYALA	4wDM	MR	7374	1939	a
	-	4wDM	MR	862x		Dsm a

a Loco kept elsewhere.

GLOUCESTERSHIRE

BRITISH WATERWAYS BOARD, SHARPNESS DOCKS
Gauge : 4'8½".   (SO 667023)

01001	REGD. No.DL1	4wDM	RH	463150	1961	
01002	DL2	0-6-0DM	WB	3151	1962	OOU

COSTAIN DOW MAC, NAAS LANE, QUEDGELEY   (Closed)
Gauge : 4'8½".   (SO 821128)

-		0-4-0DE	RH	418602	1958

GLOUCESTER CITY COUNCIL, QUEDGELEY STORE
Gauge : 4'8½".   (   )

4184		4w-4RER	GRC		1924 Pvd

MARCROFT ENGINEERING LTD, REPARCO WORKS, INDIA ROAD, GLOUCESTER
(A Division of CAIB U.K. Ltd).   (Closed)
Gauge : 4'8½".   (SO 845177)

YARD No.4856	4wDM	FH	3737	1955
-	0-4-0DE	YE	2687	1958

MINISTRY OF DEFENCE, ARMY RAILWAY ORGANISATION, ASHCHURCH DEPOT
See Section Five for full details.

INDUSTRIAL SITES

PRESERVATION SITES

DEAN FOREST RAILWAY, NORCHARD STEAM CENTRE, LYDNEY, FOREST OF DEAN
Gauge : 4'8½".   (SO 629044)

4953	PITCHFORD HALL	4-6-0	OC	Sdn		1929
5541		2-6-2T	OC	Sdn		1928
5553		2-6-2T	OC	Sdn		1928
9681		0-6-0PT	IC	Sdn		1949
2		0-4-0ST	OC	AB	2221	1946
1873	JESSIE	0-6-0ST	IC	HE	1873	1937
No.4	WILBERT REV. W. AWDRY	0-6-0ST	IC	HE	3806	1953
63.000.432 FRED WARRIOR		0-6-0ST	IC	HE	3823	1954
USKMOUTH 1		0-4-0ST	OC	P	2147	1952
(D2062) 03062		0-6-0DM		Don		1959
D2119 (03119) LINDA		0-6-0DM		Sdn		1959
D3462 (08377)		0-6-0DE		Dar		1957

No.1 LORD MARSHALL OF GORING		0-4-0DM	AB		392	1954
	-	4wDM	FH		3947	1960
2145	BASIL	0-4-0DM	HE		2145	1940
55	4210101	0-4-0DM	JF		4210101	1955
	DESMOND	0-4-0DM	JF		4210127	1957
39	CABOT	0-6-0DH	RR		10218	1965
3		0-4-0DH	S		10142	1962
2		0-4-0DH	S		10165	1964
DS 3057		4wPMR	Wkm		4254	1947
DB 965065		2w-2PMR	Wkm		7580	1956
9045		2w-2PMR	Wkm		8774	1960

GLOUCESTERSHIRE WARWICKSHIRE RAILWAY SOCIETY, TODDINGTON GOODS YARD
Gauge : 4'8½". (SP 050322)

4277		2-8-0T	OC	Sdn		1920
4936	KINLET HALL	4-6-0	OC	Sdn		1929
5080	DEFIANT	4-6-0	4C	Sdn		1939
7752		0-6-0PT	IC	NB	24040	1930
35006	PENINSULAR & ORIENTAL S.N. CO	4-6-2	3C	Elh		1941
76077		2-6-0	OC	Hor		1956
4	ROBERT NELSON No.4	0-6-0ST	IC	HE	1800	1936
	KING GEORGE	0-6-0ST	IC	HE	2409	1942
	PERCY	0-4-0ST	OC	P	1976	1939
	-	0-6-0ST	OC	WB	2655	1942
D9537		0-6-0DH		Sdn		1965
D9539	51	0-6-0DH		Sdn		1965
(D9553)	8311/34 54	0-6-0DH		Sdn		1965
	-	4wPM		Dowty		+
	-	4wDM		FH	2893	1944
D1		0-6-0DM		HC	D615	1938
21		0-4-0DM		JF	4210130	1957
19		0-6-0DH		JF	4240016	1964
TR 13	PWM 2189	2w-2PMR		Wkm	4166	1948
TR 23	PWM 4313 B52	2w-2PMR		Wkm	7516	1956

+ Converted lorry.

GREAT WESTERN RAILWAY MUSEUM, ROYAL FOREST OF DEAN, COLEFORD RAILWAY YARD, COLEFORD
Gauge : 4'8½". (SO 574107)

	-	0-4-0ST	OC	P	1893	1936

K. HARDY, BROOK HOUSE, BADGEWORTH, near CHELTENHAM
Gauge : 1'3". (SO 903193)

	-	0-4-0ST	OC	K.Hardy	1987
	-	4wBE		K.Hardy	1990

NATIONAL WATERWAYS MUSEUM, GLOUCESTER DOCKS, GLOUCESTER
Gauge : 4'8½".   (   )

-		0-4-0F	OC	AB	2126	1942

NORTH GLOUCESTERSHIRE RAILWAY CO., TODDINGTON GOODS YARD
Gauge : 2'0".   (SP 048318)

1	GEORGE B	0-4-0ST	OC	HE	680	1898 +
1091		0-8-0T	OC	Hen	15968	1918
7	JUSTINE	0-4-0WT	OC	Jung	939	1906
	PETER PAN	0-4-0ST	OC	KS	4256	1922
	ISIBUTU	4-4-0T	OC	WB	2820	1945
	YARD No.A497	4wDM		HE	6647	1967
2	DFK 538	4wDM		L	34523	1949
3	SPITFIRE	4wPM		MR	7053	1937
4		4wDM		RH	354028	1953
6		4wDM		RH	166010	1932
	IVAN	4wPM		FH	3317	1957

+ Currently under renovation elsewhere.

R.J. WASHINGTON
Gauge : 2'0".   (SO   )

-		2w-2PM	R.J.Washington	1980

WINCHCOMBE RAILWAY MUSEUM, 23 GLOUCESTER STREET, WINCHCOMBE,
near CHELTENHAM
Gauge : 2'0".   (   )

AMOS	2w-2BE	A.Fox	c1972

HAMPSHIRE

INDUSTRIAL SITES

A.W. BIGGS & SON, WATERCRESS GROWERS, DISTRICT HILL,
HURSTBOURNE PRIORS, near WHITCHURCH
Gauge : 60cm.   (SU 446462)

-		2w-2PM	R.Hutchings	OOU

D.S. PROCUREMENTS LTD, 4 ALEXANDRA CLOSE, HYTHE, SOUTHAMPTON
Gauge : 4'8½".   (   )

239		0-6-0DH	EES	8423	1963 OOU

EASTLEIGH RAILWAY PRESERVATION SOCIETY,
B.R.(MAINTAINANCE) LTD., EASTLEIGH WORKS
Gauge : 4'8½". (SU 457185)

(30828) E828   4-6-0   OC E1h   1928 Pvd

ENICHEM ELASTOMERS LTD, CHARLESTON ROAD, HARDLEY, HYTHE
Gauge : 4'8½". (SU 442058)

              -              4wDM   RH   416568 1957

ESSO PETROLEUM CO LTD, FAWLEY REFINERY
Gauge : 4'8½". (SU 452046, 453040, 462037)

    553   GREENFINCH   0-6-0DH   HE   7542 1978
    552   BLUEBIRD   0-6-0DH   HE   8998 1981
    641   REDWING   0-6-0DH   HE   8999 1981

J. WELLS & SON, FAIR OAK YARD, DUMPERS GROVE, HORTON HEATH, near EASTLEIGH
Gauge 4'8½". (SU 498170)

    BE1   4wDM   RH   221561 1943 OOU
    BE3   4wDM   RH   221564 1943 OOU
    YARD No. 766   0-4-0DM   RH   414300 1957 OOU
    ARMY 222   0-4-0DM   _(VF   5256 1945 OOU
                                          ‾(DC   2175 1945

MINISTRY OF DEFENCE, ARMY RAILWAY ORGANISATION, MARCHWOOD DEPOT.
          See Section Five for full details.

MINISTRY OF DEFENCE, NAVY DEPARTMENT
Portsmouth Dockyard
Gauge : 4'8½". (SU 642012)   RTC.

YD No.9266   4wDH   R/R S&H   7503 1967 OOU

Royal Naval Armament Depot, Bedenham, Bridgemary
Gauge : 4'8½". (SU 593035)

No.1   YD. No.26653   4wDH   BD   3730 1977
          -                  0-4-0DH   HE   9045 1980
          -                  0-4-0DH   HE   9046 1980
    211   0-4-0DM   _(VF   4860 1942
                                          ‾(DC   2168 1942

          Also one loco kept at Frater (SU 592028)

Royal Naval Armament Depot, Dean Hill
Gauge : 4'8½".   (SU 276266)

No.2	YD No. 26654	4wDH	BD	3731	1977
	-	0-4-0DM	HE	3395	1946

Also uses M.O.D., A.R.O. locos.
See Section Five for full details.

Gauge : 2'6".

P 6495		4wDM	HE	6651	1965
P 6496		4wDM	HE	6652	1965
P 13350		4wDM	HE	6659	1965
P 13351		4wDM	HE	6660	1965
	YARD No. P 26553	4wDM	HE	7495	1977

PRESERVATION SITES

EAST HAYLING LIGHT RAILWAY, HAYLING ISLAND
Gauge : 2'0".   (   )

-	0-4-0DH	S/O	AK	23	1988
-	4wDM		MR	7199	1937
-	4wDM		MR	22070	1960

MR. FARES, BEAULIEU
Gauge : 60cm.   (   )

-	4wBE	WR	D6905	1964	OOU

HAMPSHIRE NARROW GAUGE RAILWAY SOCIETY, DURLEY LIGHT RAILWAY,
"FOUR WINDS", DURLEY, BISHOPS WALTHAM
Gauge : 2'0".   (SU 522173)

	THE FELDBAHN	0-8-0T	OC	Hano	8310	1918
	CLOISTER	0-4-0ST	OC	HE	542	1891
No.2	JOSEPHINE	0-4-2ST	OC	HE	1842	1936
D.L.R.No.1	WENDY	0-4-0ST	OC	WB	2091	1919
	-	4wDM		FH	3787	1956
	AGWI PET	4wPM		MR	4724	1939
D.L.R.No.4	BRAMBRIDGE HALL	4wPM		MR	5226	1930
	-	4wDM		OK	4013	1930
5125		4wDM		OK	5125	1935
	NORDEN	0-4-0DM		OK	20777	1936
	-	0-4-0DM		OK	21160	1938
	-	4wDM		RH	392117	1956

Gauge : 1'8".

-	2w-2-2w-4BEF	GB	6132	1966

HAYDON BAILEY PRIVATE MUSEUM, JAMES WHARF, BELVEDERE ROAD, NORTHAM, SOUTHAMPTON
Gauge : 4'8½".   (SU 432123)

-		0-4-0DM	Bg	3568	1961
-		0-6-0DM	HC	D1253	1962

MARWELL'S WONDERFUL RAILWAY,
MARWELL ZOOLOGICAL PARK, COLDEN COMMON, near WINCHESTER
Gauge : 1'3".   (SU 508216)

PRINCESS ANNE	2-6-0DH	S/O	SL	75.3.87	1987

MID-HANTS RAILWAY PLC, "THE WATERCRESS LINE", ROPLEY STATION
Gauge : 4'8½".   (SU 629324)

(30120)	120	4-4-0	IC	9E	572	1899
30499		4-6-0	OC	Elh		1920
(30506)	506	4-6-0	OC	Elh		1920
31625		2-6-0	OC	Afd		1929
31806		2-6-0	OC	Bton		1926
31874	BRIAN FISK	2-6-0	OC	Woolwich		1925
34016	BODMIN	4-6-2	3C	Bton		1945
34067	TANGMERE	4-6-2	3C	Bton		1947
34105	SWANAGE	4-6-2	3C	Bton		1950
35018	BRITISH INDIA LINE	4-6-2	3C	Elh		1945
73096		4-6-0	OC	Derby		1955
76017	HERMES	2-6-0	OC	Hor		1953
3278	FRANKLYN D. ROOSEVELT					
701		2-8-0	OC	AL	71533	1944
(SLOUGH ESTATES No.3)		0-6-0ST	OC	HC	1544	1924
196	ERROL LONSDALE	0-6-0ST	IC	HE	3796	1953
601	STURDEE	2-10-0	OC	NBH	25438	1943
D3358	(08288)	0-6-0DE		Derby		1957
D5217	(25067)	Bo-BoDE		Don		1963
D5353	(27007)	Bo-BoDE		BRCW	DEL 196	1961
4		0-4-0DM		JF	22889	1939
DX 68026	72109 DB 965491	2w-2DMR		Matisa	D8 012	1972
		Rebuilt		Crewe G.R.		1980
(DS 3317)	DS 3319 "MERCURY"	2w-2PMR		Wkm	6642	1953
5	RLC/009023	2w-2PMR		Wkm	8087	1958
TR22	PWM 4312	2w-2PMR		Wkm	7515	1956

PAULTONS RAILWAY, PAULTONS PARK, OWER, near ROMSEY
Gauge : 1'3".   (SU 316167)

-	2-8-0DH	S/O	SL	RG.11.86	1987

G.W. SMITH, WAYLAND HOUSE, 31 MANOR ROAD SOUTH, WOOLSTON
Gauge : 1'6".   (SU 440113)

G.N.R. No.1	2-2-2	IC	?	c1863

T.W. SMITH, THE BUNGALOW, WHITWORTH CRESCENT, BITTERNE, SOUTHAMPTON
Gauge : 4'8½".   (SU 439140)

-		4wPM	MR	5355	1932

Gauge : 1'6".

-		4-2-2	OC	WB	1425	1893

SOUTHERN COASTCRAFTS LTD, HYTHE PIER RAILWAY
Gauge : 2'0".   (SU 423081)

-		4wRE	BE	16302	1917
-		4wRE	BE	16307	1917

HEREFORD & WORCESTER

INDUSTRIAL SITES

ALAN KEEF LTD, LEA LINE, ROSS-ON-WYE
Gauge : 4'8½".   (SO 665214)

Gauge : Metre.

YARD No. A498		4wDM		HE	6648	1967

Gauge : 3'0".

No.1		0-4-2ST	OC	KS	3024	1916	
	-	4wDH		MR	115U093	1970	

Gauge : 2'6".

10		4wDH		MR	115U094	1970	
19		4wDH		SMH	101T019	1979	

Gauge : 2'0".

	WOTO	0-4-0ST	OC	WB	2133	1924	
	-	4wDM		AK	2	1976	
1863		4w-2-4wDH	S/O	Chance	64-5031-24	1964	
	-	4wDM		MR	5243	1930	Dsm
15		4wDM		MR	5861	1934	+
	-	4wDM		MR	7057	1938	Dsm a
	-	4wDM		MR	7066	1938	Dsm
35		4wDM		MR	8683	1941	Dsm
	-	4wDM		MR	8875	1944	Dsm
4	K11141	4wDM		MR	8877	1944	Dsm
13	DIGGER	4wDM		MR	8882	1944	
No.2	K 11142	4wDM		MR	21282	1957	
No.21		4wDM		MR	21513	1955	
		4wPM		MR		c1920	
	-	4wDM		OK	7595		

-		4wDM	RH	393327	1956	Dsm
-		4wDM	RH	432664	1959	Dsm
-		4wDM	RH			Dsm b

```
+   Converted to a brake van.
a   In use as a hydraulic press.
b   Either 217973/1941 or 213853/1942.
```

Mono Rail.

-	2a-2DH	AK	M002	1989
-	2a-2DH	AK	M003	1989

BIRDS COMMERCIAL METALS LTD, LONG MARSTON DEPOT
Gauge : 4'8½".   (SP 154458)

(D2857)		0-4-0DH	YE	2816	1960
53		4wDM	FH	4016	1964
-		0-4-0DM	(VF	5267	1945
			(DC	2205	1945

Also other locos for scrap occasionally present.

H.P. BULMER LTD, CIDER MANUFACTURERS, MOORFIELDS, HEREFORD
Gauge : 4'8½".   (SO 505402)

5786		0-6-0PT	IC	Sdn		1930	Pvd +
6000	KING GEORGE V	4-6-0	4C	Sdn		1927	Pvd
(46201)	6201						
	PRINCESS ELIZABETH	4-6-2	4C	Crewe	107	1933	Pvd
1579	PECTIN	0-4-0ST	OC	P	1579	1921	Pvd
47	CARNARVON	0-6-0ST	IC	K	5474	1934	Pvd +
(D2578)	2 CIDER QUEEN	0-6-0DM		HE	6999	1968	
		Rebuild of		HE	5460	1958	
12514		0-6-0DM		HC	D1254	1962	
PWM 3767	ADRIAN	2w-2PMR		Wkm	6646	1953	Pvd +

```
+   Property of Worcester Locomotive Society Ltd.
```

MINISTRY OF DEFENCE, ARMY RAILWAY ORGANISATION, MORETON-ON-LUGG DEPOT
See Section Five for full details.

PAINTER BROS LTD, ENGINEERS, MORTIMER ROAD, HEREFORD
Gauge : 2'0".   (SO 508413)

-		4wDM	L	40407	1954
-		4wDM	LB	54181	1964 OOU

WICKHAM RAIL, BUSH BANK, SUCKLEY
(SO 735505) New WkmR railcars usually present.

PRESERVATION SITES

M. DEEM, LAMARO, ECCLES GREEN, NORTON CANON, HEREFORD
Gauge : 1'3".   (SO 387475)

101		2-4wPM	J.Taylor	c1964

DROITWICH CANAL TRUST, NORTH CLAINES
Gauge : 2'0".   (SO 853607)

-		4wDM	MR	7471 1940

R.D. HARRISON, FENCOTE OLD STATION, HATFIELD, near LEOMINSTER
Gauge : 4'8½".   (SO 601589)

TR40	PWM 4314	DB 965565	2w-2PMR	Wkm	7517 1956

Mr. HARRISON, TITLEY JUNCTION STATION, near KINGTON
Gauge : 4'8½".   (   )

	THE SHERIFF	4wDM	RH	458961 1962

HEREFORDSHIRE WATERWORKS MUSEUM TRUST, BROOMY HILL, HEREFORD
Gauge : 2'0".   (SO 497394)

-		4wDM	LB	52886 1962

W. MORRIS, BROMYARD & LINTON LIGHT RAILWAY, BROADBRIDGE HOUSE, BROMYARD
Gauge : 2'6".   (SO 657548)

-		4wDM	Bg	3406 1953

Gauge : 2'0".

	MESOZOIC	0-6-0ST	OC	P	1327	1913
	-	4wVBT		Jaywick Rly		1939 DsmT
	-	4wPM		MR	6031	1936
	-	4wDM		MR	9382	1948
1		4wDM		MR	9676	1952
2		4wDM		MR	9677	1952
No.7		4wDM		MR	20082	1953
		4wDM		MR	102G038	1972
	-	4wDM		RH	187101	1937
	-	4wDM		RH	195849	1939 a
L 10		4wDM		RH	198241	1939
	-	4wDM		RH	213848	1942
No.3	NELL GWYNNE	4wDM		RH	229648	1944
No.6	PRINCESS	4wDM		RH	229655	1944
LM 30		4wDM		RH	229656	1944
	-	4wDM		RH	246793	1947
	-	4wDH		RH	437367	1959
	-	2w-2PM		Wkm	3034	1941

a   In use as a generating unit.

OWENS BROS MOTORS, THE BUNGALOW, ROTHERWAS
Gauge : 1'3".   (SO 542373)

No.303   0-6-0PM S/O   J.Taylor   1967

JOHN QUENTIN
Gauge : 3'6½".   (   )

          SIR TOM   0-4-0ST   OC   WB   2135 1925

Gauge : 2'6".

            -                  4wDM   RH   170374 1934 Dsm

Gauge : 2'0".

            -                  4wDM   LB   52579 1961

REDDITCH STEAM LOCOMOTIVE PRESERVATION SOCIETY, DIXON'S SIDING SITE,
ENFIELD INDUSTRIAL ESTATE, REDDITCH
Gauge : 4'8½".   (SP 039682)

PLANT No.1301   0-4-0DM   JF   4100013 1948

J. SELWAY, BRANSFORD, near POWICK
Gauge : 1'6".   (   )

            -                  4-4-2   OC   Curwen   1951 +

       + Currently in storage in Walsall, West Midlands.

SEVERN VALLEY RAILWAY CO LTD, BEWDLEY & ARLEY STATION'S
Gauge : 4'8½".   (SO 793753, 800764)

       For details of locos see under Shropshire entry.

D. TURNER, "FAIRHAVEN", WYCHBOLD
Gauge : 2'0".   (SO 922660)

            -                  4wDM   MR   8600 1940

THE WOOLHOPE LIGHT RAILWAY, P.J. FORTEY, THE HORNETS NEST,
CHECKLEY, MORDIFORD, near HEREFORD
Gauge : 1'3".   (SO 608378)

   202   TREVOR   0-6-0PM S/O   J.Taylor   c1974

WEST MIDLANDS SAFARI PARK, BEWDLEY
Gauge : 1'3".   (SO 801756, 805755)

278	RIO GRANDE	2-8-0DH	S/O	SL	15.2.79	1979

A.J. WILKINSON, ROWDEN MILL STATION, near BROMYARD
Gauge : 4'8½".   (SO 627565)

D2371	(03371)	0-6-0DM	Sdn		1967
B30W	PWM 3956	2w-2PMR	Wkm	6941	1955

HERTFORDSHIRE

INDUSTRIAL SITES

CENTRAL ELECTRICITY GENERATING BOARD, RYE HOUSE POWER STATION,
ESSEX ROAD, HODDESDON
Gauge : 4'8½".   (TL 386090)

THE BLAENAVON TOTO No.6	0-4-0ST	OC	AB	1619	1919 Pvd

Privately owned locos here for restoration or storage.

McNICHOLAS CONSTRUCTION CO LTD, PLANT DEPOT, LISMIRRANE INDUSTRIAL PARK,
ELSTREE ROAD, ELSTREE
Gauge : 1'6".   (TQ 166952)

-	2w-2BE	Iso	T42	1973
-	2w-2BE	Iso	T51	1974
-	2w-2BE	Iso	T54	1974
-	2w-2BE	Iso		
-	2w-2BE	Iso		
-	2w-2BE	Iso		
ML-2-17	4wBH	Tunnequip		c1982

Locos present in yard between contracts.

WICKHAM RAIL, WARE
(TL 362140) New Wickham railcars usually present.

PRESERVATION SITES

KNEBWORTH HOUSE, KNEBWORTH PARK & WINTERGREEN RAILWAY, KNEBWORTH HOUSE, near STEVENAGE
Gauge : 1'11½". (TL 228208)

-		4wDM		MR	8717	1941 Dsm +
-		4wDM		MR	8995	1946 Dsm +
SIR TOM		4wDM		MR	40S273	1966
HORATIO		4wDM	S/O	RH	217967	1942
MAVIS		4wDM	S/O	RH	7002/0967/6	1967

+ Converted into a brake van.

C. & D. LAWSON, 11 OKELEY LANE, HIGHFIELD ESTATE, TRING
Gauge : 2'6".   (SP 913114)

	-	4wPM	L	34652	1949
No.3		4wDM	RH	297066	1950
No.4		4wDM	RH	402439	1957
No.5		4wDM	RH	432654	1959
No.6		4wDM	RH	224315	1944
No.7	ELLEN	4wDM	RH	200069	1939
No.8		4wDM	RH	244559	1946
	-	2w-2PM	Wkm	3431	1943 Dsm
	-	2w-2PM	Wkm	3578	1944 Dsm

Gauge : 2'1½".

No.1		4wDM	RH	166045	1933
No.2		4wDM	RH	247178	1947

Gauge : 2'0".

-		4wDM	RH	441944	1960

Gauge : 1'8".

-		4wDM	RH	229657	1945

Locos are currently stored elsewhere.

M. SAUL, WENGEO LANE, WARE
Gauge : 4'8½".   (TL 346147)

NEWSTEAD	0-6-0ST	IC	HE	1589	1929 +

+ Currently under renovation elsewhere.

HUMBERSIDE

BLUE CIRCLE INDUSTRIES LTD, CENTRAL WORKS, KIRTON LINDSEY
Gauge : 4'8½". (SE 950012)

	DON ATKINSON	0-4-0DH	RH	525947	1968

B.P. CHEMICALS INTERNATIONAL LTD, SALT END REFINERY, HULL
Gauge : 4'8½". (TA 165275)

No.1	802	MELVIN	PRN 22396	0-6-0DH	HE	7041	1971
No.2	800	NEVILLE	PRN 30940	0-6-0DH	HE	6971	1968

BRITISH STEEL PLC, GENERAL STEELS DIVISION
Appleby Coke Ovens
Gauge : 4'8½". (SE 917108)

5	4wRE	Schalker	1973
6	4wRE	Schalker	1973
7	4wRE	Schalker	1979 +

+ Built under licence by Starco Engineering,
   Winterton Road, Scunthorpe.

Appleby-Frodingham Works, Scunthorpe
Gauge : 4'8½". (SE 910110, 913109, 915110, 916105)

24		0-6-0ST	IC	HE	2411	1941 a Pvd
1		0-6-0DE		YE	2877	1963
5		0-6-0DE		YE	2909	1963
15		0-6-0DE		YE	2901	1963
16		0-6-0DE		YE	2902	1963
17		0-6-0DE		YE	2788	1960
25		0-6-0DE		YE	2936	1964
26		0-6-0DE		YE	2727	1959 OOU
29		0-6-0DE		YE	2938	1964
30		0-6-0DE		YE	2943	1965
31		0-6-0DE		YE	2903	1963
34		0-6-0DE		YE	2876	1963
39		0-6-0DE		YE	2790	1960 OOU
40		0-6-0DE		YE	2764	1959 OOU
42		0-6-0DE		YE	2766	1960 OOU
43		0-6-0DE		YE	2767	1960
44		0-6-0DE		YE	2768	1960
45		0-6-0DE		YE	2944	1965
46	HORNET	0-6-0DE		YE	2945	1965
47		0-6-0DE		RR	10236	1967
50		0-6-0DE		RR	10238	1967
No.51		0-6-0DE		YE	2709	1959 OOU
52	DE 2	0-6-0DE		YE	2773	1959 OOU
53		0-6-0DE		YE	2793	1961

54	DE 4	0-6-0DE	YE	2908	1963	
55		0-6-0DE	YE	2690	1959	
70		Bo-BoDE	HE	7281	1972	
71		Bo-BoDE	HE	7282	1972	
72		Bo-BoDE	HE	7283	1972	
73		Bo-BoDE	HE	7284	1972	
74		Bo-BoDE	HE	7285	1972	
75		Bo-BoDE	HE	7286	1972	
76		Bo-BoDE	HE	7287	1973	
77		Bo-BoDE	HE	7288	1973	
78		Bo-BoDE	HE	7289	1973	
79		Bo-BoDE	HE	7290	1973	
80		Bo-BoDE	HE	7474	1977	
0714/69/09		4wDH	Donelli	190/80	1980	
-		0-6-0DE	HE	7473	1976	Dsm +

+ Frame used as a snowplough.
a On loan from Rutland Railway Museum, Leicestershire.

Blast Furnace Highline, Appleby-Frodingham Works, Scunthorpe
Gauge : 4'8½". (   )

HL 1	0449-73-01	0-4-0DE	_(BD	3734	1977	
			‾(GECT	5434	1977	
HL 2	0448-73-02	0-4-0DE	_(BD	3735	1977	
			‾(GECT	5435	1977	
HL 3	0448/73/03	0-4-0DE	_(BD	3736	1977	
			‾(GECT	5436	1977	
HL 4	0448-73-04	0-4-0DE	_(BD	3737	1977	
			‾(GECT	5437	1977	
No.5	0448-73-05	0-4-0DE	_(BD	3738	1977	
			‾(GECT	5438	1977	
HL 6	0448-73-06	0-4-0DE	_(BD	3739	1977	
			‾(GECT	5439	1977	
HL 7	0449-73-07	0-4-0DE	_(BD	3740	1977	
			‾(GECT	5440	1977	

Dawes Lane Coke Ovens, Scunthorpe
Gauge : 4'8½". (SE 921118)

-		4wWE	GB	420383/1	1977
-		4wWE	GB	420383/2	1977

CIBA-GEIGY CHEMICALS LTD, PYEWIPE, GRIMSBY
Gauge : 4'8½". (TA 251115)

3793	COLONEL B	4wDM	HE	5308	1960

CONOCO LTD, HUMBER REFINERY, SOUTH KILLINGHOLME, GRIMSBY
Gauge : 4'8½". (TA 163168)

-		0-4-0DH	HE	6981	1968
M.F.P. No.1		0-4-0DM	JF	4210131	1957 OOU
M.O.P. No.8		0-4-0DM	JF	4210145	1958 OOU
-		0-6-0DH	TH	312V	1984

COURTAULDS LTD, COURTELLE DIVISION, GREAT COATES WORKS, GRIMSBY
Gauge : 4'8½".   (TA 238124)

2		4wDH	RR	10252	1966

CROXTON & GARRY LTD. MELTON DEPOT, MELTON, near HULL   (Closed)
Gauge : 4'8½".   (SE 965258)   RTC.

-		0-6-0DH	JF	4240017	1966
	THE HERBERT TURNER	0-4-0DH	RH	513139	1967

FISONS PLC, HORTICULTURE DIVISION, BRITISH MOSS PEAT WORKS, SWINEFLEET, near GOOLE
Gauge : 3'0".   (SE 719182, 752166, 768168)

02-09	S 11048		4wDM	Fisons		1976
	H 21004		4wDM	HE	7366	1974
	-		4wDM	LB	53976	1964
	-		4wDM	LB	53977	1964
	S 11211	SIMBA	4wDH	RH	432661	1959
			Rebuilt	Swanhaven, Hull		c1986
02-05	S 11042	TANIA	4wDM	RH	432665	1959
02-07	S 11210	SHEEBA	4wDM	RH	466594	1961
	S 11041		4wDM	SMH	40SD507	1978

FLIXBOROUGH WHARF LTD, FLIXBOROUGH
Gauge : 4'8½".   (SE 859147)

37		0-6-0DE	YE	2738	1959
41		0-6-0DE	YE	2765	1959 OOU

GRANT LYON EAGRE LTD
(A Division of British Steel Plc)
B.S.C. Appleby-Frodingham Works, Scunthorpe
Gauge : 4'8½".   (SE 925092)

No.2	714/24	0714/78/06	4wDM	Robel	21.12 RN5	1973
No.3	714/22	0714/78/05	4wDM	Robel	54.12-56-RT1	1966
No.4	714/26	0714/78/07	4wDM	Robel	54.12-56-RW3	1974
5		0714/69/29	4wDM	Robel 54.12-56-AA169		1978

Civil Engineers, Plant Depot, Scotter Road, Scunthorpe
Gauge : 4'8½".   (SE 871114)

-		4wDM		RH	200793	1940
-		4wDM		RH	294269	1951
-		0-4-0DE		RH	425478	1959
-		4wDM		RH		
-		4wDH	R/R	S&H	7510	1967
-		4wDM	R/R	S&H	7512	1972

Locos present in yard between contracts.

HYDRO FERTILISERS LTD, IMMINGHAM DOCKS, IMMINGHAM
Gauge : 4'8½". (TA 200157)

```
         -                   0-6-0DH      RR      10267  1967
                             Rebuilt      RFSK           1990
        20                   4wDH         TH       187V  1967
```

IMMINGHAM STORAGE CO LTD, WEST SIDE, IMMINGHAM DOCK
Gauge : 4'8½". (TA 19x16x)

```
         -                   4wDH    R/R  NNM      83506  1984
```

LINDSEY OIL REFINERY LTD, KILLINGHOLME REFINERY
Gauge : 4'8½". (TA 160176)

```
        BEAVER               0-6-0DH      AB        630  1978
        BADGER               0-6-0DH      AB        658  1980
        SPRINGBOK            4wDH         TH        212V  1969
```

NEW HOLLAND BULK SERVICES, NEW HOLLAND
Gauge : 4'8½". (TA 082243)

```
       166                   0-6-0DH      S       10166  1963
```

SCUNTHORPE SLAG LTD, SANTON WORKS, SCUNTHORPE
Gauge : 4'8½". (SE 923123)

```
       E/1/4                 0-6-0DE      YE       2661  1958 OOU
```

STEPNEY TUNNELLING LTD, PLANT DEPOT, GROVEHILL, BEVERLEY
Gauge : 48cm. (TA 050394)

```
         -                   2w-2BE       Iso        T1  1972
         -                   2w-2BE       Iso        T8  1972
        3                    2w-2BE       Iso        T9  1972
         -                   2w-2BE       Iso       T10  1972
         -                   2w-2BE       Iso       T53  1974
        N2 J/N 16290         2w-2BE       Iso
        L 2                  2w-2BE      (Iso?)
```

        Locos present in yard between contracts.

TIOXIDE (UK) LTD, PYEWIPE WORKS, GRIMSBY
Gauge : 4'8½". (TA 254113)

```
        4      DEREK         0-4-0DM      RH     375713  1954
       (5)                   4wDM         RH     412429  1957 Dsm
        6                    0-4-0DM      RH     414303  1957
        7                    4wDM         RH     421418  1958
```

WANSFORD TROUT FARM, near DRIFFIELD
Gauge : 2'0".   (   )

-		2w-2PM			
-		2w-2PM			
-		2w-2PM			
-		2w-2PM			OOU

WILLIAM BLYTH, FAR INGS TILERIES, BARTON-ON-HUMBER
Gauge : 2'0".   (TA 023233)

IVOR	4wDM	MR	8678	1941	
-	4wDM	RH	223692	1943	OOU

PRESERVATION SITES

CLEETHORPES MINIATURE RAILWAY, MARINE EMBANKMENT, CLEETHORPES
Gauge : 1'2¼".   (TA 321073)

-	2-8-0GasH	S/O SL	7217	1972	
800	2-8-0GasH	S/O SL	15.5.78	1978	

R. DALE, HAWERBY HALL FARM, WOLD NEWTON, near LOUTH
Gauge : 2'0".   (TF 256982)

-	4wDM	MR	9264	1947

HUMBERSIDE LOCOMOTIVE PRESERVATION GROUP, DAIRYCOATES DEPOT, HULL
Gauge : 4'8½".   (TA 068269)

(30777)	777 SIR LAMIEL	4-6-0	OC	NB	23223	1925
34053	SIR KEITH PARK	4-6-2	3C	Bton		1947
42859		2-6-0	OC	Crewe	5981	1930
45163		4-6-0	OC	AW	1204	1935
(45305)	ALDERMAN A.E. DRAPER					
5305		4-6-0	OC	AW	1360	1937
(61306)	1306 MAYFLOWER	4-6-0	OC	NBQ	26207	1948
No.8		0-4-0ST	OC	AB	2369	1955
	ALBERT	4wDM		RH	275882	1950
N.C.B. 11 1963		4wDH		TH/S	134C	1964
17/502		4wDH		TH	265V	1976

KINGSTON-UPON-HULL CORPORATION, TRANSPORT MUSEUM, 36 HIGH STREET, HULL
Gauge : 3'0".   (TA 102284)

1	0-4-0Tram	OC	K	T56	1882

MUSEUM OF ARMY TRANSPORT, FLEMINGATE, BEVERLEY
Gauge : 4'8½".   (TA 041392)

	WOOLMER	0-6-0ST	OC	AE	1572	1910
	GAZELLE	0-4-2WT	IC	Dodman		1893
ARMY 92	WAGGONER	0-6-0ST	IC	HE	3792	1953
	RORKE'S DRIFT	0-4-0DM		⌐(EE	847	1934
				⌐(DC	2047	1934
ARMY 110		4wDM		RH	411319	1958
ARMY 9035		2w-2PMR		Wkm	8195	1958

Gauge : 2'6".

2		4wPM	MR	3849	1927
WD 767139|	2w-2PM	Wkm	3282	1943

Gauge : 60cm.

WD 2182		4wPM	MR	461	1917
No.4	LOD 758228	4wDM	MR	8667	1941
LO 3009	LOD 758028	4wDM	MR	8855	1943
-		4wDM	MR	22209	1964 +

+ Currently under renovation at John Appleton Engineering,
  Leiston, Suffolk.

ISLE OF MAN

INDUSTRIAL SITES

CIVIL AVIATION AUTHORITY, LAXEY
Gauge : 3'6".   (SC 432847)

-	4wDMR	Wkm	7642	1958
-	4wDMR	Wkm	10956	1976

PRESERVATION SITES

J. EDWARDS, BALLAKILLINGAN HOUSE, CHURCHTOWN, near RAMSEY
Gauge : 3'0".   (SC 425945)

No.14	THORNHILL	2-4-0T	OC	BP	2028	1880 Pvd

GROUDLE GLEN RAILWAY LTD, LHEN COAN, GROUDLE GLEN
Gauge : 2'0".   (SC 418786)

	SEA LION	2-4-0T	OC	WB	1484	1896
		Rebuilt		BNFL		1987
No.1	DOLPHIN	4wDM		HE	4394	1952
No.2	WALRUS	4wDM		HE	4395	1952

ISLE OF MAN RAILWAY SOCIETY, CASTLETOWN GOODS YARD
Gauge : 3'0".   (SC 268680)

No.7	(TYNWALD)		2-4-0T	OC	BP		2038	1880	Dsm

ISLE OF MAN RAILWAYS, DEPARTMENT OF TOURISM & TRANSPORT
Isle of Man Steam Railway
Gauge : 3'0".   Locos are kept at :-

Douglas   (SC 374754, 375755)
Port Erin   (SC 198689)

No.4		LOCH	2-4-0T	OC	BP		1416	1874	
No.5		(MONA)	2-4-0T	OC	BP		1417	1874	OOU +
No.6		(PEVERIL)	2-4-0T	OC	BP		1524	1875	OOU
No.8		(FENELLA)	2-4-0T	OC	BP		3610	1894	OOU +
No.9		(DOUGLAS)	2-4-0T	OC	BP		3815	1896	OOU +
No.10		(G.H. WOOD)	2-4-0T	OC	BP		4662	1905	Dsm
No.11		MAITLAND	2-4-0T	OC	BP		4663	1905	
No.12		HUTCHINSON	2-4-0T	OC	BP		5126	1908	
No.13		KISSACK	2-4-0T	OC	BP		5382	1910	
19			0-4-0+4DMR		WkB/Dundalk			1950	
20			0-4-0+4DMR		WkB/Dundalk			1951	
		-	4wDM		MR		22021	1959	OOU
		-	2w-2PMR		Wkm		5763	1950	OOU
		-	2w-2PMR		Wkm		7442	1956	OOU

+   In store for Isle of Man Railway Society.

Manx Electric Railway, Ramsey Electric Railway Museum, Ramsey Station
Gauge : 3'0".   (SC 454943)

	-		4wPM	S/O	FH		2027	1937	
No.23			4w-4wWE		IOMT			1900	Pvd +

+   Now unmotorised.

Port Erin Railway Museum, Strand Road, Port Erin
Gauge : 3'0".   (SC 198689)

No.1		SUTHERLAND	2-4-0T	OC	BP		1253	1873	
		MANNIN	2-4-0T	OC	BP		6296	1926	
No.4	15	CALEDONIA	0-6-0T	OC	D		2178	1885	

Snaefell Mountain Railway, Laxey
Gauge : 3'6".   (SC 432847)

	-		4wPMR		Wkm		5864	1951	OOU

ISLE OF WIGHT

WIGHT LOCOMOTIVE SOCIETY, ISLE OF WIGHT STEAM RAILWAY, HAVEN STREET STATION
Gauge : 4'8½".   (SZ 556898)

(32640)	(W1)	NEWPORT					
	IWC 11		0-6-0T	IC	Bton		1878
(32646)	(W8)	FRESHWATER					
	FYN 2		0-6-0T	IC	Bton		1876
W24	CALBOURNE		0-4-4T	IC	9E	341	1891
38	AJAX		0-6-0T	OC	AB	1605	1918
37	INVINCIBLE		0-4-0ST	OC	HL	3135	1915
D2059	(03059)	EDWARD	0-6-0DM		Don		1959
D2554	(05001)		0-6-0DM		HE	4870	1956
	-		4wDMR		Bg/DC	1647	1927 Dsm
2059	TIGER		0-4-0DH		NB	27415	1954
DS 3320	PWM 3766		2w-2PMR		Wkm	6645	1953

KENT

BLUE CIRCLE INDUSTRIES LTD, SWANSCOMBE WORKS
Gauge : 4'8½".   (TQ 601753)

No.5		4wDH	S	10020	1959

BOWATERS UNITED KINGDOM PAPER CO LTD, SITTINGBOURNE WORKS
Gauge : 4'8½".   (TQ 920667)

(D3763)	08596	0-6-0DE	Derby		1959

B.P. OIL REFINERY LTD.
Bitumen Terminal, Isle of Grain, Rochester
Gauge : 4'8½".   (TQ 864761)

No.24	F-35	0-6-0DH	HE	6950	1967

Grain Refinery, Rochester
Gauge : 4'8½".   (TQ 864758

WIN 4820-0008	MAN OF KENT	0-6-0DH	TH	294V	1981
WIN 4820-0009	KENTISH MAID	0-6-0DH	TH	295V	1981

DEPARTMENT OF THE ENVIRONMENT, LYDD GUN RANGES, LYDD, ROMNEY MARSH
Gauge : 60cm.   (TR 033198)

L 5329	ANN		4wDM	RH	191646	1938	
L.R.No.2	LOD 758366	AD40	4wDM	RH	202000	1940	OOU
RTT/767149			2w-2PM	Wkm	3151	1942	
RTT/767150			2w-2PM	Wkm	3152	c1943	
RTT/767151			2w-2PM	Wkm	3153	c1943	
RTT/767152			2w-2PM	Wkm	3154	c1943	Dsm
RTT/767154			2w-2PM	Wkm	3156	c1943	OOU
RTT/767155			2w-2PM	Wkm	3157	c1943	Dsm
RTT/767156			2w-2PM	Wkm	3158	c1943	
RTT/767157			2w-2PM	Wkm	3159	c1943	Dsm
RTT/767158			2w-2PM	Wkm	3160	c1943	OOU
RTT/767159			2w-2PM	Wkm	3161	c1943	
RTT/767160			2w-2PM	Wkm	3162	c1943	
RTT/767161			2w-2PM	Wkm	3234	1943	Dsm
RTT/767162			2w-2PM	Wkm	3235	1943	
RTT/767163			2w-2PM	Wkm	3236	1943	
RTT/767164			2w-2PM	Wkm	3237	1943	Dsm
RTT/767165			2w-2PM	Wkm	3238	1943	
RTT/767166			2w-2PM	Wkm	3239	1943	Dsm
RTT/767167			2w-2PM	Wkm	3240	1943	OOU
RTT/767169			2w-2PM	Wkm	3242	1943	OOU
RTT/767170			2w-2PM	Wkm	3243	1943	Dsm
RTT/767171			2w-2PM	Wkm	3163	c1943	OOU
RTT/767172			2w-2PM	Wkm	3164	c1943	
RTT/767173			2w-2PM	Wkm	3165	c1943	OOU
RTT/767174			2w-2PM	Wkm	3166	c1943	OOU
RTT/767175			2w-2PM	Wkm	3167	c1943	
RTT/767176			2w-2PM	Wkm	3168	c1943	OOU
RTT/767177			2w-2PM	Wkm	3169	c1943	Dsm
RTT/767178			2w-2PM	Wkm	3170	c1943	
RTT/767179			2w-2PM	Wkm	3171	c1943	Dsm
RTT/767180			2w-2PM	Wkm	3149	1942	OOU
RTT/767181			2w-2PM	Wkm	3150	1942	Dsm
RTT/767182			2w-2PM	Wkm	2522	1938	OOU
RTT/767184			2w-2PM	Wkm	2555	1939	Dsm
RTT/767189			2w-2PM	Wkm	2561	1939	Dsm
RTT/767190			2w-2PM	Wkm	2562	1939	Dsm

Also uses M.O.D., A.R.O. locos; for full details see Section Five.

FOSTER YEOMAN QUARRIES LTD, GRAIN REFINERY SITE, ROCHESTER
Gauge : 4'8½".   (TQ 875743)

(D3817	08650)	55	0-6-0DE	Hor	1959

INDEPENDENT SEA TERMINALS, RIDHAM DOCK
Gauge : 4'8½".   (TQ 918684)

(D3225)	08157	1020	0-6-0DE	Dar	1955

MANSTON LOCOMOTIVE PRESERVATION SOCIETY,
POWERGEN RICHBOROUGH POWER STATION
Gauge : 4'8½". (TR 334620)

```
34070   MANSTON          4-6-2     3C  Bton             1947 Pvd
```

MEDWAY PORTS AUTHORITY, CHATHAM DOCKYARD
Gauge : 4'8½". (TQ 777698)

```
YARD No.12228 ALLINGTON CASTLE 0-4-0DH      HE          6975  1968
```

QUEENBOROUGH ROLLING MILLS LTD, QUEENBOROUGH WHARF SCRAPYARD,
ISLE OF SHEPPEY
Gauge : 4'8½". (TQ 896716, 911716, 912719)

```
(D2027   03027)  18        0-6-0DM    Sdn               1958
 15                        0-4-0DM    Bicester          c1955 00U
         ELEMENTARY        0-6-0DH    EEV       D1229   1967
                  Rebuilt             HE        8900    1977
 6685    BIG JOHN          0-6-0DH    HE        6685    1968
 16                        0-4-0DM    (VF       5257    1945 00U
                                      (DC       2176    1945
```

Also other locos occasionally present for resale,
and used in yards as required.

REDLAND BRICKS LTD
Funton Works, Sheerness Road, Lower Halstow, near Sittingbourne
Gauge : 2'0". (TQ 875677)

```
         -              2w-2BE      Red(F)            1979
```

Otterham, near Rainham
Gauge : 2'0". (TQ 829666)

```
         -              4wBE        Red(F)            c1956
         -              4wBE        Red(W)            c1974
```

REED PAPER & BOARD (U.K.) LTD, EMPIRE PAPER MILLS, GREENHITHE
Gauge : 4'8½". (TQ 595753)

```
         BATMAN         0-4-0DE      RH        512842  1965
```

THE RUGBY GROUP PLC, ROCHESTER WORKS, HALLING
Gauge : 4'8½". (TQ 704650)

```
 16                    6wDH         RR        10275   1969
 15                    4wDH         TH        186V    1967
```

SHEERNESS STEEL CO, SHEERNESS, ISLE OF SHEPPEY
Gauge : 4'8½".   (TQ 912747)

| (D3201) | 08133 | 1 | 0-6-0DE | Derby | 1955 |
| (D3286) | 08216 | 2 | 0-6-0DE | Derby | 1956 |

TRANSMANCHE-LINK, CHANNEL TUNNEL CONTRACTORS, TUNNEL CONSTRUCTION SITE,
SHAKESPEARE CLIFF, near DOVER
Gauge : 900mm.   (TR 296393)

RA1		4wBE/WE		HE	9162	1987
RA2		4wBE/WE		HE	9163	1987
RA3		4wBE/WE		HE	9164	1988
RA4		4wBE/WE		HE	9165	1988
RA5		4wBE/WE		HE	9166	1988
RA6		4wBE/WE		HE	9167	1988
RA7		4wBE/WE		HE	9168	1988
RA8		4wBE/WE		HE	9169	1988
RA9		4wBE/WE		HE	9170	1988
RA10		4wBE/WE		HE	9171	1988
RA11		4wBE/WE		HE	9172	1988
RA12		4wBE/WE		HE	9173	1988
RA13		4wBE/WE		HE	9400	1988
RA14		4wBE/WE		HE	9401	1988
RA15		4wBE/WE		HE	9402	1988
RA16		4wBE/WE		HE	9403	1988
RR1		4wBE/WE	R/A	HE	9158	1987
RR2		4wBE/WE	R/A	HE	9159	1987
RR3		4wBE/WE	R/A	HE	9160	1987
RR4		4wBE/WE	R/A	HE	9161	1987
RR5		4wBE/WE	R/A	HE	9177	1988
RR6		4wBE/WE	R/A	HE	9178	1988
RR7		4wBE/WE	R/A	HE	9179	1988
RR8		4wBE/WE	R/A	HE	9180	1988
RR9		4wDH	R/A	HE	9282	1988
RR10		4wDH	R/A	HE	9283	1988
RR18				HE		
RR19				HE		
RS 101		4wDH		RFS	101L	1989
RS 102		4wDH		RFS	L102	1989
RS 103		4wDH		RFS	L103	1989
RS 104		4wDH		RFS	L104	1989
RS 105		4wDH		RFS	L105	1989
RS 106		4wDH		RFS	L106	1989
RS 107		4wDH		RFS	L107	1990
-		4wDM		Schöema	4415	1980
-		4wDM		Schöema	4418	1981
-		4wDM		Schöema	5000	1989
-		4wDM		Schöema	5001	1989
-		4wDM		Schöema	5002	1989
-		4wDM		Schöema	5004	1989
-		4wDM		Schöema	5005	1989
-		4wDM		Schöema	5006	1989
-		4wDM		Schöema	5007	1989
-		4wDM		Schöema	5008	1989
-		4wDM		Schöema	5009	1989
-		4wDM		Schöema	5010	1989
-		4wDM		Schöema	5011	1989
-		4wDM		Schöema	5012	1989
-		4wDM		Schöema	5013	1989

-		4wDM	Schöema	5014	1989	
-		4wDM	Schöema	5015	1989	
-		4wDM	Schöema	5059	1989	
-		4wDM	Schöema	5060	1989	
-		4wDM	Schöema	5061	1989	
-		4wDM	Schöema	5062	1989	
-		4wDM	Schöema	5063	1989	
-		4wDM	Schöema	5064	1989	
-		4wDM	Schöema	5087	1989	
-		4wDM	Schöema	5088	1989	
-		4wDM	Schöema	5089	1989	
-		4wDM	Schöema	5090	1989	
-		4wDH	SMH	101T018	1979	
-		4wDH	SMH	101T020	1979	

PRESERVATION SITES

A.J.R. BIRCH & SON LTD, HOPE FARM, SELLINDGE, near ASHFORD
Gauge : 4'8½".   (TR 119388)

(31065)	65	0-6-0	IC	Afd		1896 +
34028	EDDYSTONE	4-6-2	3C	Bton		1946
	ST THOMAS	0-6-0ST	OC	AE	1971	1927
41		4wDM		FH	3885	1958
N.C.B. 8 1962		4wDH		TH/S	120C	1962

   + Frame and wheels stored at an unknown location.

BREDGAR & WORMSHILL LIGHT RAILWAY, "THE WARREN", SWANTON STREET, BREDGAR, near SITTINGBOURNE
Gauge : 750mm.   (TQ 873585)

105		0-6-0WT	OC	Hen	29582	1956

Gauge : 2'6".

6		0-4-0T	OC	La Meuse	3355	1929

Gauge : 2'0".

	KATIE	0-6-0WT	OC	Jung	3872	1931
	VOLUNTEER	0-6-0ST	OC	P	2050	1944
	BRONHILDE	0-4-0WT	OC	Sch	9124	1927
	-	0-6-0DH		BD	3768	1983
	-	0-6-0DH		BD	3771	1983
J.G.F. 4		4wDM		L	41545	1955
	-	4wDM		MR	5877	1935
	-	4wDM		MR	8606	1941
	-	4wDM		MR	8704	1942
6	STURRY	4wDM		RH	349061	1953
E1		4wBE		Riordan	T6664	1967

CHATHAM DOCKYARD HISTORIC TRUST, CHATHAM DOCKYARD
Gauge : 4'8½".   (TQ 758689)

YARD No.361 AJAX		0-4-0ST	OC	RSHN	7042	1941
YARD No.562 ROCHESTER CASTLE		4wDM		FH	3738	1935
-		0-4-0DE		YE	2856	1961

DOVER TRANSPORT MUSEUM, DOVER PUMPING STATION, CONNAUGHT ROAD, DOVER
Gauge : 2'0".   (   )

-	4wPM	FH	3116	1946
-	4wDM	MR	8730	1941
-	4wDM	RH		

+ Either 283871/1950 or 444193/1960

EAST KENT LIGHT RAILWAY SOCIETY, SHEPHERDSWELL
Gauge : 4'8½".   (TR 258483)

ST DUNSTAN	0-6-0ST	OC	AE	2004	1927
MINNIE	0-6-0ST	OC	FW	358	1878
-	0-4-0DM		JF	4160002	1952

TED HAMER, near CANTERBURY
Gauge : 2'0".   (   )

8	ELSA	0-6-0WT	OC	OK	7116	1913

HEWITTS FARM, CHELSFIELD, ORPINGTON
Gauge : 2'0".   (TQ 488635)

-	0-4-0DM	S/O	Bg	3024	1939

KENT COUNTY COUNCIL, CANTERBURY HERITAGE CENTRE, STOUR STREET, CANTERBURY
Gauge : 4'8½".   (TQ 146577)

(INVICTA)	0-4-0	OC	RS	24	1830

NORTH DOWNS STEAM RAILWAY CO LTD
Chatham Dockyard
Gauge : 4'8½".   (TQ 758689)

R39	THALIA	0-4-0DM	(RSHN	7816	1954
			+(DC	2503	1954

Stone Lodge Farm Park, Cotton Lane, Stone, near Dartford
Gauge : 4'8½".   (TQ 562745)

No.3	C.E.G.B. No.13						
	NORTH DOWNS		0-6-0T	OC	RSH	7846	1955
No.12	TOPHAM		0-6-0ST	OC	WB	2193	1922
No.6	PRINCESS MARGARET		0-4-0DM		AB	376	1948
249	ESL 118A/ESL 118B	4w-4+4-4wRE			BRCW		1932
		Rebuilt			Acton		1961
No.4			0-4-0DH		JF	4220008	1959
No.2	BURT		4wDM		MR	9019	1950
No.5	CRABTREE		4wDM		RH	338416	1953
No.1	SCOTTIE		4wDM		RH	412427	1957
	PAXMAN		0-4-0DE		RH	412718	1958
No.9	TELEMON	R40	0-4-0DM		(VF	D295	1955
					(DC	2568	1955
No.8	OCTANE		0-4-0DE		YE	2686	1958

ROMNEY HYTHE & DYMCHURCH RAILWAY, NEW ROMNEY
Gauge : 1'3".   (TR 074249)

1	GREEN GODDESS	4-6-2	OC	DP	21499	1925
2	NORTHERN CHIEF	4-6-2	OC	DP	21500	1925
3	SOUTHERN MAID	4-6-2	OC	DP	22070	1926
4	THE BUG	0-4-0T	OC	Krauss	8378	1926
5	HERCULES	4-8-2	OC	DP	22071	1926
7	TYPHOON	4-6-2	OC	DP	22073	1926
8	HURRICANE	4-6-2	OC	DP	22074	1926
10	DOCTOR SYN	4-6-2	OC	YE	2295	1931
No.12	JOHN SOUTHLAND	4w-4wDM		TMA	6143	1983
14		4w-4wDM		TMA	2336	1989
	-	4wDH		MR	7059	1938
PW2	SCOOTER	2w-2PM		RHDR		1949
PW3	REDGAUNTLET	4wPM		AK		1977

SITTINGBOURNE & KEMSLEY LIGHT RAILWAY LTD, SITTINGBOURNE & KEMSLEY
Gauge : 4'8½".   (TQ 905643, 920662)

No.1		0-4-0F	OC	AB	1876	1925	Pvd
No.4		0-4-0ST	OC	HL	3719	1928	Pvd
	BEAR	0-4-0ST	OC	P	614	1896	Pvd

Gauge : 2'6".

	PREMIER	0-4-2ST	OC	KS	886	1905
	LEADER	0-4-2ST	OC	KS	926	1905
	MELIOR	0-4-2ST	OC	KS	4219	1924
	UNIQUE	2-4-0F	OC	WB	2216	1923
	ALPHA	0-6-2T	OC	WB	2472	1932
	TRIUMPH	0-6-2T	OC	WB	2511	1934
	SUPERB	0-6-2T	OC	WB	2624	1940
	VICTOR	4wDM		HE	4182	1953
	EDWARD LLOYD	0-4-0DM		RH	435403	1961

<u>SOUTH EASTERN STEAM CENTRE</u> (Closed)
Gauge : 4'8½".

| 4902 | S 13003 S | | | 4w-4wRER | Elh | | 1949 + |

+ Currently stored elsewhere.

<u>TENTERDEN RAILWAY CO LTD, (KENT & EAST SUSSEX RAILWAY)</u>
Gauge : 4'8½".   Locos are kept at :-

Rolvenden Station	(TQ 865328)
Tenterden Station	(TQ 882336)
Wittersham Road Station	(TQ 866288)

30065	MAUNSELL			0-6-0T	OC	VIW	4441	1943
(30070)	WAINWRIGHT	21		0-6-0T	OC	VIW	4433	1943
(31556)	1556							
	PRIDE OF SUSSEX			0-6-0T	IC	Afd		1909
(32650)	SUTTON	No.10		0-6-0T	IC	Bton		1876
32670	BODIAM	No.3		0-6-0T	IC	Bton		1872
32678	8			0-6-0T	IC	Bton		1880
No.15	HASTINGS			0-6-0ST	IC	HE	469	1888
No.26	LINDA			0-6-0ST	IC	HE	3781	1952
No.23	WD191	HOLMAN F. STEPHENS		0-6-0ST	IC	HE	3791	1952
No.25	NORTHIAM			0-6-0ST	IC	HE	3797	1953
No.24	WILLIAM H. AUSTEN			0-6-0ST	IC	HE	3800	1953
14	CHARWELTON			0-6-0ST	IC	MW	1955	1917
No.19	376			2-6-0	OC	Nohab	1163	1919
No.12	MARCIA			0-4-0T	OC	P	1631	1923
No.27	ROLVENDEN			0-6-0ST	IC	RSH	7086	1943
No.10	GERVASE			0-4-0VBT	VCG	S	6807	1928
		Rebuild of		0-4-0ST		MW	1472	1900 a
(W20W)	20			4w-4wDMR		AEC		1940
D2023	FAITH	46	5	0-6-0DM		Sdn		1958
(D2024)		47	4	0-6-0DM		Sdn		1958
(D9504)	506			0-6-0DH		Sdn		1964
(D9525)				0-6-0DH		Sdn		1965
	-			Bo-BoDE		BTH		1932
42				0-6-0DM		HE	4208	1948
No.1				0-4-0DE		RH	423661	1958
	-			2w-2PM		NLP		
	TITAN			0-4-0DM		(VF	D140	1951
						(DC	2274	1951
900312	YORK	21		2w-2PMR		Wkm		1931
(900923)				2w-2PMR		Wkm	400	1931 Dsm
	-			2w-2PMR		Wkm	473	1931 DsmT
(900393)				2w-2PMR		Wkm	673	1932 DsmT
TR 1	6872			2w-2PMR		Wkm	6872	1954
DB 965042				2w-2PMR		Wkm	6950	1955 +
9043				2w-2PMR		Wkm	6965	1955
7438				2w-2PMR		Wkm	7438	1956

+ Currently in storage at D.Wickham & Co, Ware, Herts.
a Carries plate S 6710.

TUNBRIDGE WELLS & ERIDGE RAILWAY PRESERVATION SOCIETY LTD,
ERIDGE STATION, ERIDGE ROAD, ERIDGE GREEN, TUNBRIDGE WELLS
Gauge : 4'8½".   (   )

		0-4-0DM		(RSHN	7924	1959
-				(DC	2591	1959

W.M.G. HEDGES & SON, STEAM ENGINEERS, CHATHAM DOCKYARD
Gauge : 4'8½".   (TQ 758689)

-	0-4-0ST	OC	AB	2352	1954

Works with steam locos occasionally present for restoration.

WOODLANDS RAILWAY
Gauge : 1'3".   (   )

PAM	4wPM		C.Mace	1982
SIMON	4wPM	S/O	C.Mace	1985
PERCY	4wPM	S/O	C.Mace	1989

LANCASHIRE

<u>INDUSTRIAL SITES</u>

BLACKPOOL TRANSPORT SERVICES LTD, BLUNDELL STREET DEPOT & WORKS, BLACKPOOL
Gauge : 4'8½".   (SD 307350)

441	D801 CTP	4wDM	R/R	Bruff		1986
440	YFV 577Y	4wDM	R/R	Unimog	12/983	1982

BOROUGH OF PRESTON, PRESTON DOCKS
Gauge : 4'8½".   (SD 504295)

ENERGY	4wDH	RR	10281	1968
ENTERPRISE	4wDH	RR	10282	1968
PROGRESS	4wDH	RR	10283	1968

BRITISH FUEL CO, BLACKBURN COAL CONCENTRATION DEPOT, WHALLEY BANKS
Gauge : 4'8½".   (SD 677275)

D2272	ALFIE	0-6-0DM	(RSH	7914	1958
			(DC	2616	1958
2588		0-4-0DM	(RSH	7921	1957
			(DC	2588	1957

CASTLE CEMENT (RIBBLESDALE) LTD, CLITHEROE WORKS, WEST BRADFORD ROAD,
CLITHEROE
Gauge : 4'8½".   (SD 749434)

11	-	0-6-0DH	EEV	D1137	1966
No.10		0-6-0DH	GECT	5396	1975
No.9		0-6-0DH	GECT	5401	1975

CENTRILINE LTD, UNIT 481, WALTON SUMMIT EMPLOYMENT CENTRE,
RANGLET ROAD, BAMBER BRIDGE, PRESTON
Gauge : 1'6".   (   )

OC LC 05	4wBH	Tunnequip
OC LC 06	4wBH	Tunnequip
OC LC 07	4wBH	Tunnequip

IMPERIAL CHEMICAL INDUSTRIES LTD, MOND DIVISION, HILLHOUSE WORKS, THORNTON
Gauge : 4'8½".   (SD 345435)

1	2w-2DH	R/R	TH	V326	1988
2	2w-2DH	R/R	TH	V327	1988

LANFINA BITUMEN LTD, PRESTON WORKS, DOCK ESTATE, PRESTON
Gauge : 4'8½".   (SD 508298)

-	4wDM	FH	3906	1959

LEYLAND DAF LTD, CENTURION WAY, LEYLAND
Gauge : 4'8½".   (SD 544237)

5	0-4-0DM	JF	4210108	1955

MINISTRY OF DEFENCE, ROYAL ORDNANCE FACTORY, CHORLEY
Gauge : 4'8½".   (SD 562203)

18238	R.O.F. CHORLEY No.3	0-4-0DH	JF	4220021	1962
18242	R.O.F. CHORLEY No.4	0-4-0DH	JF	4220022	1962

NATIONAL POWER, PADIHAM POWER STATION, BURNLEY
(A Division of C.E.G.B.)
Gauge : 4'8½".   (SD 783334)

No.1	0-4-0DH	AB	473	1961	
No.2	0-4-0DH	AB	474	1961	00U
No.2	Bo-BoWE	HL	3872	1936	00U
-	0-4-0DH	JF	4210001	1949	00U

NUCLEAR ELECTRIC,
Heysham Power Station
(A Division of C.E.G.B.)
Gauge : 4'8½".   (SD 401599)

No.1	LANCASTER	0-6-0F	OC	AB	1572	1917	
No.2		0-4-0F	OC	AB	1950	1928	OOU
	DOUG TOTTMAN -	Bo-BoBE		RSHN	7284	1945	
		Rebuilt		Kearsley			
				Power Station		1982	

Uranium Fuel Centre, Springfields Factory, Salwick, near Preston
(A Division of C.E.G.B.)
Gauge : 4'8½".   (SD 468317)

4301 A		0-4-0DM	HC	D628	1943
4301 A		0-4-0DM	HC	D629	1945

R.O. HODGSON LTD, WARTON ROAD, CARNFORTH
Gauge : 4'8½".   (SD 499709)

(D2196)	03196				
40	JOYCE GLYNIS	0-6-0DM	Sdn		1961

WINFIELD SHOE CO LTD, HAZEL MILL,
BLACKBURN ROAD, ACRE, HASLINGDEN, ROSSENDALE
Gauge : 4'8½".   (SD 78x25x)

-	0-6-0DE	HC	D1075	1959

PRESERVATION SITES

EAST LANCASHIRE RAILWAY PRESERVATION SOCIETY, ROSSENDALE YARD, RAWTENSTALL
Gauge : 4'8½".   (SD 807224)

-	4wDM	RH	.	312432	1951

FLEETWOOD LOCOMOTIVE CENTRE LTD, WYRE DOCK, FLEETWOOD
Gauge : 4'8½".   (SD 335468)

4979	WOOTTON HALL	4-6-0	OC	Sdn		1930
45491		4-6-0	OC	Derby		1943
	ALEXANDER	0-4-0ST	OC	AB	1865	1926
26		0-6-0ST	OC	AE	1810	1918
	-	0-4-0ST	OC	P	737	1899
	-	0-4-0DM		HC	D1031	1956
No.5	21811	0-4-0DH		HE	7161	1970
	-	0-6-0DE		YE	2743	1959

FRONTIERLAND, MORECAMBE
Gauge : 1'4".   (SD 428641)

    1865   4w-4-4wPM S/O A.Herschell MH-304-6-86 c1959

R.B.A. HOLDEN, 32 VICTORIA STREET, OSWALDTWISTLE, ACCRINGTON
Gauge : 2'0".   (   )

                  -                    4wPM   Lancs Tanning   1958

LYTHAM MOTIVE POWER MUSEUM, DOCK ROAD, LYTHAM
Gauge : 4'8½". (SD 381276)

    68095   No.42   0-4-0ST   OC Cowlairs   1887
   RIBBLESDALE No.3   0-4-0ST   OC HC   1661   1936
          PENICUIK   0-4-0ST   OC HL   3799   1935
   HODBARROW No.6   SNIPEY   0-4-0CT   IC N   4004   1890
   GARTSHERRIE No.20   0-4-0ST   OC NB   18386   1908
          HOTTO   4wPM   H   965   1930

Gauge : 1'10¾".

   No.10   JONATHAN   0-4-0ST   OC HE   678   1898
              -                 4wDM   HE   2198   1940

PADIHAM RAILWAY SOCIETY, PADIHAM POWER STATION SITE,
BLACKBURN ROAD, near BURNLEY
Gauge : 4'8½".   (SD 783334)

              -                 0-4-0T   OC Lewin   1863
    3052   CAR No.90   4w-4wRER   MC   1932

PLEASURE BEACH RAILWAY, SOUTH SHORE, BLACKPOOL
Gauge : 1'9".   (SD 305332)

    4472   MARY LOUISE   4-6-2DH   S/O HC   D578   1933
    4473   CAROL JEAN   4-6-4DM   S/O HC   D579   1933
    6200   THE PRINCESS ROYAL 4-6-2DH   S/O HC   D586   1935

RIO GRANDE EXPRESS, ZOOLOGICAL GARDENS, EAST PARK DRIVE, BLACKPOOL
Gauge : 1'3".   (SD 335362)

    279   7   2-8-0DH   S/O SL   7219   1972

Gauge : 4'8½".   (SD 496708)

5538		2-6-2T	OC	Sdn		1928
(30850)	850 LORD NELSON	4-6-0	4C	Elh		1926
44871	SOVEREIGN	4-6-0	OC	Crewe		1945
(45407)	5407	4-6-0	OC	AW	1462	1937
(46441)	6441	2-6-0	OC	Crewe		1950
48151		2-8-0	OC	Crewe		1942
(52322)	1300	0-6-0	IC	Hor	420	1896
(60007)	No. 4498 SIR NIGEL GRESLEY	4-6-2	3C	Don	1863	1937
70000	BRITANNIA	4-6-2	OC	Crewe		1951
No.1	GLENFIELD	0-4-0CT	OC	AB	880	1902
	JOHN HOWE	0-4-0ST	OC	AB	1147	1908
No.1969	JANE DARBYSHIRE	0-4-0ST	OC	AB	1969	1929
2134	CORONATION	0-4-0ST	OC	AB	2134	1942
	HURRICANE	0-4-0ST	OC	AB	2230	1947
No.4	BRITISH GYPSUM	0-4-0ST	OC	AB	2343	1953
29	ELIZABETH	0-4-0ST	OC	AE	1865	1922
	PITSFORD	0-6-0ST	OC	AE	1917	1923
No.11	SIRAPITE	4wWT	G	AP	6158	1906
RENISHAW IRONWORKS No.6		0-6-0ST	OC	HC	1366	1919
65		0-6-0ST	OC	HC	1631	1929
231K22	LA FRANCE	4-6-2	4CC	La Loire		1913
		Rebuilt				1948
	CALIBAN	0-4-0ST	OC	P	1925	1937
3		0-4-0ST	OC	P	2084	1948
	JOYCE	4wT	VCG	S	7109	1927
7	GASBAG	4wVBT	VCG	S	8024	1929
01.1104		4-6-2	3C	Sch	11360	1940
No.7	CHLOE	0-4-0ST	IC	SS	1435	1863
BARROW STEEL CO. No.17		0-4-0ST	IC	SS	1585	1865
	-	0-6-0F	OC	WB	3019	1952
	LINDSAY	0-6-0ST	IC	WCI		1887
D2381	03381	0-6-0DM		Sdn		1961
(D3290)	08220	0-6-0DE		Derby		1956
D5500	(31018)	A1A-A1ADE	BT		71	1957
	TRENCHARD	0-4-0DM		AB	401	1956
No.5		0-4-0DM		Bg	3027	1939
	NEW JERSEY	Bo-BoDE		GEU	30483	1949
7049	TOM ROLT	0-6-0DM		HE	2697	1944
	-	0-6-0DH		HE	7017	1971
	-	0-4-0DH		RR	10206	1965
	ESKDALE	0-6-0DE		YE	2718	1958
433/13		0-4-0DE		YE	2857	1961

Other locos are occasionally present for repairs.

Gauge : 2'2".

No.1		4wDH	HE	8972	1979 +
No.2		4wDH	HE	8970	1979 +

+ Stored for private owner.

Gauge : 1'3".

18	GEORGE THE FIFTH	4-4-2	OC	BL	18	1911
22	PRINCESS ELIZABETH	4-4-2	OC	BL	22	1914
5751	PRINCE WILLIAM	4-6-2	OC	G&S	9	1946

					Lane	1956
	ROYAL ANCHOR		4w-4wDH		Lane	1956
	THE CUB		4-4wDM		Minirail	1954
D511	DOCTOR DIESEL		4w-4wDE		Minirail/SL	1969

WEST LANCASHIRE LIGHT RAILWAY, STATION ROAD, HESKETH BANK, near PRESTON
Gauge : 2'0".   (SD 448229)

No.3	IRISH MAIL		0-4-0ST	OC	HE	823	1903
No.9			0-6-0WTT	OC	KS	2405	1915
21	No.35 UTRILLAS		0-4-0WT	OC	OK	2378	1907
No.22	No.34 MONTALBAN		0-4-0WT	OC	OK	6641	1913
No.1	CLWYD		4wDM		RH	264251	1951
No.2	TAWD		4wDM		RH	222074	1943
No.4			4wPM		FH	1777	1931
No.5			4wDM		RH	200478	1940
No.7			4wDM		MR	8992	1946
No.8			4wDM		HE	4478	1953 +
No.10			4wDM		FH	2555	1946
No.12			4wDM		MR	7955	1945 Dsm a
No.16	20 L22		4wDM		RH	202036	1941
No.18	U192 T.R.A.No.13 TRENT		4wDM		RH	283507	1949
No.19			4wPM		L	10805	1939
No.20			4wPM		Bg	3002	1937
No.21			4wDM		HE	1963	1939
No.25			4wDM		RH	297054	1950
No.26	8		4wDM		MR	11223	1963
No.27	No.31 MILL REEF		4wDM		MR	7371	1939
No.29	31 RED RUM		4wDM		MR	7105	1936
No.33			4wPM		Bg	2095	1936 b
No.36			4wDM		RH	339105	1953
No.38			0-4-0DM		HC	(DM750	1949?)
	-		4wDM		FH	3916	1959

+ Plate reads 4480/1953.
a Converted into a brake van.
b Currently under renovation by P. Siddington,
  20, Alvaston Road, Derby.

LEICESTERSHIRE

INDUSTRIAL SITES

BARDON HILL GROUP PLC, BARDON HILL GRANITE QUARRY, COALVILLE
Gauge : 4'8½".   (SK 446129)

No.59	DUKE OF EDINBURGH	6wDH	RR	10273	1968

BRITISH COAL
See Section Four for full details.

BRUSH ELECTRICAL MACHINES LTD, TRACTION DIVISION, LOUGHBOROUGH
Gauge : 4'8½".   (SK 543207)

PLANT No.11079 SPRITE   0-4-0DH   HC   D1341 1966

          New BT locos occasionally present.

CASTLE CEMENT (KETTON) LTD, KETTON WORKS
Gauge : 4'8½".   (SK 982056, 987057)

No.1   0-4-0DH   JF   4220007 1960 OOU
No.4   0-6-0DH   JF   4240012 1961
         -                    0-6-0DH   TH   293V 1980

DAVY MORRIS LTD, NORTH ROAD, LOUGHBOROUGH
Gauge : 4'8½".   (SK 537209)   RTC.

    1   4wDM   MR   2026 1920 OOU

E.C.C. CONSTRUCTION MATERIALS LTD, CROFT GRANITE DIVISION, CROFT
Gauge : 4'8½".   (SP 517960)

    EX 256   MSC 0256   0-4-0DH   JF   4220016 1962
    MS 2137 EDWIN   0-4-0DH   RH   437363 1960 OOU
    MS 5482 KATHRYN   0-6-0DH   RR   10256 1966

POWER GEN, CASTLE DONINGTON POWER STATION
(A Division of C.E.G.B.)
Gauge : 4'8½".   (SK 433284)

         POWERGEN 1   0-4-0ST   OC RSHN   7817 1954
         POWERGEN 2   0-4-0ST   OC RSHN   7818 1954
No.1   0-4-0DH   AB   415 1957 OOU
No.2   0-4-0DH   AB   416 1957
   3   0-4-0DE   RH   420142 1958

REDLAND AGGREGATES LTD,
BARROW-UPON-SOAR RAIL-LOADING TERMINAL
Gauge : 4'8½".   (SK 587168)

  (D2324)   0-6-0DM   ⌐(RSH   8183 1961
                                            ⌐(DC   2705 1961
  (D2867) DIANE   0-4-0DH   YE   2850 1961
    503   NELLY   4wDH   ASEA   0C0488 1982
           JO   4wDH   DeDietrich   89134 1988

STANTON PLC, HOLWELL WORKS, ASFORDBY HILL, MELTON MOWBRAY
Gauge : 4'8½". (SK 725199)

```
19        4493/45              4wDH     S         10019  1960
36        4493/46              4wDH     S         10036  1960
```

VIC BERRY LTD, RAILWAY VEHICLE DIVISION, WESTERN BOULEVARD, LEICESTER
Gauge : 4'8½". (SK 580035)

```
(D2069)   03069                0-6-0DM   Don                1959
(D2397)   03397                0-6-0DM   Don                1961
          -                    0-4-0DM   Bg/DC     2724     1963
          -                    0-6-0DH   EEV       D1227    1967
          -                    0-6-0DH   EEV       D1228    1967
1483                           4wDM      HE        5306     1958
HF 0591                        0-4-0DM   JF        4210112  1956  OOU
HF 0793   M 793    DOBBIN      0-4-0DM   JF        4210133  1957
```

Also other locos for scrap occasionally present.

PRESERVATION SITES

REV. E.R. BOSTON (Deceased), BOSTONS LODGE, CADEBY RECTORY, MARKET BOSWORTH
Gauge : 4'8½". (SK 426024)

```
V47                           0-4-0ST   OC  P       2012   1941
```

Gauge : 2'0".

```
No.2                          0-4-0WT   OC  OK      7529   1914
          PIXIE               0-4-0ST   OC  WB      2090   1919
          L.A.W.R.            0-4-0PM   S/O Bg      1695   1928
          -                   4wDM          HC      D558   1930
          NEW STAR            4wPM          L       4088   1931
87004                         4wDM          MR      2197   1923
87009                         4wDM          MR      4572   1929
          -                   4wPM          MR      5038   1930  Dsm
          -                   4wDM          MR      5853   1934
42                            4wDM          MR      7710   1939
20                            4wDM          MR      8748   1942
          -                   4wPM          OK      4588   1932
87008                         4wDM          RH      179870 1936
85051                         4wDM          RH      404967 1957
```

F.R. COOMBES, WASH LANE, RAVENSTONE, near COALVILLE
Gauge : 4'8½". (SK 409142)

```
          -                   4wPM          FH      2895   1944
```

GREAT CENTRAL RAILWAY (1976) LTD, LOUGHBOROUGH (G.C.) STATION & ROTHLEY
Gauge : 4'8½".   (SK 543194)

5224		2-8-0T	OC	Sdn		1924
6990	WITHERSLACK HALL	4-6-0	OC	Sdn		1948
34039	BOSCASTLE	4-6-2	3C	Bton		1946
35005	CANADIAN PACIFIC	4-6-2	3C	Elh		1941
35025	BROCKLEBANK LINE	4-6-2	3C	Elh		1948
(45593)	5593 KOLHAPUR	4-6-0	3C	NBQ	24151	1935
47406		0-6-0T	IC	VF	3977	1926
48305		2-8-0	OC	Crewe		1943
(61264)	1264	4-6-0	OC	NB	26165	1947
(62660)	506					
	BUTLER-HENDERSON	4-4-0	IC	Gorton		1920
68088		0-4-0T	IC	Dar	1205	1923
69523		0-6-2T	IC	NB	22600	1921
92212		2-10-0	OC	Sdn		1959
68009		0-6-0ST	IC	HE	3825	1954
No.3		0-4-0ST	OC	HL	3581	1924
	-	0-4-0ST	OC	P	1963	1938
7597	ZEBEDEE	0-6-0T	OC	RSHN	7597	1949
D3101		0-6-0DE		Derby		1955
D4067	MARGARET ETHEL - THOMAS ALFRED NAYLOR					
		0-6-0DE		Dar		1961
9019	(55019)	Co-CoDE	_(EE	2924	1961
	ROYAL HIGHLAND FUSILIER			⌐(VF	D576	1961
M51616		2-2w-2w-2DHR		Derby C&W		1959
M51622		2-2w-2w-2DHR		Derby C&W		1959
	BARDON	0-4-0DM		AB	400	1956
D4279	ARTHUR WRIGHT	0-4-0DM		JF	4210079	1952
No.1	QWAG	4wDM		RH	371971	1954
	-	2w-2PMR		Wkm	(693	1932?)DsmT

LEICESTERSHIRE COUNTY COUNCIL
Leicestershire Museum of Technology,
Abbey Meadows, Corporation Road, Leicester
Gauge : 4'8½".   (SK 589067)

	MARS II	0-4-0ST	OC	RSHN	7493	1948

Gauge : 2'0".

-	4wPM		FH	(1776	1931?)
-	4wPM		MR	5260	1931
-	4wDM		RH	223700	1943 Dsm

The Industrial Adventure, Former Snibston Mine, Ashby Road, Coalville
Gauge : 4'8½".   (SK 420144)

No.2	0-4-0F	OC	AB	1815	1924
-	0-4-0ST	OC	BE	314	1906
CADLEY HILL No.1	0-6-0ST	IC	HE	3851	1962
-	0-6-0DH		HE	6289	1966

MARKET BOSWORTH LIGHT RAILWAY, SHACKERSTONE STATION, MARKET BOSWORTH
Gauge : 4'8½".   (SK 379066)

	THE KING		0-4-0WT	OC	EB		1906
	WALESWOOD		0-4-0ST	OC	HC	750	1906
N.C.B. 11			0-4-0ST	OC	HE	1493	1925
No.7			0-4-0ST	OC	P	2130	1951
	RICHARD III		0-6-0T	OC	RSHN	7537	1949
No.4			0-6-0T	OC	RSHN	7684	1951
	-		4wVBT	G	Shackerstone		1983
		Rebuilt from	4wDM		RH	235513	1945
	LINDA		0-4-0ST	OC	WB	2648	1941
LAMPORT No.3			0-6-0ST	OC	WB	2670	1942
2			0-6-0ST	OC	WB	3059	1953
(D2245)	11215		0-6-0DM		(RSH	7864	1956
					(DC	2577	1956
(M50397)			2-2w-2w-2DMR		PR		1957 +
No.2	NANCY		4wDM		RH	263001	1949
	-		0-6-0DM		RH	347747	1957
	-		4wDM		RH	393304	1956
	-		0-4-0DE		RH	423657	1958
2	48		0-6-0DM		RSH	7697	1953
9033			2w-2PMR		Wkm	6857	1954

+   Now unmotorised.

RUTLAND RAILWAY MUSEUM, COTTESMORE IRON ORE MINES SIDINGS,
ASHWELL ROAD, COTTESMORE, near OAKHAM
Gauge : 4'8½".   (SK 887137)

	(FIREFLY)	0-4-0ST	OC	AB	776	1896	
B.S.C. No.2		0-4-0ST	OC	AB	1931	1927	
	DRAKE	0-4-0ST	OC	AB	2086	1940	
	SIR THOMAS ROYDEN	0-4-0ST	OC	AB	2088	1940	
	SWORDFISH	0-6-0ST	OC	AB	2138	1941	
	SALMON	0-6-0ST	OC	AB	2139	1942	
	DORA	0-4-0ST	OC	AE	1973	1927	
	RHOS	0-6-0ST	OC	HC	1308	1918	
COAL PRODUCTS No.6		0-6-0ST	IC	HE	2868	1943	
		Rebuilt		HE	3883	1963	
65		0-6-0ST	IC	HE	3889	1964	
YARD No.440	SINGAPORE	0-4-0ST	OC	HL	3865	1936	
	UPPINGHAM	0-4-0ST	OC	P	1257	1912	
	ELIZABETH	0-4-0ST	OC	P	1759	1928	
8		0-4-0ST	OC	P	2110	1950	
(D9518)	No.7 9312/95	0-6-0DH		Sdn		1964	
(D9520)	45 R.R.M. No.17	0-6-0DH		Sdn		1964	
(D9521)	No.3 9312/90	0-6-0DH		Sdn		1964	
D9555		0-6-0DH		Sdn		1965	
	-	0-4-0DM		AB	352	1941	
	-	0-4-0DH		AB	499	1965	
	-	0-6-0DH		EEV	D1200	1967	Dsm
	-	0-6-0DH		EEV	D1231	1967	
	PHOENIX	4wDM		FH	3887	1958	
NORWOOD COKING PLANT No.1		0-4-0DH		HE	6688	1968	
3		0-4-0DH		NB	27656	1957	

	-	4wDM	RH	305302	1951	
	-	4wDM	RH	306092	1950	
	21 90 02	0-4-0DH	RH	504565	1965	
	BETTY	0-4-0DH	RR	10201	1964	
No.5	2444/20	0-6-0DE	YE	2670	1958	
No.28		0-6-0DE	YE	2791	1962	
	1382	0-6-0DE	YE	2872	1962	

WELLAND VALLEY VINTAGE TRACTION CLUB, GLEBE ROAD, MARKET HARBOROUGH
Gauge : 3'0".   (SP 742868)

KETTERING FURNACES No.8   0-6-0ST   OC MW   1675 1906

> Currently under renovation at a private address.

LINCOLNSHIRE

<u>INDUSTRIAL SITES</u>

GEORGE FISCHER (LINCOLN) LTD, IRONFOUNDERS, NORTH HYKEHAM
Gauge : 4'8½". (SK 933672)   RTC.

-	2-2wBE	WR	T8070	1979	OOU

H.M. DETENTION CENTRE, NORTH SEA CAMP, FREISTON, near BOSTON
Gauge : 2'0". (TF 397422)   RTC.

-	4wDM	L	10994	1939	Dsm
-	4wDM	L	33650	1949	OOU
-	4wDM	L	33651	1949	Dsm
-	4wDM	LB	51917	1960	Dsm
-	4wDM	LB	55413	1967	OOU
-	4wDM	LB	56371	1970	OOU

> One transferred to Prison Service College, Warwickshire.

PORT OF BOSTON AUTHORITY, BOSTON DOCKS
Gauge : 4'8½". (TF 329431)

(D3871)	08704			
	PORT OF BOSTON	0-6-0DE	Hor	1960

PRESERVATION SITES

N. BANKS, TYSDALE FARM, COMMON WAY, TYDD ST MARY
Gauge : 2'0".   (TF 444187)

-		4wDM	OK	6931	1937
-		4wDM	OK	7734	1938

FULSTOW STEAM CENTRE, CARPENTERS ROW, MAIN STREET, FULSTOW, near LOUTH
Gauge : 4'8½".   (TF 337976)

-		0-4-0ST	OC	P	1749	1928
AGECROFT No.1		0-4-0ST	OC	RSHN	7416	1948
No.1	FULSTOW	0-4-0ST	OC	RSHN	7680	1950

T. HALL, NORTH INGS FARM, DORRINGTON, near RUSKINGTON
Gauge : 2'0".   (TF 098527)

-		4wDM	Clay Cross		+
-		4wPM	L	962	c1930
-		4wDM	L	29890	1946
No.6	4	4wDM	MR	7403	1939
3		4wDM	MR	7493	1940
	INGRID	4wDM	OK		c1932
-		4wDM	RH	172901	1935
No.1		4wDM	RH	371937	1956
-		4wDM	RH	375701	1954
No.4		4wDM	RH	421433	1959

+   Constructed from parts supplied by Listers in 1961 or 1973.

LINCOLN CITY COUNCIL, LINCOLN CENTRAL STATION, LINCOLN
Gauge : 4'8½".   (SK 975709)

D3167	(08102)	0-6-0DE	Derby		1955

LINCOLNSHIRE COUNTY COUNCIL,
MUSEUM OF LINCOLNSHIRE LIFE, BURTON ROAD, LINCOLN
Gauge : 4'8½".   (SK 972723)

-	4wDM	RH	463154	1961

Gauge : 2'6".

-	4wPM	RP	52124	1918

Gauge : 2'3".

-	4wDM	RH	192888	1939

Gauge : 2'0".

-	4wDM	RH	421432	1959

LINCOLNSHIRE RAILWAY MUSEUM, BURGH LE MARSH
Gauge : 4'8½".   (   )

| | | 0-6-0ST | OC | HC | 1604 | 1928 |

LYSAGHTS SPORTS & SOCIAL CLUB, HOLTON LE MOOR
Gauge : 2'6".   (TF 094973)

| | CANNONBALL | 4wDM | S/O | RH | 175403 | 1935 |

J. SCHOLES, STAMFORD
Gauge : 2'0".   (   )

		4wDM		HE	7120	1969
LOD 758227	-	4wDM		MR	8813	1943
	-	4wDM		MR	8820	1943

Locos stored at private location.

J.H.P. WRIGHT, FENLAND AIRFIELD, HOLBEACH ST, JOHNS
Gauge : 1'9".   (TF 333179)

| | - | 4-6-2DM | S/O | HC | D611 | 1938 | Dsm |
| 6203 | QUEEN ELIZABETH | 4-6-2DM | S/O | HC | D612 | 1938 | OOU |

GREATER LONDON

INDUSTRIAL SITES

ARTHUR GUINNESS, SON & CO (PARK ROYAL) LTD, PARK ROYAL BREWERY
Gauge : 4'8½".   (TQ 195828)

| (D3030) | 08022 | LION | 0-6-0DE | Derby | 1953 |
| (D3074) | 08060 | UNICORN | 0-6-0DE | Dar | 1953 |

BOVIS CONSTRUCTION LTD, c/o LONDON UNDERGROUND LTD, ALDWYCH STATION
Gauge : 4'8½".   (TQ 818300)

| | - | 2-2wBER | Track Supplies | NP/L023 | 1985 |

BRITISH COAL, WEST DRAYTON FORWARD LANDSALE DEPOT,
off TAVISTOCK ROAD, WEST DRAYTON
Gauge : 4'8½".   (TQ 054802)

| | - | 0-6-0DH | HE | 5590 | 1964 | OOU |

BRITISH GYPSUM LTD, ERITH
Gauge : 4'8½".   (TQ 507789)   RTC.

-		4wDM	RH	512572	1965	OOU

BRITISH RAILWAYS BOARD, EASTERN REGION, STRATFORD DIESEL REPAIR DEPOT
Gauge : 4'8½".   (TQ 383849)

Industrial locos occasionally present for overhaul.

COAL MECHANISATION (TOLWORTH) LTD, TOLWORTH COAL CONCENTRATION DEPOT
Gauge : 4'8½".   (TQ 198656)

(D2246)	BLUEBELL	0-6-0DM	(RSH	7865	1956
			(DC	2578	1956
D2310		0-6-0DM	(RSH	8169	1960
			(DC	2691	1960

DAY & SON, BRENTFORD TOWN GOODS DEPOT, TRANSPORT AVENUE, BRENTFORD
Gauge : 4'8½".   (TQ 166778)

(12049)	0-6-0DE	Derby	1948

THE DELTIC 9000 FUND, c/o BRITISH RAIL, SELHURST DEPOT
Gauge : 4'8½".   (TQ 333678)

D9000	(55022) ROYAL SCOTS GREY	Co-CoDE	(EE	2905	1960 Pvd
			(VF	D557	1960
D9016	(55016) GORDON HIGHLANDER	Co-CoDE	(EE	2921	1961 Pvd
			(VF	D573	1961

Privately owned locos kept here in between appearances at
Open Days, Exhibitions, etc.

DOCKLANDS LIGHT RAILWAY, POPLAR DEPOT
Gauge : 4'8½".   (TQ 376806)

	SOOTY	4wDM		Wkm	11622	1986
F540 YCK		4wDM	R/R	Permaquip		1989 a

a Property of Balfour Beatty Ltd (Contractor on site).

600 FERROUS FRAGMENTISERS LTD, SCRUBS LANE, WILLESDEN
Gauge : 4'8½".   (TQ 210832)

(D2018	03018)	600	2	0-6-0DM	Sdn		1958

FORD MOTOR CO LTD, DAGENHAM
Gauge : 4'8½".   (TQ 496825, 499827)

(D2267)	No.01		0-6-0DM	_(RSH	7897	1958
				⁻(DC	2611	1958
(D2280)	No.2 P 1381 C		0-6-0DM	_(RSH	8098	1960
				⁻(DC	2657	1960
(D2051)	No.4		0-6-0DM	Don		1959
L/DH/6	060-6-0DH			HC	D1396	1967
GT/PL/1	P 260 C		4wDM	Robel	21 11 RK1	

G.L. PLANT LTD, PLANT DEPOT, DRAKE ROAD, MITCHAM (Member of Gleeson Group)
Gauge : 2'0".   (TQ 281673)

-		0-4-0BE	WR	N7639	1973 00U

J. MURPHY & SONS LTD, PLANT DEPOT, HIGHGATE ROAD, KENTISH TOWN
Gauge : 2'0".   (TQ 287855)

-	4wBE	CE	B1534A	1971
JMLM 16	4wBE	CE	B1534B	1977
-	4wBE	CE	B1547B	1977
-	4wBE	CE	B3070A	1983
-	4wBE	CE	B3070B	1983
-	4wBE	CE	B3070C	1983
-	4wBE	CE	B3070D	1983
-	0-4-0BE	WR	M7550	1972
-	4wBE	WR	N7605	1973
-	4wBE	WR	N7606	1973
-	4wBE	WR	N7607	1973
-	0-4-0BE	WR	N7617	1973
-	4wBE	WR	N7620	1973
-	4wBE	WR	N7621	1973

Locos present in yard between contracts.

Gauge : 1'6".

-	0-4-0BE	WR	M7548	1972

LONDON UNDERGROUND LTD
Gauge : 4'8½". Locos are kept at :-

Acton Works, Bollo Lane	(TQ 196791)
Amersham Station	(SP 964982)
Chalfont & Latimer Station	(SU 996976)
Cockfosters Depot	(TQ 288962)
Ealing Common Depot, Uxbridge Road	(TQ 189802)
Golders Green Depot, Finchley Road	(TQ 253875)
Hainault Depot	(TQ 450918)
Hammersmith Depot	(TQ 234787)
Highgate Depot (Closed)	(TQ 279886)
Lillie Bridge Depot	(TQ 250782)
Morden Depot	(TQ 255680)
Neasden Depot	(TQ 206858)
New Cross Depot	(TQ 360778)
Northfields Depot, Northfields Avenue	(TQ 167789)
Northumberland Park Depot	(TQ 34x88x)
Rickmansworth Station	(TQ 057946)
Stonebridge Park Depot	(TQ 192845)
Upminster Depot	(TQ 570871)
Wembley Park Signal Engineers Depot	(TQ 194864)
West Ruislip Depot	(TQ 094862)

Locos may also be found temporarily in depots/sidings at

Aldgate, Arnos Grove, Barking, Brixton (Underground), Edgware,
Edgware Road, Elephant & Castle (Underground), Farringdon, High
Barnet, High Street Kensington, London Road, Loughton, Moorgate,
Parsons Green, Queens Park, South Harrow, Stanmore, Triangle
Sidings, Uxbridge, Walthamstow (Underground), Wembley Park
Depot, White City, Woodford.

L 11	4w-4wRE	MC		1931/1932 OOU
	Rebuilt	Acton		1964
L 15	4w-4wBE/RE	MC		1970
L 16	4w-4wBE/RE	MC		1970
L 17	4w-4wBE/RE	MC		1970
L 18	4w-4wBE/RE	MC		1970
L 19	4w-4wBE/RE	MC		1970
L 20	4w-4wBE/RE	MC		1964
L 21	4w-4wBE/RE	MC		1964
L 22	4w-4wBE/RE	MC		1965
L 23	4w-4wBE/RE	MC		1965
L 24	4w-4wBE/RE	MC		1965
L 25	4w-4wBE/RE	MC		1965
L 26	4w-4wBE/RE	MC		1965
L 27	4w-4wBE/RE	MC		1965
L 28	4w-4wBE/RE	MC		1965
L 29	4w-4wBE/RE	MC		1965
L 30	4w-4wBE/RE	MC		1965
L 31	4w-4wBE/RE	MC		1965
L 32	4w-4wBE/RE	MC		1965
L 33	4w-4wBE/RE	Acton		1962 OOU
L 35	4w-4wBE/RE	GRCW		1938
L 36	4w-4wBE/RE	GRCW		1938
L 37	4w-4wBE/RE	GRCW		1938 OOU
L 38	4w-4wBE/RE	GRCW		1938
L 39	4w-4wBE/RE	GRCW		1938 OOU
L 44	4w-4wBE/RE	Don	DON L44	1973
L 45	4w-4wBE/RE	Don	DON L45	1974

L 46	4w-4wBE/RE	Don	DON L46	1974	
L 47	4w-4wBE/RE	Don	DON L47	1974	
L 48	4w-4wBE/RE	Don	DON L48	1974	
L 49	4w-4wBE/RE	Don	DON L49	1974	
L 50	4w-4wBE/RE	Don	DON L50	1974	
L 51	4w-4wBE/RE	Don	DON L51	1974	
L 52	4w-4wBE/RE	Don	DON L52	1974	
L 53	4w-4wBE/RE	Don	DON L53	1974	
L 54	4w-4wBE/RE	Don	DON L54	1974	
L 55	4w-4wBE/RE	RYP		1951	
L 56	4w-4wBE/RE	RYP		1951	
L 58	4w-4wBE/RE	RYP		1951	
L 59	4w-4wBE/RE	RYP		1951	
L 60	4w-4wBE/RE	RYP		1951	
L 61	4w-4wBE/RE	RYP		1952	
L 62	4w-4wBE/RE	MC		1985	
L 63	4w-4wBE/RE	MC		1985	
L 64	4w-4wBE/RE	MC		1985	
L 65	4w-4wBE/RE	MC		1985	
L 66	4w-4wBE/RE	MC		1985	
L 67	4w-4wBE/RE	MC		1986	
L 130	4w-4RE	MC		1934	c
	Rebuilt	Acton		1967	
L 131	4w-4RE	MC		1934	c
	Rebuilt	Acton		1967	
L 132	4w-4wRE	Cravens		1960	d
	Rebuilt	Derby		1987	
L 133	4w-4wRE	Cravens		1960	d
	Rebuilt	Derby		1987	
L 134	4w-4RE	M		1927	
	Rebuilt	Acton		1967	
L 135	4w-4RE	MC		1934	
	Rebuilt	Acton		1967	
L 141	2-2w-2w-2RE	MC		1938	OOU
	Rebuilt	Acton		1973	
L 142	2-2w-2w-2RE	MC		1938	OOU
	Rebuilt	Acton		1973	
L 143	2-2w-2w-2RE	MC		1938	OOU
	Rebuilt	Acton		1973	
L 144	2-2w-2w-2RE	MC		1940	OOU
	Rebuilt	Acton		1975	
L 145	2-2w-2w-2RE	MC		1938	OOU
	Rebuilt	Acton		1975	
L 146	2-2w-2w-2RE	MC		1938	
	Rebuilt	Acton		1976	
L 147	2-2w-2w-2RE	MC		1938	
	Rebuilt	Acton		1976	
L 148	2-2w-2w-2RE	MC		1938	
	Rebuilt	Acton		1977	
L 149	2-2w-2w-2RE	MC		1938	
	Rebuilt	Acton		1977	
L 150	2-2w-2w-2RE	MC		1938	a
	Rebuilt	Acton		1978	
L 151	2-2w-2w-2RE	MC		1938	a
	Rebuilt	Acton		1978	
L 152	2-2w-2w-2RE	MC		1940	OOU
	Rebuilt	Acton		1978	
L 153	2-2w-2w-2RE	MC		1940	OOU
	Rebuilt	Acton		1978	
ESL 107	4w-4-4-4wRE	BRCW/Met.Amal		1903	OOU
	Rebuilt	Acton		1939	

No./ID	Name	Type	Builder	Works No.		Date	Notes
ESL 117		4w-4-4-4wRE	BRCW/Met.Amal			1903	OOU
		Rebuilt	Acton			1940	
4416		2w-2-2-2wRE	GRCW			1939	c Pvd
		Rebuilt	Acton			1971	
4417		2w-2-2-2wRE	GRCW			1939	c Pvd
		Rebuilt	Acton			1971	
TCC 1		4w-4wRE	MC			1939	
		Rebuilt	Acton			1978	
TCC 5		4w-4wRE	MC			1938	
		Rebuilt	Acton			1978	
PC 857		4w-4wRER	MC			c1939	b
		Rebuilt	Acton			1980	
PC 858		4w-4wRER	MC			1939	b
		Rebuilt	Acton			1980	
PC 859		4w-4wRER	MC			1939	b
		Rebuilt	Acton			1981	
12	SARAH SIDDONS	4w-4wRE	VL			1922	
A723 LNW	TMM 774	4wDM R/R Unimog 424131	10 088213	12/985		1982	
A456 NWX	L84	4wDM R/R Unimog 414121.10.101335		1035/83		1983	
C622 EWT	L85	4wDM	R/R Unimog			1986	
DL 81		0-6-0DH	RR	10278		1968	OOU
DL 82		0-6-0DH	RR	10272		1967	
DL 83		0-6-0DH	RR	10271		1967	
40/174		2w-2PMR	Wkm	7702		1957	
40/175		2w-2PMR	Wkm	7703		1957	
40/219		2w-2PMR	Wkm	8822		1961	Dsm
40/233		2w-2PMR	Wkm	9522		1963	Dsm
40/234	N	2w-2PMR	Wkm	9523		1963	
40/245		2w-2PMR	Wkm	9813		1965	
40/246		2w-2PMR	Wkm	9814		1965	

a Converted to weedkilling train.
b Converted to Personnel Carriers, no longer powered.
c Pilot Motor Cars.
d Track Recording Pilot Motor Cars.

PLASSER RAILWAY MACHINERY (G.B.) LTD, DRAYTON GREEN ROAD, WEST EALING
Gauge : 4'8½".   (TQ 161809)

	Type	Builder
-	4wDM	Plasser
-	4wDM	Plasser

POST OFFICE (LONDON) RAILWAY
Gauge : 2'0".   Locos are kept at :-

King Edward Building, St Pauls	(TQ)
Mount Pleasant Parcels Office, Clerkenwell	(TQ 311823)
New Western District Office, Rathbone Place	(TQ 296814)

No./ID	Type	Builder	Works No.	Date	Notes
-	4wRE	EE	601	1926	Pvd
-	4wRE	EE	652	1926	+
1	4wBE	EE	702	1926	
2	4wBE	EE	703	1926	
3	4wBE	EE	704	1926	
BREAKDOWN No.1	4wRE	EE			a
WAGON CAR DEPOT BREAKDOWN No.2	4wRE	EE			a
BREAKDOWN No.3	4wRE	EE			a
66	2w-2-2-2wRE	(EE	3335	1962	
		(EES	8314	1962	

169		2w-2RE	HE	9134	1982 d
170		2w-2RE	HE	9134	1983 d
501		2w-2-2-2wRE	GB	420461/1	1980
502		2w-2-2-2wRE	GB	420461/2	1980
503		2w-2-2-2wRE	GB	420461/3	1980
504		2w-2-2-2wRE	(GB	420461/4	1980
			(HE	9103	1980
505		2w-2-2-2wRE	(GB	420461/5	1980
			(HE	9104	1980
506		2w-2-2-2wRE	(GB	420461/6	1980
			(HE	9105	1980
507		2w-2-2-2wRE	(GB	420461/7	1980
			(HE	9106	1980
508	GREAT WEST EXPRESS	2w-2-2-2wRE	(GB	420461/8	1980
			(HE	9107	1980
509		2w-2-2-2wRE	(GB	420461/9	1981
			(HE	9108	1981
510		2w-2-2-2wRE	(GB	420461/10	1981
			(HE	9109	1981
511		2w-2-2-2wRE	(GB	420461/11	1981
			(HE	9110	1981
512		2w-2-2-2wRE	(GB	420461/12	1981
			(HE	9111	1981
513		2w-2-2-2wRE	(GB	420461/13	1981
			(HE	9112	1981
514	CAPITAL EXPRESS	2w-2-2-2wRE	(GB	420461/14	1981
			(HE	9113	1981
515		2w-2-2-2wRE	(GB	420461/15	1981
			(HE	9114	1981
516		2w-2-2-2wRE	(GB	420461/16	1981
			(HE	9115	1981
517		2w-2-2-2wRE	(GB	420461/17	1981
			(HE	9116	1981
518		2w-2-2-2wRE	(GB	420461/18	1981
			(HE	9117	1981
519		2w-2-2-2wRE	(GB	420461/19	1981
			(HE	9118	1981
520		2w-2-2-2wRE	(GB	420461/20	1981
			(HE	9119	1981
521		2w-2-2-2wRE	(GB	420461/21	1981
			(HE	9120	1981
522		2w-2-2-2wRE	(GB	420461/22	1981
			(HE	9121	1981
523		2w-2-2-2wRE	(GB	420461/23	1981
			(HE	9122	1981
524		2w-2-2-2wRE	(GB	420461/24	1981
			(HE	9123	1981
525		2w-2-2-2wRE	(GB	420461/25	1981
			(HE	9124	1981
526		2w-2-2-2wRE	(GB	420461/26	1981
			(HE	9125	1981
527		2w-2-2-2wRE	(GB	420461/27	1982
			(HE	9126	1982
528		2w-2-2-2wRE	(GB	420461/28	1982
			(HE	9127	1982
529		2w-2-2-2wRE	(GB	420461/29	1982
			(HE	9128	1982
530		2w-2-2-2wRE	(GB	420461/30	1982
			(HE	9129	1982
531		2w-2-2-2wRE	(GB	420461/31	1982
			(HE	9130	1982

No.	Name	Type	Builder	Works No.	Year		Notes
532	THE LONDON FLYER	2w-2-2-2wRE	(GB	420461/32	1982		
			(HE	9131	1982		
533	GREAT·EAST EXPRESS	2w-2-2-2wRE	(GB	420461/33	1982		
			(HE	9132	1982		
534		2w-2-2-2wRE	(GB	420461/34	1982		
			(HE	9133	1982		
752		2w-2-2-2wRE	EE	752	1930	OOU	c
	-	2w-2-2-2wRE	EE	753	1930		b
754		2w-2-2-2wRE	EE	754	1930	OOU	e
755		2w-2-2-2wRE	EE	755	1930		
756		2w-2-2-2wRE	EE	756	1930		
759		2w-2-2-2wRE	EE	759	1930	OOU	c
760		2w-2-2-2wRE	EE	760	1930		
761		2w-2-2-2wRE	EE	761	1930		
762		2w-2-2-2wRE	EE	762	1930		
763		2w-2-2-2wRE	EE	763	1930	OOU	c
793		2w-2-2-2wRE	EE	793	1930	OOU	c
795		2w-2-2-2wRE	EE	795	1930	OOU	c
797		2w-2-2-2wRE	EE	797	1930	OOU	e
799		2w-2-2-2wRE	EE	799	1930	OOU	c
801		2w-2-2-2wRE	EE	801	1930		
802		2w-2-2-2wRE	EE	802	1930	OOU	c
804		2w-2-2-2wRE	EE	804	1930	OOU	c
805		2w-2-2-2wRE	EE	805	1930		
806		2w-2-2-2wRE	EE	806	1930		
810		2w-2-2-2wRE	EE	810	1930	OOU	
811		2w-2-2-2wRE	EE	811	1930		
812		2w-2-2-2wRE	EE	812	1930		
813		2w-2-2-2wRE	EE	813	1930	OOU	c
814		2w-2-2-2wRE	EE	814	1930		
815		2w-2-2-2wRE	EE	815	1930		
816		2w-2-2-2wRE	EE	816	1930	OOU	c
817		2w-2-2-2wRE	EE	817	1930	OOU	c
818		2w-2-2-2wRE	EE	818	1930	OOU	c
819		2w-2-2-2wRE	EE	819	1930		
820		2w-2-2-2wRE	EE	820	1931	OOU	c
	-	2w-2-2-2wRE	EE	821	1931		b
822		2w-2-2-2wRE	EE	822	1931	OOU	c
823		2w-2-2-2wRE	EE	823	1931	OOU	e
824		2w-2-2-2wRE	EE	824	1931		
826		2w-2-2-2wRE	EE	826	1931	OOU	c
827		2w-2-2-2wRE	EE	827	1931		
828		2w-2-2-2wRE	EE	828	1931	OOU	f
830		2w-2-2-2wRE	EE	830	1931	OOU	c
925		2w-2-2-2wRE	EE	925	1936	OOU	c
926		2w-2-2-2wRE	EE	926	1936		
928		2w-2-2-2wRE	EE	928	1936		
929		2w-2-2-2wRE	EE	929	1936	OOU	e
930		2w-2-2-2wRE	EE	930	1936	OOU	f
931		2w-2-2-2wRE	EE	931	1936		
932		2w-2-2-2wRE	EE	932	1936	OOU	c

+ Converted to a battery carrier.
a Converted to a wagon.
b Converted to a passenger car.
c Stored in a disused tunnel at Rathbone Place.
d Spare power units. May be found running in 501 to 534
  or stored in Mount Pleasant Depot.
e Stored in a siding under Wimpole Street.
f Stored in a siding under High Holborn.

PURFLEET DEEP WHARF & STORAGE CO LTD, ERITH WHARF
Gauge : 4'8½".   (TQ 517779)   RTC.

1	THETIS	0-4-0DM	_(RSHN	7815	1954	OOU
			‾(DC	2502	1954	
	PRIAM	0-4-0DM	_(VF	D293	1955	OOU
			‾(DC	2566	1955	

TAYLOR WOODROW PLANT CO LTD, PLANT DEPOT, GREENFORD
(Part of Taylor Woodrow Group of Companies)
Gauge : 2'0".   (TQ 126826)

280537		4wBE	GB	420253	1970
	Rebuilt	WR		1983
6		4wBE	WR	6502	1962
	Rebuilt	WR		10102	1983
-		4wBE	WR	6503	1962
	Rebuilt	WR		10104	1983
4951		4wBE	WR	6504	1962
	Rebuilt	WR		10106	1983
8		4wBE	WR	6505	1962
	Rebuilt	WR		10105	1983

        Locos present in yard between contracts.

WALSH BROS (TUNNELLING) LTD, PLANT DEPOT, 9 BARNARD HILL, MUSWELL HILL
Gauge : 1'6".   (   )

| - | 4wBE | CE |

WARD FERROUS METALS LTD, THAMES ROAD, SILVERTOWN
Gauge : 4'8½".   (TQ 417801)

| SUSAN | 4wDH | TH | 176V | 1966 |

PRESERVATION SITES

FLYING SCOTSMAN SERVICES, SOUTHALL M.P.D., SOUTHALL
Gauge : 4'8½".   (TQ 133798)

35028	CLAN LINE	4-6-2	3C	Elh		1948
(60103)	No.4472					
	FLYING SCOTSMAN	4-6-2	3C	Don	1564	1923
	LORD LEVERHULME	0-4-0DM		AB	388	1953

G.W.R. PRESERVATION GROUP, SOUTHALL M.P.D., SOUTHALL
Gauge : 4'8½".   (TQ 133798)

2885		2-8-0	OC	Sdn		1938
4110		2-6-2T	OC	Sdn		1936
9682		0-6-0PT	IC	Sdn		1949
68078		0-6-0ST	IC	AB	2212	1946
2 •	WILLIAM MURDOCH	0-4-0ST	OC	P	2100	1949
	BIRKENHEAD	0-4-0ST	OC	RSHN	7386	1948
A.E.C. No.1		4wDM		AEC		1938
ARMY 251						
	FRANCIS BAILY OF THATCHAM	0-4-0DM		RH	390772	1956

KEW BRIDGE STEAM MUSEUM, GREEN DRAGON LANE, BRENTFORD
Gauge : 2'0".   (TQ 188780)

	LILLA	0-4-0ST	OC	HE	554	1891
2		4wDM		L	44052	c1958

KINGS COLLEGE, STRAND
Gauge : 1'3".   (TQ 308808)

	PEARL	2-2-2WT	IC	B(C)	1860

LONDON REGIONAL TRANSPORT, LONDON TRANSPORT MUSEUM, COVENT GARDEN
Gauge : 4'8½".   (TQ 303809)

	-	4wWT	G	AP	807	1872
23		4-4-0T	OC	BP	710	1866
5	JOHN HAMPDEN	Bo-BoRE		VL		1922
4248		4w-4RER		GRC		1924
11182		2-2w-2w-2RER		MC		1939
22679		2w-2-2-2wRER		MC		1952 +

+   Stored at L.U.L. Ealing Common Depot.

MUSEUM OF LONDON, LONDON WALL, LONDON EC 2
Gauge : 2'0".   (TQ 322816)

-	4w Atmospheric Car	c1865

NORTH WOOLWICH OLD STATION MUSEUM, PIER ROAD, NORTH WOOLWICH
Gauge : 4'8½".   (TQ 433798)

35010	BLUE STAR	4-6-2	3C	Elh		1942 +
45293		4-6-0	OC	AW	1348	1936 +
No.14	DOLOBRAN	0-6-0ST	IC	MW	1762	1910 a
No.15	RHYL 8310/41	0-6-0ST	IC	MW	2009	1921 a
No.229		0-4-0ST	OC	N	2119	1876
	-	0-6-0ST	OC	P	2000	1942
No.29		0-6-0ST	IC	RSHN	7667	1950
	-	4wDM		FH	(3294	1948?)
54256		2w-2-2-2wRER		BRCW		1939 +

+ Not on public display.
a Stored at Custom House.

S.C. ROBINSON, 47 WAVERLEY GARDENS
Gauge : 2'0".   (TQ 187831)

-	4wDM	RH	209429	1942

ROYAL AIR FORCE MUSEUM, HENDON
Gauge : 2'0".   (   )

A.M.W. No. 165	4wDM	RH	194784	1939

Currently under overhaul at R.A.F. Stanbridge.

SCIENCE MUSEUM, SOUTH KENSINGTON
Gauge : 5'0".   (TQ 268793)

"PUFFING BILLY"	4w	VCG	Wm.Hedley	1827-1832 +

+ Incorporates part of loco of same name built c1814.

Gauge : 4'8½".

4073	CAERPHILLY CASTLE	4-6-0	4C	Sdn		1923
BAUXITE No.2		0-4-0ST	OC	BH	305	1874
	THE AGENORIA	0-4-0	VC	Foster Rastrick		1829
	"SANS PAREIL"	0-4-0	VC	Hackworth		1829
	ROCKET	0-2-2	OC	RS	19	1829
	DELTIC	Co-CoDE		EE	2007	1955
No.13		4wRE		MP/BP		1890 a
3327		4w-4RER		M		1929

a Actual identity unknown. Plates from No.1 fitted to
   another loco of same batch.

GREATER MANCHESTER

<u>INDUSTRIAL SITES</u>

<u>A.B.B. BRITISH WHEELSET LTD, TRAFFORD PARK WORKS,</u>
<u>ASHBURTON ROAD WEST, TRAFFORD PARK, MANCHESTER</u>
Gauge : 4'8½".   (SJ 779971)

10	0-4-0DM	(RSH	7810	1954
		(DC	2513	1954
14	0-4-0DM	(RSH	8089	1959
		(DC	2654	1959

<u>ADAMS PEAT PRODUCTS, FOUR LANE ENDS MILL, FOUR LANE ENDS,</u>
<u>off ASTLEY ROAD, IRLAM</u>
Gauge : 2'0".   (SJ 699954)

-	4wDM	L	7954	1936

<u>ASHTON CANAL CARRIERS, HANOVER STREET NORTH, GUIDE BRIDGE</u>
Gauge : 2'0".   (SJ 920978)

-	4wDM	HE	2820	1943
-	4wDM	HE	6012	1960
-	4wDM	MR	22031	1959
-	4wDM	RH	200761	1941

<u>BRITISH COAL</u>
See Section Four for full details.

<u>BRITISH FUEL CO, BROADWAY DEPOT, HUNT LANE, CHADDERTON</u>
Gauge : 4'8½".   (SD 903054)   RTC.

-	4wDH	TH/S	107C	1961 OOU

<u>FOREMOST, CROXDEN HORTICULTURAL PRODUCTS LTD, TWELVE YARDS ROAD, IRLAM</u>
Gauge : 2'0".   (SJ 715966)

-	4wDM	AK	No.5	1979
51651	4wDM	LB	51651	1960

<u>HOLLANDS MOSS PEAT CO, CHAT MOSS, IRLAM</u>
Gauge : 2'0".   (SJ 708963)

8	4wDM	L	37170	1951

J.C. GILLESPIE CIVIL ENGINEERING LTD, PLANT DEPOT, SORBY ROAD, IRLAM
Gauge : 1'6".   (SJ 720933)

-		2w-2BE	Iso	T79	1975	
-		0-4-0BE	WR	F7117	1966	
		Rebuilt	WR	10142	1985	
-		0-4-0BE	WR			

J.F. DONELON & CO LTD, PLANT DEPOT, CROWN LANE, HORWICH, near BOLTON
Gauge : 750mm.   (SD 628110)

	-	4wDH	Ageve	898	1980	
1		4wDH	Ageve			
	-	4wDH	Ruhr	3909	1969	
	-	4wDH	Ruhr	3920	1969	

Gauge : 2'0".

	T/E 319	4wBE	CE	5238	1966	
	T/E 320	4wBE	CE	5868/3	1971	
10	T/E 331	4wBE	CE	B0182B	1974	
	T/E 332	4wBE	CE	B0182A	1974	
	T/E 378	4wBE	CE	5481	1968	
	-	4wBE	CE	5378	1967	Dsm

Gauge : 1'6".

	T/E 300		4wBE	CE			
	T/E 301		0-4-0BE	WR	N7608	1974	
9	T/E 303		4wBE	CE	B0151	1973	
5	T/E 304		0-4-0BE	WR	N7609	1974	
11	T/E 306		0-4-0BE	Donelon			
15	T/E 307		0-4-0BE	Donelon			
12	T/E 333		4wBE	CE	B0171B	1974	
7	T/E 363		4wBE	CE	B0171C	1974	
3	T/E 364		4wBE	CE	B0171A	1974	
2	T/E 372		4wBE	CE	B1570	1978	
13	T/E 390		0-4-0BE	Donelon			
16	T/E 405	JFD 50	4wBE	CE	B0171D	1974	
	T/E 425		0-4-0BE	Donelon			
	T/E 429		4wBE	CE	5911C	1972	
	T/E 430		0-4-0BE	Donelon			
	T/E 435		0-4-0BE	Donelon			
	T/E 471		4wBE	CE	B2200A	1979	

Locos present in yard between contracts.

JOSE K. HOLT GORDON LTD, SCRAP MERCHANTS, CHEQUERBENT, WESTHOUGHTON
Gauge : 4'8½".   (SD 674062)

-	0-6-0ST	OC	AE	1600	1912	OOU
HARRY	0-6-0ST	IC	HC	1776	1944	Pvd

KILROE CIVIL ENGINEERING LTD, PLANT DEPOT, LOMAX STREET, RADCLIFFE
Gauge : 2'0".   (SD 784068)

	1511	4wBE	CE	5382	1966	
	1520	4wBE	CE	B0445	1975	
	2143	4wBE	CE	B3611	1989	
S10		4wBE	WR			

Gauge : 1'6".

	1291	4wBE	CE			+
S12	1293	4wBE	CE			+
	1294	4wBE	CE			+
S11	1517	4wBE	CE	B0122	1973	
	-	4wBE	CE			+
	-	4wBH	Kilroe			
S1		0-4-0BE	WR			a
S2		0-4-0BE	WR			a
	1293	0-4-0BE	WR			a
	1300	0-4-0BE	WR			a
	1508	0-4-0BE	WR			a
	1514	0-4-0BE	WR			a
	1518	0-4-0BE	WR			a
	2019	0-4-0BE	WR			a

+ Original locos included 5858/1971, B0105.A & B/1973 and
   B0172/1973.
a Original locos included 6600/1962, C6711/1963, N7611/1972
   and N7612 to N7614/1973
Locos present in yard between contracts.

MANCHESTER SHIP CANAL CO LTD,
MODE WHEEL SHED & WORKSHOPS, MODE WHEEL ROAD (SOUTH), WEASTE, SALFORD
& BARTON DOCK LOCO SHED, TRAFFORD PARK
Gauge : 4'8½".   (SJ 791955, 798977)

4002	0-6-0DE	HC	D1076	1959	OOU
D2	0-6-0DM	HC	D1187	1960	
D3	0-6-0DM	HC	D1188	1960	
D6	0-6-0DM	HC	D1191	1960	
D11	0-6-0DM	HC	D1255	1962	
CE 9123	4wDM	Robel 54-12-65 RR1			
DH23	4wDH	RR	10226	1965	
DH24	4wDH	RR	10227	1965	
DH26	4wDH	RR	10229	1965	

      See also entry under Cheshire.

NOBELS EXPLOSIVES CO LTD, ROBURITE WORKS, SHEVINGTON, near WIGAN
(Subsidiary of I.C.I. Ltd).
Gauge : 2'0".   (SD 543075)

3	4wDM	RH	381704	1955	
4	4wDM	RH	260716	1949	Dsm
5	4wDM	RH	280866	1949	Dsm
6	4wDM	RH	280865	1949	
7	4wDM	RH	260719	1948	
-	4wDM	RH	304439	1950	Dsm
-	4wDM	RH	381705	1955	

NORTH WEST WATER AUTHORITY, EASTERN DIVISION, ASHTON WORKS, DUKINFIELD
Gauge : 2'0".   (SJ 932973)

	CHAUMONT	4wDH	HU	LX1002	1968

PARKFIELD GROUP PLC, HORWICH WORKS, near BOLTON
Gauge : 4'8½".   (SD 637110)

(D2066)	03066	0-6-0DM	Don		1959
(D2094)	03094	0-6-0DM	Don		1960

PIKROSE & CO LTD, WINGROVE & ROGERS DIVISION, DELTA ROAD, AUDENSHAW
New WR locos under construction, and locos for repair, usually present.

POWELL DUFFRYN WAGON CO LTD, HEYWOOD WORKS, GREEN LANE, HEYWOOD
Gauge : 4'8½".   (SD 868102)

2		4wDH	S	10003	1959

POWER GEN, CARRINGTON POWER STATION, PARTINGTON
(A Division of C.E.G.B.)
Gauge : 4'8½".   (SJ 727933, 731934)

3	0-6-0DH	HE	8976	1979
No.2	0-6-0DH	HE	8977	1980

QUEGHAN CONSTRUCTION CO, PLANT DEPOT, STAMPSTONE STREET, OLDHAM
Gauge : 1'6".   (SJ   )

-	0-4-0BE	WR	7662	1975
	Rebuilt	WR	10132	1984
-	0-4-0BE	WR	7663	1975
	Rebuilt	WR	10133	1984

Locos present in yard between contracts.

RIVERTOWER CONSTRUCTION LTD,
PLANT DEPOT, UNIT 7, MORT LANE INDUSTRIAL ESTATE, TYLDESLEY
Gauge : 1'6"/2'0". (SD 712026)

432/33	4wBE	CE	5481	1968	
432/36	4wBE	CE	5628	1969	Dsm
-	4wBE	CE	5640	1969	
-	4wBE	CE	5940C	1972	
-	4wBE	CE	5940D	1972	
-	4wBE	CE	5961A	1972	
-	4wBE	CE	B0119D	1973	
-	4wBE	CE	B0142A	1973	
001	4wBE	CE			

		4wBE	CE		
		4wBE	CE		

At least eleven locos are now owned by this company.
Some more may be out on contract work.

Locos present in yard between contracts.

TRAFFORD PARK ESTATES CO, 'B' JUNCTION, THIRD AVENUE, TRAFFORD PARK
Gauge : 4'8½".   (SJ 794970, SJ 789971)

(D3538	08423)	0-6-0DE	Derby		1958
(D3836	08669)	0-6-0DE	Crewe		1960
	-	6wDH	TH	180V	1967
	-	4wDH	WB	3208	1961
	R.A. LAWDAY	0-6-0DE	YE	2878	1963

VERNON & ROBERTS LTD, AQUEDUCT WORKS, PEEL STREET, STALYBRIDGE
Gauge : 4'8½".   (SJ 957983)

BELLA	4wDM	RH	186309	1937	OOU

PRESERVATION SITES

C.P. BLACKHAM, 36 FRESHFIELD ROAD, HEATON MERSEY, STOCKPORT
Gauge : 1'3".   (   )

BLUE PACIFIC	4-6-0VB	OC	NL.Guinness	c1935

Loco is kept at private premises in Heaton Moor,
with no public access.

EAST LANCASHIRE RAILWAY PRESERVATION SOCIETY, BURY TRANSPORT MUSEUM, CASTLECROFT ROAD, BURY
Gauge : 4'8½".   (SD 803109)

7229		2-8-2T	OC	Sdn		1935	
45337		4-6-0	OC	AW	1392	1937	
46428		2-6-0	OC	Crewe		1948	
47298		0-6-0T	IC	HE	1463	1924	+
73156		4-6-0	OC	Don		1956	
76079		2-6-0	OC	Hor		1957	
80097		2-6-4T	OC	Bton		1954	
92207	MORNING STAR	2-10-0	OC	Sdn		1959	
	-	0-4-0ST	OC	AB	1927	1927	
32	GOTHENBURG	0-6-0T	IC	HC	680	1903	
No.70	PHOENIX	0-6-0T	IC	HC	1464	1921	
8	SIR ROBERT PEEL	0-6-0ST	IC	HE	3776	1952	
193	SHROPSHIRE	0-6-0ST	IC	HE	3793	1953	
E.L.R. No.1		0-6-0T	OC	RSHN	7683	1951	

```
   D335    (40135)              1Co-Co1DE    _(EE          3081  1961
                                             ‾(VF          D631  1961
  (D345)   40145                1Co-Co1DE    _(EE          3091  1961
                                             ‾(VF          D641  1961
   D832    ONSLAUGHT            B-BDH        Sdn                 1960
  D1041    WESTERN PRINCE       C-CDH        Crewe               1962
 (D2587)                        0-6-0DM      HE            7180  1969
                                Rebuild of   HE            5636  1959
  D2767    76.067               0-4-0DH      NB           28020  1960
                                Rebuilt      AB                  1968
 (D2774)                        0-4-0DH      NB           28027  1960
                                Rebuilt      AB                  1968
 (D2956)   01003                0-4-0DM      AB             398  1956
  D5054    (24054)              Bo-BoDE      Crewe               1959
  D7076                         B-BDH        BPH           7980  1963
 (D7612)   25901                Bo-BoDE      Derby               1966
 (D7659)   25909                Bo-BoDE      BP            8069  1966
  D9531                         0-6-0DH      Sdn                 1965
RDB 975003 (Sc 79998) GEMINI    4w-4wBER     Derby C&W/Cowlairs 1958
           BENZOLE              4wDM         FH            3438  1950
           WINFIELD             4wDM         MR            9009  1948
           JEAN                 0-4-0DH      RR           10204  1965
           -                    4wDH         S            10175  1964
```

+ Currently under renovation at Kirby, Merseyside.

GREATER MANCHESTER SCIENCE & RAILWAY MUSEUM, LIVERPOOL ROAD, MANCHESTER
Gauge : 5'6".   (SJ 831978)

```
   3157                         4-4-0    IC  VF            3064  1911
```

Gauge : 4'8½".

```
  44806    MAGPIE               4-6-0    OC  Derby               1944
 (58926)   1054                 0-6-2T   IC  Crewe         2979  1888
           LORD ASHFIELD        0-6-0F   OC  AB            1989  1930
AGECROFT No.3                   0-4-0ST  OC  RSHN          7681  1951
           "NOVELTY"            2-2-0VBWT VC  Science Mus.       1929 +
 (27001)   ARIADNE  1505        Co-CoWE      Gorton        1066  1954
           -                    4wBE         EE            1378  1944
      1    -                    Bo-BoWE      HL            3682  1927
           -                    0-4-0DM      JF         4210074  1952
```

+ Replica of original loco built 1829, by Braithwaite & Ericsson

Gauge : 3'6".

```
   2352                         4-8-2+2-8-4T  4C  BP        6639  1930
```

Gauge : 3'0".

```
No.3       PENDER               2-4-0T   OC  BP            1255  1873 +
```

+ Loco is sectioned to show moving parts.

HAIGH RAILWAY, HAIGH HALL COUNTRY PARK, WIGAN
Gauge : 1'3".   (SD 599087)

| 7204 | HAIGH HALL | 2-4-2 | OC | Guest | | 14 | 1954 |
| 15 | W. BROGAN B.E.M. | 0-6-0DM | | Guest | | 15 | 1959 |

MOSELEY INDUSTRIAL TRAMWAY MUSEUM, MANOR SCHOOL,
NORTHDOWNS ROAD, CHEADLE, STOCKPORT
Gauge : 2'6".   (SJ 864871)

| "No.27" | RTT/767032 | 2w-2PM | | Wkm | 3564 | | Dsm |

Gauge : 2'0".

No.1	THE LADY D	4wDM		MR	8934	1944	
2	MOSELEY	4wDM		RH	177639	1936	
3		4wDM		MR	8878	1944	
"No.4"		4wDM	S/O	RH	229647	1943	
5		4wDM		RH	223667	1943	
6		4wPM		MR	9104	1941	
6	NEATH ABBEY	4wDM		RH	476106	1964	
7		4wDM		MR	8663	1941	
"No.8"		4wPM		KC		c1926	
9		4wPM		MR	4565	1928	
"No.10"		4wDM		MR	7522	1948	
"No.11"	ALD HAGUE	4wDM		FH	3465	1954	
"No.12"		4wDM		L	38296	1952	Dsm
No.13		4wDM		MR	11142	1960	
14	KNOTHOLE WORKER	4wDM		MR	22045	1959	
"No.15"	NICK THE GREEK	4wDM		MR	8937	1944	
"No.19"		4wDM		MR	7512	1938	
"No.20"		4wBE		GB	2345	1950	
"No.21"		4wDM		MR	8669	1941	
"No.22"		4wDM		L	8022	1936	
23		4wDM		L	52031	c1960	a
"No.25"		4wDM		HE	4758	1954	
"No.28"	RTT/767094	2w-2PM		Wkm			Dsm
81.03	L.C.W.W. 18	4wDM		HE	6299	1964	
–		4wPM		L	3834	1931	
–		4wDM		LB	52885	1962	
–		4wDM		RH			+

+ Either 195846/1939 or 221610/1943.
a Plate reads 25031.

Gauge : 1'6".

| "No.16" | | 4wBE | | GB | 2960 | 1959 | Dsm |
| "No.17" | | 4wBE | | GB | 420172 | 1969 | Dsm |

RED ROSE LIVE STEAM GROUP, ASTLEY GREEN COLLIERY MUSEUM, ASTLEY, TYLDESLEY
Gauge : 4'8½".   (SJ 705998)

| – | | 0-4-0ST | OC | AE | 1563 | 1908 | |
| – | | 4wDM | | RH | 244580 | 1946 | |

Gauge : 2'0".

-	4wDM	MR	11218	1962
-	4wDM	RH	422573	1958

SALFORD METROPOLITAN DISTRICT COUNCIL, GEORGE THORNS RECREATION CENTRE, LIVERPOOL ROAD, IRLAM
Gauge : 4'8½". (SJ 722943)

-	0-4-0F	OC	P	2155	1955

MERSEYSIDE

<u>INDUSTRIAL SITES</u>

BRITISH COAL
> See Section Four for full details.

DAVE WILLIAMS, HAULAGE CONTRACTOR, BIRKENHEAD DOCKS
Gauge : 4'8½". (   )

WABANA	0-4-0DM	⌐(RSH	7814	1953	00U
		└(DC	2500	1953	
W.H.SALTHOUSE	0-4-0DM	⌐(RSH	7923	1959	00U
		└(DC	2590	1959	
PEGASUS	0-4-0DM	⌐(VF	D99	1949	00U
		└(DC	2270	1949	
-	0-4-0DM	⌐(VF	D138	1951	00U
		└(DC	2272	1951	

FORD MOTOR CO LTD, HALEWOOD FACTORY, LIVERPOOL
Gauge : 4'8½". (SJ 450845)

No.1	A12	0-4-0DH	YE	2807	1960
No.2	A36	0-4-0DH	YE	2675	1961
No.3	A16	0-4-0DH	YE	2679	1962

J.M.R.(SALES) LTD, BIRKDALE TRADING ESTATE, LIVERPOOL ROAD, BIRKDALE, SOUTHPORT
Gauge : 1'3". (SD 329146)

PRINCESS ANNE	4-6wDE	S/O	Barlow		1962

NORWEST HOLST PLANT LTD, BRIDGE HOUSE, DUNNINGSBRIDGE ROAD, BOOTLE
Gauge : 2'0".   (   )

		4wBE	CE	B2228A	1980
		4wBE	CE	B2228B	1980
		4wBE	WR	556801	1988

PILKINGTON FLAT GLASS LTD, GLASS MANUFACTURERS,
COWLEY HILL WORKS, ST. HELENS
Gauge : 4'8½". (SJ 514965)

		0-4-0DE	YE	2626	1956 OOU
		0-4-0DE	YE	2781	1960 OOU

SHEPPERD FRAGMENTISERS LTD, ST. HELENS JUNCTION SCRAPYARD, ST. HELENS
Gauge : 4'8½". (SJ 534932)

		0-4-0DM	JF	4210085	1953 OOU

U.G. GLASS CONTAINERS LTD, PEASLEY GLASSWORKS,
PEASLEY CROSS LANE, ST. HELENS
Gauge : 4'8½". (SJ 519945)

No.1	PEASLEY	0-4-0DE	YE	2653	1957
No.2	PEASLEY	0-4-0DE	YE	2730	1958

PRESERVATION SITES

KNOWSLEY SAFARI PARK, KNOWSLEY HALL, near PRESCOT
Gauge : 1'3". (SJ 460936)

4498	SIR NIGEL GRESLEY	4-6w-2DE	S/O	Barlow		1950

LAKESIDE MINIATURE RAILWAY, MARINE LAKE, SOUTHPORT
Gauge : 1'3". (SD 331174)

	RED DRAGON	4-4-2	OC	G.Walker	1984 +
4468	DUKE OF EDINBURGH	4-6-2+4-4DE	S/O	Barlow	1948
No.2510	PRINCE CHARLES	4-6-2+4-4DE	S/O	Barlow	1954
GOLDEN JUBILEE 1911-1961		4-6w+4-4DE		Barlow	1963
		2-6-4PM	S/O	S.Battison	1958
No.1		4-6-0DM	S/O	Jubilee Min Rly	1987 a
	PRINCESS ANNE	6w-6DH		SL	1971
14		2w-2PM		G.Walker	1985

    + Incorporates parts of BL 15/1909.
    a In store.

MERSEYSIDE METROPOLITAN COUNTY COUNCIL, MERSEYSIDE COUNTY MUSEUM,
WILLIAM BROWN STREET, LIVERPOOL
Gauge : 4'8½".   (SJ 349908)

	LION	0-4-2	IC	Todd,Kitson	
				& Laird	1838
M.D. & H.B. No.1		0-6-0ST	OC	AE	1465	1904
No.3		4w-4wRER		BM		1892

SOUTHPORT PIER RAILWAY, SOUTHPORT
Gauge : 60cm.   (SD 335176)

ENGLISH ROSE	4w-4DH	SL	23	1973

SOUTHPORT RAILWAY CENTRE, DERBY ROAD MOTIVE POWER DEPOT, SOUTHPORT
Gauge : 4'8½".   (SD 341170)

5193		2-6-2T	OC	Sdn		1934	
	EFFICIENT	0-4-0ST	OC	AB	1598	1918	
	LUCY	0-6-0ST	OC	AE	1568	1909	a
5	CECIL RAIKES	0-6-4T	IC	BP	2605	1885	
	FITZWILLIAM	0-6-0ST	IC	HE	1954	1939	
NS 8812		0-6-0ST	IC	HE	3155	1944	
9		0-6-0ST	IC	HE	3855	1954	
	WHITEHEAD	0-4-0ST	OC	P	1163	1908	
	HORNET	0-4-0ST	OC	P	1935	1937	
	-	0-4-0ST	OC	P	1999	1941	
5		0-6-0ST	OC	P	2153	1954	
AG-2	AGECROFT No.2	0-4-0ST	OC	RSHN	7485	1948	
	ST.MONANS	4wVBT	VCG	S	9373	1947	
(D2149	03149) D2148	0-6-0DM		Sdn		1960	
(D2593)	D2595	0-6-0DM		HE	7179	1969	
		Rebuild of		HE	5642	1959	
D2870		0-4-0DH		YE	2677	1960	
(D5081)	24081	Bo-BoDE		Crewe		1960	
M28361M		4w-4wRER		Derby C&W		c1939	
No.1		4wBE		GB	2000	1945	
	PERSIL	0-4-0DM		JF	4160001	1952	
B.I.C.C. No.46		0-4-0DH		NB	27653	1956	
	SEFTON	4wDH		TH	123V	1963	
STANLOW No.4		0-4-0DH		TH	160V	1966	

          a   Under renovation at Cammell Laird Ltd.

Gauge : 2'0".

12	4wDM	MR	11258	1964

WIRRAL BOROUGH COUNCIL, SHORE ROAD MUSEUM, BIRKENHEAD
Gauge : 4'8½".   (   )

   M28690M IVOR T. DAVIES G.M. 4w-4wRER   Derby C&W   c1939

          Currently in store at B.R. Birkenhead Central Depot.

NORFOLK

BRITISH SUGAR CORPORATION LTD
Cantley Factory
Gauge : 4'8½".   (TG 386034)

1	0-6-0DM	RH	304468	1950
2	0-6-0DM	RH	395301	1956 OOU

Saddlebow Factory, South Lynn, Kings Lynn
Gauge : 4'8½".   (TF 609178)

-	0-4-0DM	RH	281266	1950
-	0-4-0DM	RH	327974	1954

COSTAIN DOW MAC, ATLAS WORKS, LENWADE
Gauge : 4'8½".   (TF 114179)

D2118	0-6-0DM	Sdn		1959

DOW CHEMICAL CO LTD, CROSSBANK ROAD, KINGS LYNN
Gauge : 4'8½".   (TF 613215)

(D2997 07013)	0-6-0DE	RH	480698	1962

MAY GURNEY & CO LTD, PLANT DEPOT, TROWSE, NORWICH
Gauge : 1'6".   (   )

-	2w-2BE	Iso
-	2w-2BE	Iso

Locos present in yard between contracts.

MAYER NEWMAN & CO LTD, LONGWATER WORKS, LONGWATER TRADING ESTATE,
DEREHAM ROAD, COSTESSEY, NORWICH
Gauge : 4'8½".   (TG 159110)

-	4wDM	RH	466625	1962 OOU

REDLAND AGGREGATES LTD, TROWSE STATION, NORWICH
Gauge : 4'8½".   (TG 243071)

-	4wDH	TH/S	177C	1967

PRESERVATION SITES

ALAN BLOOM, BRESSINGHAM LIVE STEAM MUSEUM, BRESSINGHAM HALL, near DISS
Gauge : 4'8½". (TM 080806)

(30102)	GRANVILLE	0-4-0T	OC	9E	406	1893
(32662)	662	0-6-0T	IC	Bton		1875
(41966)	80 THUNDERSLEY	4-4-2T	OC	RS	3367	1909
(42500)	2500	2-6-4T	3C	Derby		1934
(46100)	6100 ROYAL SCOT	4-6-0	3C	Derby		1930
(46233)	6233 DUCHESS OF SUTHERLAND					
		4-6-2	4C	Crewe		1938
(62785)	490	2-4-0	IC	Str	836	1894
70013	OLIVER CROMWELL	4-6-2	OC	Crewe		1951
	-	0-4-0F	OC	AB	1472	1916
6841	WILLIAM FRANCIS					
		0-4-0+0-4-0T	4C	BP	6841	1937
141R73	TOM PAINE	2-8-2	OC	Lima	8939	1945 +
No.1		0-4-0T	OC	N	4444	1892
No.25		0-4-0ST	OC	N	5087	1896
377	KING HAAKON 7	2-6-0	OC	Nohab	1164	1919
	MILLFIELD	0-4-0CT	OC	RSHN	7070	1942
5865	PEER GYNT	2-10-0	OC	Schichau		1944

+ Carries incorrect worksplate.

Gauge : 1'11".   Nursery Line.

No.1643	BRONLLWYD	0-6-0WT	OC	HC	1643	1930
No.316	GWYNEDD	0-4-0STT	OC	HE	316	1883
No.994	GEORGE SHOLTO	0-4-0ST	OC	HE	994	1909
No.1	EIGIAU	0-4-0WT	OC	OK	5668	1912
	A.W.A. 12120	4wDM		MR	22210	1964

Gauge : 1'3".   Waveney Valley Railway.

No.1662	ROSENKAVALIER	4-6-2	OC	Krupp	1662	1937
No.1663	MäNNERTREU	4-6-2	OC	Krupp	1663	1937
4472	FLYING SCOTSMAN	4-6-2	3C	WP.Stewart		1976
	IVOR	4wPH		Bressingham		1979

BURE VALLEY RAILWAY LTD, THE BROADLAND LINE, AYLSHAM
Gauge : 1'3".   (TG 197265)

6	SAMSON	4-8-2	OC	DP	22072	1926 +
11	BLACK PRINCE	4-6-2	OC	Krupp	1664	1937 +
9	WINSTON CHURCHILL	4-6-2	OC	YE	2294	1931 +
	-	4w-4wDH		BVR		1989

+ On loan from Romney, Hythe & Dymchurch Railway, Kent.

BURGH HALL BYGONE VILLAGE, FLEGGBURGH (BURGH ST MARGARETS),
near GREAT YARMOUTH
Gauge : 4'8½". (TG 450140)

1928	JOHN GLADDEN	2-6-4T	OC	Nohab	2229	1953
1212		4w-4DMR		EK		1958
No.1		0-4-0DM		JF	20337	1934
LNER 338	900338	2w-2PM		Wkm	(626?)	

Gauge : 2'0".

SEZELA No.4		0-4-0T	OC	AE	1738	1915
	-	4wDM	S/O	MR	9869	1953
1	THE ROYAL TOURNAMENT	4wDH		SL		1986
	BANK OF SCOTLAND	4wDH		SL		1986

G.T. CUSHING, STEAM MUSEUM, THURSFORD GREEN, THURSFORD, near FAKENHAM
Gauge : 1'10¾". (TF 980345)

	CACKLER	0-4-0ST	OC	HE	671	1898

PETER HILL, CAISTER CASTLE, near YARMOUTH
Gauge : 4'8½".   (TG 504123)

42	RHONDDA	0-6-0ST	IC	MW	2010	1921

R.& A. JENKINS, FRANSHAM STATION, GREAT FRANSHAM, near SWAFFHAM
Gauge : 4'8½".   (TF 888135)

	-	0-6-0ST	OC	HC	1208	1916
	-	4wDM		RH	398611	1956

KILVERSTONE WILDLIFE PARK, THETFORD
Gauge : 1'8".   (TL 890840)

	ROBIN HOOD	4-6-4DH	S/O	HC	D570	1932
4472	FLYING SCOTSMAN	4-6-2DM	S/O	HC	D582	1933

M. MAYES
Gauge : 4'8½".   (   )

	-	0-4-0VBT	OC	Cockerill	2525	1907

Loco under renovation at secret location.

NORTH NORFOLK RAILWAY CO LTD
Gauge : 4'8½".   Locos are kept at :-

		Sheringham	(TG 156430)				
		Weybourne	(TG 118419)				

(61572)	8572	4-6-0	IC	BP		6488	1928
(65462)	564	0-6-0	IC	Str			1912
	HARLAXTON	0-6-0T	OC	AB		2107	1941
ED4	EDMUNDSONS	0-4-0ST	OC	AB		2168	1943
	WISSINGTON	0-6-0ST	IC	HC		1700	1938
1982	RING HAW	0-6-0ST	IC	HE		1982	1940
3809		0-6-0ST	IC	HE		3809	1954
	PONY	0-4-0ST	OC	HL		2918	1912
1970	JOHN D.HAMMER	0-6-0ST	OC	P		1970	1939
12		0-6-0T	OC	RSH		7845	1955
2370		0-6-0F	OC	WB		2370	1929
No.4	"BIRCHENWOOD"	0-6-0ST	OC	WB		2680	1944
D5386	(27066)	Bo-BoDE		BRCW	DEL	229	1962
12131		0-6-0DE		Dar			1952
(E79960)		2w-2DMR		WMD		1265	1958
E79963		2w-2DMR		WMD		1268	1958
3052	CAR No.291	4w-4wRER		MC			1932 +
N.C.B. 10 1963		0-4-0DH		EES		8431	1963
	RANSOMES	4wDM		RH		466629	1962
M. & G. N. No.1		2w-2PMR		Wkm		1521	1934 Dsm
No.2		2w-2PMR		Wkm		1522	1934 Dsm
960243	THE BUG	2w-2PMR		Wkm		1642	1934 Dsm

+ Now unmotorised.

ST. JOHN FOTI, WELLE MANOR HALL, NEW ROAD, UPWELL
Gauge : 2'0".   (   )

B.L.R. No.6		4wDM		L	37658	1952
	-	4wDM	S/O	MR	9774	1953 +

+ Stored at Norfolk Punch factory.

STRUMPSHAW HALL STEAM MUSEUM, STRUMPSHAW HALL, near ACLE
Gauge : 1'11½". (TF 345065)

No.6	GINETTE MARIE	0-4-0WT	OC	Jung	7509	1937
No.1		4wDM	S/O	MR	7192	1937

Gauge : 1'3".

2	CAGNEY	4-4-0	OC	P.McGarigle	1902

ERIC WALKER, WOLFERTON STATION, near KINGS LYNN
Gauge : 4'8½". (TF 661286)

960220		2w-2PMR	Wkm	1933

WENSUM VALLEY RAILWAY, COUNTY SCHOOL STATION, NORTH ELMHAM, near DEREHAM
Gauge : 4'8½".   (TF 990227)

		0-4-0DH	RH	497753	1963
-					

YAXHAM LIGHT RAILWAYS, YAXHAM STATION YARD, YAXHAM, near DEREHAM
Gauge : 1'11½".   (TG 003102)

No.16	ELIN	0-4-0ST	OC	HE	705	1899
Y.P.L.R. No.1		0-4-0VBT	VCG	Potter		1970
No.2	RUSTY	4wDM		L	32801	1948
No.3	PEST	4wDM		L	40011	1954
No.4	GOOFY	4wDM		OK	7688	c1936
No.5		4wDM		OK	7728	1938
No.6	COLONEL LOD/758097	4wDM		RH	202967	1940
No.7		4wDM		RH	170369	1934
No.9		4wDM		RH	202969	1940
No.10	OUSEL	4wDM		MR	7153	1937
No.13		4wDM		MR	7474	1940
No.14	LOD 758375	4wDM		RH	222100	1943
No.15	EPV 785	2-2wDM		Potter		1977
	DOE 3982	4wDM		FH	3982	1962

NORTHAMPTONSHIRE

INDUSTRIAL SITES

BRITISH STEEL TUBES, CORBY WORKS, CORBY
(Subsidiary of British Steel Plc)
Gauge : 4'8½".   (SP 909899)

B.S.C. 1		0-6-0DH	EEV	D1049	1965	+
B.S.C. 2		0-6-0DH	GECT	5395	1974	OOU
B.S.C. 3		0-6-0DH	GECT	5365	1972	
D 27		0-6-0DH	EEV	5358	1971	
D 32		0-6-0DH	GECT	5388	1973	OOU
33		0-6-0DH	GECT	5394	1974	OOU a
D 35		0-6-0DH	GECT	5407	1975	

+ Carries plate GECT 5408/1975 in error.
a Carries plate EEV D1052/1965 in error.

PRESERVATION SITES

BILLING AQUADROME LTD, BILLING, near NORTHAMPTON
Gauge : 2'0".   (SP 808615)

006		4wDH	S/O	AK	14	1984

CORBY DISTRICT COUNCIL, EAST CARLTON COUNTRY HERITAGE PARK, CORBY
Gauge : 4'8½".   (   )

-		0-6-0ST	OC	HL	3827 1934

DAVENTRY BOROUGH COUNCIL, NEW STREET RECREATION GROUND, DAVENTRY
Gauge : 4'8½".   (SP 574624)

"CHERWELL"		0-6-0ST	OC	WB	2654 1942

IRCHESTER NARROW GAUGE RAILWAY TRUST, IRCHESTER COUNTRY PARK, IRCHESTER
Gauge : Metre.   (SP 904659)

	CAMBRAI	0-6-0T	OC	Corpet	493	1888
No.85	BANSHEE	0-6-0ST	OC	P	1870	1934
No.86		0-6-0ST	OC	P	1871	1934
	THE ROCK	0-4-0DM		HE	2419	1941
ND 3647		4wDM		MR	22144	1962
ND 3645		4wDM		RH	211679	1941

Gauge : 3'0".

-		0-6-0DM	RH	281290	1949
(ED 10)		4wDM	RH	411322	1958

Gauge : 2'0".

-		4wDM	MR	3797	1926
		Rebuild of MR		1367	

NORTHAMPTON STEAM RAILWAY, PITSFORD STATION SITE
Gauge : 4'8½".   (SP 735667)

3862		2-8-0	OC	Sdn		1942
7283	YVONNE	0-4-0VBT	OC	Cockerill	2945	1920
45	COLWYN	0-6-0ST	IC	K	5470	1933
-		0-4-0ST	OC	P	2104	1950 +
(D67)	45118	1Co-Co1DE		Crewe		1962
D5185	25035					
	CASTELL DINAS BRAN	Bo-BoDE		Dar		1963
(D5401)	(27056)	Bo-BoDE		BRCW	DEL 244	1962
4002	S 13004 S	4w-4wRER		Elh		1949
No.1		4wDM		RH	275886	1949
764		0-4-0DM		RH	319286	1953
	REDLAND	0-4-0DH		TH	146C	1964
		Rebuild of 0-4-0DM		JF	4210018	1950

+ Actually built 1948 but plates are dated as shown.

NORTHAMPTONSHIRE IRONSTONE RAILWAY TRUST LTD, HUNSBURY HILL SITE
Gauge : 4'8½".   (SP 735584)

	TRYM	0-4-0ST	OC	HE	287	1883	
No.14	BRILL	0-4-0ST	OC	MW	1795	1912	
9365	BELVEDERE	4wVBT	VCG	S	9365	1946	
	MUSKETEER	4wVBT	VCG	S	9369	1946	
	HYLTON	4wDH		FH	3967	1961	
16		0-4-0DM		HE	2087	1940	
	EXPRESS	4wDM		RH	235511	1945	Dsm
39	SPITFIRE	4wDM		RH	242868	1946	
	-	4wDM		RH	299100	1950	
	-	4wDM		RH	386875	1955	

Gauge : Metre.

No.87	8315/87	0-6-0ST	OC	P	2029	1942	
A 16 W		2w-2PMR		Wkm	6887	1954	+

    + Loco is dual gauge fitted for 4'8½" and Metre.

Gauge : 2'0".

	-	4wPM	L	14006	1940	
T 8		4wDM	MR	8731	1941	Dsm
22		4wDM	MR	8756	1942	
T 13		4wDM	MR	8969	1945	Dsm
T 11		4wDM	MR	9711	1952	

OVERSTONE SOLARIUM LIGHT RAILWAY, OVERSTONE SOLARIUM, SYWELL
Gauge : 2'0".   (SP 819654)   RTC.

-	4wDM	S/O	MR	8727	1941	OOU

RUSHDEN HISTORICAL TRANSPORT SOCIETY, THE OLD STATION, RUSHDEN
Gauge : 4'8½".   (SP 957672)

SIR VINCENT	4wWT	G	AP	8800	1917
-	0-4-0DM		AB	363	1942

Gauge : 2'0".

2	81 A 26	4wDM	MR	7333	1938

WICKSTEED PARK LAKESIDE RAILWAY, KETTERING
Gauge : 4'8½".   (SP 879773)

-	0-4-0ST	OC	AB	2323	1952	Pvd

Gauge : 2'0".   (SP 883770)

LADY OF THE LAKE	0-4-0DM	S/O	Bg	2042	1931
KING ARTHUR	0-4-0DH	S/O	Bg	2043	1931
CHEYENNE	4wDM	S/O	MR	22224	1966

K. WOOLMER, 15 BAKERS LANE, WOODFORD, KETTERING
Gauge : 2'0".   (SP 969769)

-	4wDM	L	36743	1951	Pvd

NORTHUMBERLAND

INDUSTRIAL SITES

AYLE COLLIERY CO LTD, AYLE EAST DRIFT, ALSTON
Gauge : 2'6".   (NY 728498)

6/41	0-4-0BE	WR	6133	1959	Dsm
6/43	4wBE	WR	6297	1960	Dsm
6/44	0-4-0BE	WR	6595	1962	Dsm
6/44	0-4-0BE	WR	C6710	1963	Dsm
6/46	0-4-0BE	WR	6593	1962	Dsm

Gauge : 2'0".

-	4wBE		CE	5667	1969	
EL 16	4wBE		CE	5667	1969	Dsm
UG 10	4wBE		CE	5667	1969	
-	4wBE	FLP	GB	2382	1951	Dsm
-	4wDM	FLP	HE	3496	1947	Dsm
-	4wDM	FLP	HE	4569	1956	OOU
-	0-4-0DM	FLP	HE	4991	1955	OOU
	Rebuilt 4wDM	FLP	Ayle		1977	
LE/12/75 P17271	4wBE		WR	C6765	1963	
-	0-4-0BE		WR	P7664	1975	
-	0-4-0BE		WR	P7731	1975	

Gauge : 1'10".

-	4wBE	WR	5655	1956	Dsm

Gauge : 1'7½".

-	0-4-0BE	WR	D6754	1964	Dsm

BLENKINSOPP COLLIERY LTD, CASTLE DRIFT, GREENHEAD
Gauge : 2'0".   (NY 665645)

-	0-4-0BE	WR	8079	1980	OOU

BRITISH COAL
        See Section Four for full details.

DEPARTMENT OF THE ENVIRONMENT, REDESDALE RANGES MILITARY TARGET RAILWAY
Gauge : 2'6".   (NT 827016)

L4135	2w-2PM	Wkm	3174	1943
L4134	2w-2PM	Wkm	3175	1943
-	2w-2PM	Wkm	3245	1943

E.R.S. MINING DEVELOPMENTS (N.E.) LTD, WHITTLE MINE, NEWTON-ON-THE-MOOR
Gauge : 2'0".   (NU 175062)

-	4wBE	(EE	2527	1957
		(RSHN	7086	1957
-	4wBE	(EE	2474	1958
		(RSHN	7942	1958
-	4wDM	SMH	104063G	1976

Battery locos used underground.

NORTHUMBRIAN PEAT, WARK FOREST, STONEHAUGH
Gauge : 2'0".   (NY 777730)

-	4wDM	L	26366	1944

PRESERVATION SITES

R. CANT, SLAGGYFORD LIGHT RAILWAY, THE ISLAND, SLAGGYFORD, near ALSTON
Gauge : 2'0".   (NY 680523)

3236	0-4-0DM	S/O Bg	3236	1947
-	4wDM	HE	2577	1942

HEATHERSLAW LIGHT RAILWAY CO LTD, HEATHERSLAW MILL, near COLDSTREAM
Gauge : 1'3".   (NT 933385)

THE LADY AUGUSTA	0-4-2	OC	B.Taylor	1989
CLIVE	6wDM		N.Smith	1989

JOHN MOFFITT, "HUNDAY", NATIONAL TRACTOR & FARM MUSEUM, NEWTON, STOCKSFIELD
Gauge : 2'6".   (NZ 042652)

YARD No.85 HUNDAY	0-4-0DM	HE	2250	1940
LOT No.3 9/56	0-4-0DM	HE	2252	1940

J. PARKER, SLAGGYFORD STATION
Gauge : 4'8½".   (NY 676524)

No.13	THE BARRA	0-4-0ST	OC	HL	3732	1928

WANSBECK DISTRICT COUNCIL,
WOODHORN MINING MUSEUM, WOODHORN COLLIERY, ASHINGTON
Gauge : 2'6".   (NZ 287884)

		4wDM	RH	256314	1949

NOTTINGHAMSHIRE

INDUSTRIAL SITES

A.B. & P.G. KETTLE, VEHICLE BREAKERS, QUARRY LANE, MANSFIELD
Gauge : 2'6".   (   )

P 9261	0-4-0DM	HE	2254	1940

Kept elsewhere at a private location.

BRITISH COAL
See Section Four for full details.

BRITISH GYPSUM LTD, RUSHCLIFFE WORKS, EAST LEAKE
Gauge : 4'8½".   (SK 553280)   RTC.

	4wDM	RH	236364	1946
	4wDM	RH	398616	1956

J.R. CLARK, PLANT DEALER, WIGSLEY WOOD SCRAPYARD
Gauge : 4'8½".   (SK 845708)

A 1092	4wDM	RH	224354	1945 OOU
SF 364	4wDM	RH	338415	1953 OOU

ENGLISH GLASS LTD, HARWORTH
Gauge : 4'8½".   (SK 624909)

	4wDM	RH	299099	1950
161 301	4wDM	RH	458959	1961

NATIONAL POWER, STAYTHORPE POWER STATION, near NEWARK
(A Division of C.E.G.B.)
Gauge : 4'8½".   (SK 763538, 764534)

No.5A	0-4-0DE	RH	420137	1958
No.1B	0-4-0DE	RH	421435	1958
No.2B	0-4-0DE	RH	449754	1961

Andrew Barclay 473 of 1961 at National Power, Padiham Power Station, Lancashire, on 22nd May 1985. (J.A. Foster)

Brook Victor 308 of 1968 at Royal Ordnance Bishopton, Strathclyde, on 10th August 1990. (J.A. Foster)

Brush/Bagnall 3072 of 1954 at BSC Velindre, West Glamorgan, on 5th
September 1988.   (K.A. Scanes)

Clayton B0457 of 1974 at Wincilate Ltd, Aberllefeni, Gwynedd, on 5th
September 1989.   (J.A. Foster)

Clayton B1551B of 1977 (of Fairclough Civil Engineering) on hire to Kilroe
Civil Engineering at the Don Valley Sewer Contract, Phase 4, Sheffield, South
Yorkshire, on 3rd March 1990.   (J.A. Foster)

Deutz 57138 of 1960 at Bord Na Mona, Tionnsca Abhainn Einne, County
Mayo, on 12th May 1988.   (D. Allison)

Drewry 2567 of 1955, at Butterley Ltd, Ripley, Derbyshire, on 15th April 1976. (A.J. Booth)

English Electric 8449 of 1965 (with Thomas Hill 278v of 1978 alongside) at Blue Circle Ltd, Westbury, Wiltshire, on 30th May 1990. (A.J. Booth)

John Fowler 4240010 of 1960 at Castle Cement Ltd, Newcastle, Tyne & Wear, on 16th September 1990. (J.A. Foster)

Greenwood & Batley 420452 of 1979, at Monckton Coke & Chemical Co Ltd, Royston, West Yorkshire, on 27th August 1983. (A.J. Booth)

Hibberd 3831 of 1958 of the Ffestiniog Railway Co at Minffordd, Gwynedd, on 13th September 1989. (J.A. Foster)

Thomas Hill 177c of 1967 at Redland Aggregates Ltd, Norwich, Norfolk, on 20th August 1988. (K.A. Scanes)

Hudswell Clarke D1279 of 1963, at C.F. Booth Ltd, Rotherham, South Yorkshire, on 28th October 1990. Ex-BR 12071 is in the left background.

(A.J. Booth)

Hunslet 7183 of 1970, at Anglesey Aluminium Metal Ltd, Holyhead,
Gwynedd, on 13th September 1990. (A.J. Booth)

Hunslet 8936 of 1980 at Bord Na Mona, near Littleton Power Station, County Tipperary, on 14th May 1988.   (D. Allison)

Alan Keef 21 of 1987 at Nobels Explosives Co Ltd, Ardeer Works, Strathclyde, on 22nd June 1987.   (K.A. Scanes)

Lister 55413 of 1967, at HM North Sea Camp Open Prison, Boston, Lincolnshire, on 1st June 1990. (A.J. Booth)

Motor Rail 8678 of 1941, at William Blyth Ltd, Barton on Humber, Humberside, on 6th May 1990. The regular driver stands alongside. (A.J. Booth)

Moyse 1464 of 1979 at Stockton Haulage Ltd, Middlesbrough, Cleveland, on 25th September 1990. (J.A. Foster)

North British 27644 of 1959 at Tees Storage Co Ltd, Middlesbrough, Cleveland, on 22nd March 1989. (J.A. Foster)

Plasmor Ltd of Knottingley, West Yorkshire, built this loco themselves in 1972. Photographed on 13th June 1990. (A.J. Booth)

Ruston & Hornsby 174532 of 1936 at L & P Peat Products Ltd, Nutberry Works, Eastriggs, Dumfries & Galloway, on 8th May 1985. (K.A. Scanes)

Ruston & Hornsby 459517 of 1961 at MOD Ashchurch, Gloucestershire, on 14th September 1990. (D.R. Colley)

Schoma 4017 of 1974 of British Rail's Departmental Stock at Georgemas Junction, Highland, on 6th May 1988. (R. Bryant)

Schweizerische Locomotiv-und-Maschinenfabrik 925 of 1895 on the Snowdon Mountain Railway, Llanberis, Gwynedd, on 6th September 1989.

(J.A. Foster)

Sentinels 10108 of 1963 and 10186 of 1964 at RMC Industrial Minerals Ltd, Peak Dale, Derbyshire, on 27th March 1990. (D.R. Colley)

Wingrove & Rogers 6717 of 1963 at United Engineering Steels Ltd, Cable Street Mills, Wolverhampton, West Midlands, on 25th September 1984.

(A.J. Booth)

Yorkshire Engine Co 2748 of 1959, formerly of British Coal, Littleton Colliery, at Peak Rail Ltd, Darley Dale, Derbyshire, on 25th December 1989.

(B. Cuttell)

Yorkshire Engine Co 2819 of 1960, at Mostyn Docks & Trading Ltd, Clwyd, on 16th August 1982.   (A.J. Booth)

Ex-BR 03037 (Swindon 1959) at British Coal Opencast Executive, Oxcroft Disposal Point, Clowne, Derbyshire, on 6th May 1989.   (J.A. Foster)

Ex-BR 08133 (Derby 1955) at Sheerness Steel Co, Sheerness, Kent, on 17th
October 1985.   (K.A. Scanes)

Ex-BR D9529, seen here heading an IRS special train at Corby Quarries on
17th May 1980, is now preserved at the Nene Valley Railway, Cambridge-
shire.   (A.J. Booth)

NOTTINGHAM SLEEPER CO LTD, ALPINE INDUSTRIAL PARK,
JOCKEY LANE, ELKESLEY, RETFORD
Gauge : 4'8½".   (   )

	THE TYKE		4wDM	FH	2914	1944	OOU

THE PERMANENT WAY EQUIPMENT CO LTD, 1 GILTWAY, GILTBROOK, NOTTINGHAM
New Permaquip locos under construction, and locos for repair usually
present.

POWER GEN. HIGH MARNHAM POWER STATION, near NEWARK
(A Division of C.E.G.B.)
Gauge : 4'8½". (SK 802712)

	-	0-4-ODH	AB	441	1959

SEVERN TRENT WATER AUTHORITY
Stoke Bardolph Sewage Works, Nottingham
Gauge : 2'0".   (SK 633421)

U168	4wDM	SMH	40SD529	1984

Trent River Management Division, Scarrington Road Depot, West Bridgford
Gauge : 2'0".   (SK 585385)

U84	4wDM	RH	7002/0567/6	1967

Locos, used on river bank work, etc., are here as required
for repair.

PRESERVATION SITES

J. CRAVEN, EAST VIEW, MAIN STREET, WALESBY, NEWARK
Gauge : 4'8½".   (SK 684707)

TR 11	PWM 2187	A155W	2w-2PMR	Wkm	4164	1948

Gauge : 3'0".

W6/2-2		4wDM	RH	418770	1957
	-	2w-2PMR	Wkm	9673	1964

GRAND CENTRAL DINER, GREAT NORTHERN WAY, LONDON ROAD, NOTTINGHAM
Gauge : 4'8½".   (SK 580392)

	-	0-4-OST	OC	P	1555	1920	Pvd

NEWARK & SHERWOOD DISTRICT COUNCIL,
MILLGATE FOLK MUSEUM, 48 MILLGATE, NEWARK
Gauge : 4'8½".   (SK 794537)

-		0-6-0ST	IC HC		1682	1937

A.J. WILSON, 6 TRENTDALE ROAD, CARLTON, NOTTINGHAM
Gauge : 2'0".   (SK 607408)

THE WASP		2-2wPM	Wilson	1969

OXFORDSHIRE

INDUSTRIAL SITES

A.R.C. SOUTHERN, LINCH HILL, STANTON HARCOURT, near WITNEY
Gauge : 2'0".   (SP 410044)

73	BERLIN	0-4-0WT	OC	Freud	73	1901	Pvd

> Currently in store at A.R.C. Southern, Swindon Plant,
> Ermin Street, Stratton St Margaret, near Swindon, Wiltshire,
> (SU 181865). Loco to be displayed at local shows, etc.

HERRING BROS., BUCKINGHAM ROAD, near BICESTER
Gauge : 4'8½".   (   )

6984	OWSDEN HALL	4-6-0	OC	Sdn	1948	Pvd

LANSDOWN INTERNATIONAL FACILITIES LTD, MILTON FREIGHT TERMINAL,
MILTON TRADING ESTATE, DIDCOT
Gauge : 4'8½".   (SU 495916)

HUNTER		0-6-0DH	HE	7276	1972

MINISTRY OF DEFENCE, ARMY RAILWAY ORGANISATION
Bicester Central Workshops.
See Section Five for full details.

Bicester Military Railway
See Section Five for full details.

PRESERVATION SITES

BLENHEIM PALACE RAILWAY, WOODSTOCK
Gauge : 1'3".   (SP 444163)

-	6w-4PM	G&S		1960

CHOLSEY & WALLINGFORD RAILWAY, ST. JOHNS ROAD, WALLINGFORD
Gauge : 4'8½".   (SU 600891)

No.3 (D3190)	THAMES 08123	0-4-0ST	OC	AB	2315	1951
	GEORGE MASON	0-6-0DE		Derby		1955
	CARPENTER	0-4-0DM		FH	3270	1948
304470	IRIS	0-4-0DM		RH	304470	1951
A144	PWM 2176	2w-2PMR		Wkm	4153	

COTSWOLD WILD LIFE PARK, BURFORD
Gauge : 2'0".   (SP 237084)

No.3		0-4-0DH	S/O	AK	17	1985
001		4wDM	S/O	MR	9976	1954

GREAT WESTERN SOCIETY, DIDCOT RAILWAY CENTRE
Gauge : 7'0¼".   (SU 524906)

	FIRE FLY	2-2-2	IC	F.F.P.		1989 +

+ Not yet completed.

Gauge : 4'8½".

1338		0-4-0ST	OC	K	3799	1898
1340	TROJAN	0-4-0ST	OC	AE	1386	1897
1363		0-6-0ST	OC	Sdn	2377	1910
1466		0-4-2T	IC	Sdn		1936
3650	BRIAN	0-6-0PT	IC	Sdn		1939
3738		0-6-0PT	IC	Sdn		1937
3822		2-8-0	OC	Sdn		1940
4144		2-6-2T	OC	Sdn		1946
4942	MAINDY HALL	4-6-0	OC	Sdn		1929
5029	NUNNEY CASTLE	4-6-0	4C	Sdn		1934
5051	EARL BATHURST / DRYSLLWYN CASTLE	4-6-0	4C	Sdn		1936
5322		2-6-0	OC	Sdn		1917
5572		2-6-2T	OC	Sdn		1929
5900	HINDERTON HALL	4-6-0	OC	Sdn		1931
6023	KING EDWARD II	4-6-0	4C	Sdn		1930
6024	KING EDWARD I	4-6-0	4C	Sdn		1930
6106		2-6-2T	OC	Sdn		1931
6697		0-6-2T	IC	AW	985	1928
6998	BURTON AGNES HALL	4-6-0	OC	Sdn		1949
7202		2-8-2T	OC	Sdn		1934
7808	COOKHAM MANOR	4-6-0	OC	Sdn		1938

71000	DUKE OF GLOUCESTER	4-6-2	3C	Crewe		1954	
2	"PONTYBEREM"	0-6-0ST	OC	AE	1421	1900	
No.5		0-4-0WT	OC	GE		1857	
No.1	BONNIE PRINCE CHARLIE	0-4-0ST	OC	RSH	7544	1949	
(D1010)	D1035						
	WESTERN YEOMAN	C-CDH		Sdn		1962	
D7018		B-BDH		BPH	7912	1961	
(W22W)	No.22	Bo-BoDMR		AEC		1941	
(5)		0-4-0PM		Derby C&W		1960	
D26		0-6-0DH		HE	5238	1962	
(A 21 W)		2w-2PMR		Wkm	6892	1954	DsmT
(B 37 W	PWM 3963)	2w-2PMR		Wkm	6948	1955	DsmT

Gauge : 2'6".

(822)	THE EARL	0-6-0T	OC	BP	3496	1902

R. HILTON, "POPLARS", NORTH MORETON, DIDCOT
Gauge : 3'0".   (SU 552904)

No.1	(ED 10)	0-4-0ST	OC	WB	1889	1911

Gauge : 1'11½".

	KIDBROOKE	0-4-0ST	OC	WB	2043	1917

SHROPSHIRE

<u>INDUSTRIAL SITES</u>

MINISTRY OF DEFENCE, ARMY RAILWAY ORGANISATION, DONNINGTON DEPOT
      See Section Five for full details.

NORTH WEST WATER AUTHORITY, LLANFORDA HALL WATERWORKS, OSWESTRY
Gauge : 2'0".   (SJ 277295)   RTC.

81 A 07	81.01	4wDM	RH	496038	1963	OOU
81 A 06	8101	4wDM	RH	496039	1963	OOU

<u>PRESERVATION SITES</u>

BRITISH RAIL, TELFORD CENTRAL STATION, TELFORD
Gauge : 4'0".   (SJ 704092)

-		4wG	TU	1987 Pvd

CAMBRIAN RAILWAYS SOCIETY LTD, OSWESTRY
Gauge : 4'8½".   (SJ 294297)

	-	0-6-0ST	OC	AB	885	1900
	-	0-4-0ST	OC	BP	1827	1879
	NORMA	0-6-0ST	IC	HE	3770	1952
	ADAM	0-4-0ST	OC	P	1430	1916
	OLIVER VELTOM	0-4-0ST	OC	P	2131	1951
	-	4wVBT	VCG	S	9374	1947
	-	4wDM		FH	3057	1946
	-	4wDM		FH	3541	1952
No.1		0-4-0DM		HC	D843	1954
	-	0-6-0DM		HE	3526	1947

HAMPTON LOADE MINIATURE RAILWAY, HAMPTON LOADE STATION
Gauge : 1'3".   (SO 744863)

4472	THE FLYING SCOTSMAN	4-6-2PH	S/O	Art	c1972
	THE LEDBURY FLYER	2w-2BER		J.Taylor	1976 Dsm +
1935	SILVER JUBILEE	4-6-4PE	S/O	?, Pengam	1935

+   Currently under renovation elsewhere.

IRONBRIDGE GORGE MUSEUM TRUST LTD, COALBROOKDALE MUSEUM
Gauge : 4'8½".   (SJ 667048, 668047)

5		0-4-0ST	OC	Coalbrookdale		c1865
	-	0-4-0VBT	VCG	S	6155	1925
	Rebuilt from	0-4-0ST	OC	MW		
	-	0-4-0VBT	VCG	S	6185	1925
	Rebuilt from 6	0-4-0ST	OC	Coalbrookdale		c1865

Gauge : 3'0".

	-	4wG	OC	GKN	1990

SEVERN VALLEY RAILWAY CO LTD
Gauge : 4'8½".   Locos are kept at :-

Arley, Hereford & Worcester	(SO 800764)
Bewdley, Hereford & Worcester	(SO 793753)
Bridgnorth	(SO 715926)
Hampton Loade	(SO 744863)
Highley	(SO 749831)

813		0-6-0ST	IC	HC	555	1901
1501		0-6-0PT	OC	Sdn		1949
2857		2-8-0	OC	Sdn		1918
4150		2-6-2T	OC	Sdn		1947
4566		2-6-2T	OC	Sdn		1924
4930	HAGLEY HALL	4-6-0	OC	Sdn		1929
5164		2-6-2T	OC	Sdn		1930
5764		0-6-0PT	IC	Sdn		1929
6960	RAVENINGHAM HALL	4-6-0	OC	Sdn		1944
(7325)	9303	2-6-0	OC	Sdn		1932

Number	Name	Type		Builder	Works No.	Date	
7714		0-6-0PT	IC	KS	4449	1930	
7802	BRADLEY MANOR	4-6-0	OC	Sdn		1938	
7812	ERLESTOKE MANOR	4-6-0	OC	Sdn		1939	
7819	HINTON MANOR	4-6-0	OC	Sdn		1939	
34027	TAW VALLEY	4-6-2	3C	Bton		1946	
42968		2-6-0	OC	Crewe		1934	
43106		2-6-0	OC	Dar	2148	1951	
(45000)	5000	4-6-0	OC	Crewe	216	1934	
45110	R.A.F. BIGGIN HILL	4-6-0	OC	VF	4653	1935	
(45690)	5690 LEANDER	4-6-0	3C	Crewe	288	1936	
		Rebuilt		Derby		1973	
45699	GALATEA	4-6-0	3C	Crewe		1936	
46443		2-6-0	OC	Crewe		1950	
46521		2-6-0	OC	Sdn		1953	
47383		0-6-0T	IC	VF	3954	1926	
(48773)	8233	2-8-0	OC	NB	24607	1940	
(61994)	3442 THE GREAT MARQUESS	2-6-0	3C	Dar	1761	1938	
75069		4-6-0	OC	Sdn		1955	
78019		2-6-0	OC	Dar		1954	
80079		2-6-4T	OC	Bton		1954	
	THE LADY ARMAGHDALE	0-6-0T	IC	HE	686	1898	
MW 2047	WARWICKSHIRE	0-6-0ST	IC	MW	2047	1926	
600	GORDON	2-10-0	OC	NB	25437	1943	
No.4		0-4-0ST	OC	P	1738	1928	
D1013	WESTERN RANGER	C-CDH		Sdn		1962	
D1062	WESTERN COURIER	C-CDH		Crewe		1963	
D3022	(08015)	0-6-0DE		Derby		1953	
3586	(08471)	0-6-0DE		Crewe		1958	
D7633	(25283 25904)	Bo-BoDE		BP	8043	1965	
12099		0-6-0DE		Derby		1952	
	SILVER SPOON	0-4-0DM		RH	281269	1950	
	-	0-4-0DM		RH	319290	1953	
11509	ALAN	0-4-0DM		RH	414304	1957	
D2961		0-4-0DE		RH	418596	1957	
(PT 2P)		2w-2PMR		Wkm	1580	1934	
PWM 3189		2w-2PMR		Wkm	5019	1948	DsmT
PWM 3774		2w-2PMR		Wkm	6653	1953	Dsm
DB 965054		2w-2PMR		Wkm	7577	1957	DsmT
. (PT 1P TP 49P)		2w-2PMR		Wkm	7690	1957	
(9021)	6	2w-2PMR		Wkm	8085	1958	

TELFORD HORSEHAY STEAM TRUST, THE OLD LOCO SHED, HORSEHAY, TELFORD
Gauge : 4'8½".   (SJ 675073)

Number	Name	Type		Builder	Works No.	Date
5619		0-6-2T	IC	Sdn		1925
	PETER	0-6-0ST	OC	AB	782	1896
3		0-4-0ST	OC	P	1990	1940
	F.T. CLAMP	4wVBT	VCG	S	9535	1952
1	TOM	0-4-0DH		NB	27414	1954
D2959		4wDM		RH	382824	1955
(TR36)	PWM 2786 A14W	2w-2PMR		Wkm	6885	1954

Gauge : 2'0".

Number	Name	Type		Builder		Date
	THOMAS	4wVBT	VCG	Kierstead Ltd/AK		1979
1		4wPM		D.Skinner		c1975

SOMERSET

INDUSTRIAL SITES

A.R.C. (SOUTHERN) LTD, WHATLEY QUARRY, near FROME
Gauge : 4'8½".   (ST 733479)

3	2323		4wDH	TH	152V	1965 OOU
No.4			4wDH	TH	200V	1968
	PRIDE OF WHATLEY		6wDH	TH	V325	1987

COURTAULDS LTD, FILMS, BRIDGWATER
Gauge : 4'8½".   (ST 309382)

D2133		0-6-0DM	Sdn	1960

DELTA CIVIL ENGINEERING COMPANY LTD, PLANT DEPOT, WYLDS ROAD, BRIDGWATER
Gauge : 2'0".   (   )

-		4wBH	Decon	1982
-		4wBH	Decon	1986
LOX 005		4wBE	WR	
LOX 006		4wBE	WR	
LOX 007		4wBE	WR	
LOX 008		4wBE	WR	
LOX 009		4wBE	WR	

Gauge : 1'6".

EL2		0-4-0BE	WR	N7641	1974
EL4		0-4-0BE	WR	N7615	1973
EL5		0-4-0BE	WR	N7616	1973

Locos present in yard between contracts.

FOSTER YEOMAN QUARRIES LTD, MEREHEAD STONE TERMINAL,
TORR WORKS, SHEPTON MALLET
Gauge : 4'8½".   (ST 693426)

(D3044	08032)				
33	MENDIP	0-6-0DE	Derby		1954
44	WESTERN YEOMAN II	Bo-BoDE	GM	798083-1	1980
	-	0-6-0DE	YE	2641	1957

P.C. VALLINS, BLACKMOOR VALE NURSERY, TEMPLECOMBE
Gauge : 2'0".   (ST 706230)

-		4wDM	Jung	5869	
-		4wPM	L	9256	1937
20		4wPM	L	18557	1942

ROYAL ORDNANCE PLC, EXPLOSIVES DIVISION, BRIDGWATER
Gauge : 4'8½".   (ST 33x41x)

```
P 5104  R.O.F.BRIDGWATER No.1  0-4-0DH      AB        578  1972
  6321  R.O.F.BRIDGWATER No.2  0-4-0DH      AB        579  1972
```

Gauge : 2'6".   (ST 33x42x   RTC.

```
 3582   A.   125              4wBE   FLP GB  1877  1943 OOU
 3584   C.                    4wBE   FLP GB  1699  1940 OOU
```


TAUNTON CIDER COMPANY LTD, SILK MILL, NORTON FITZWARREN, near TAUNTON
Gauge : 4'8½".   (ST 196256)

```
          -                  4wDM   R/R Unimog          1986
```


PRESERVATION SITES

CRICKET ST. THOMAS WILDLIFE PARK, near CHARD
Gauge : 1'3".   (ST 376086)

```
    SAINT THOMAS           4w-4wDH      G&S           1953
```


EAST SOMERSET RAILWAY CO LTD, WEST CRANMORE RAILWAY STATION,
SHEPTON MALLET
Gauge : 4'8½".   (ST 664429)

```
 6634                        0-6-2T   IC  Sdn              1928
 47493                       0-6-0T   IC  VF       4195    1927
 69023   JOEM                0-6-0T   IC  Dar      2151    1951
 75029   THE GREEN KNIGHT    4-6-0    OC  Sdn              1954
 92203   BLACK PRINCE        2-10-0   OC  Sdn              1959
 1398    LORD FISHER         0-4-0ST  OC  AB       1398    1915
 1719    LADY NAN            0-4-0ST  OC  AB       1719    1920
 32110                       0-6-0T   IC  Bton             1877
 4101                        0-4-0CT  OC  D        4101    1901
 W38     PWM 3764            2w-2PMR      Wkm      6643    1953
```

Gauge : 3'6".

```
          -                  2w-2PM       Ford             1938
```


ALAN GARTELL, YENSTONE, near TEMPLECOMBE
Gauge : 2'0".   (   )

```
    TEMPLECOMBE            4wDM      L        42494   1956
          -               4wDM      LB       51989   1960
          -               4wDM      LB       53726   1963
          -               4wDM      LB       55070   1966
          -               4wDM      Eclipse
```

C. HEAL, UNITY FARM CARAVAN SITE, COAST ROAD, BERROW
Gauge : 1'6".   (ST 295538)

125		4w-4wPM	A.R.Deacon		1975

MENDIP DISTRICT COUNCIL, WELSHMILL ADVENTURE PLAYGROUND, FROME
Gauge : 4'8½".   (ST 778486)

-		4wVBT	VCG S	9387	1948

SOMERSET & AVON RAILWAY COMPANY LTD, MELLS ROAD, near KILMERSDON
Gauge : 4'8½".   (ST 719510)

1		4wDH	TH/S	133C	1963 +
2		4wDH	TH/S	136C	1964

+   Carries plate 133V in error.

Gauge : 2'0".

2201	4wDM	FH	2201	1939
-	4wDM	FH		

SOMERSET & DORSET RAILWAY MUSEUM TRUST, WASHFORD STATION
Gauge : 4'8½".   (ST 044412)

53808	88	2-8-0	OC	RS	3894	1925
	ISABEL	0-6-0ST	OC	HL	3437	1919
No.1		0-4-0F	OC	WB	2473	1932
24		4wDM		RH	210479	1941 +

+   Currently under overhaul at M.R.C. Derbys.
    Carries plate RH 306089.

Gauge : 2'0".

-	4wDM	L	42319	1956

SOMERSET COUNTY COUNCIL ?, CHILDRENS PLAYGROUND, WANSTROW
Gauge : 4'8½".   (ST 712417)

(D3003)	0-6-0DE	Derby		1952

WESTONZOYLAND ENGINE GROUP, WESTONZOYLAND PUMPING STATION, near BRIDGWATER
Gauge : 2'0".   (ST 340328)

	-	4wDM	L	34758	1949
87030	30	4wDM	MR	40S310	1968

WEST SOMERSET RAILWAY CO
Gauge : 4'8½".   Locos are kept at :-

	Bishops Lydeard		(ST 164290)			
	Minehead		(SS 975463)			
	Williton Goods Yard		(ST 085416)			

3205		0-6-0	IC	Sdn		1946
3850		2-8-0	OC	Sdn		1942
4160		2-6-2T	OC	Sdn		1948
4561		2-6-2T	OC	Sdn		1924
5542		2-6-2T	OC	Sdn		1928 +
6412	THE FLOCKTON FLYER	0-6-0PT	IC	Sdn		1934
7820	DINMOR MANOR	4-6-0	OC	Sdn		1950
No.20	JENNIFER	0-6-0T	OC	HC	1731	1942
D2205	6 45 11223	0-6-0DM		(VF	D212	1953
				(DC	2486	1953
D2271		0-6-0DM		(RSH	7913	1957
				(DC	2615	1957
D2994	(07010)	0-6-0DE		RH	480695	1962
D7017		B-BDH		BPH	7911	1961
(D9500)	No.1 9312/92	0-6-0DH		Sdn		1964
D9526		0-6-0DH		Sdn		1964
D9551	50 8311/29 29	0-6-0DH		Sdn		1965
E 50341		2-2w-2w-2DMR		GRC		1957
W 50413		2-2w-2w-2DMR		PR		1957
(W)50414		2-2w-2w-2DMR		PR		1957
(E)51118		2-2w-2w-2DMR		GRC		1957
(E)51485		2-2w-2w-2DMR		Cravens		c1959
No.57		0-6-0DH		RR	10214	1964
	PWM 5671	2w-2PMR		Wkm	8502	1960
A37W		2w-2PMR		Wkm		

Gauge : 2'0".

808.		2w-2-2-2wRE	EE	808	1931

Built 1931 but originally carried plates dated 1930.

+   Currently under restoration at Priorswood Trading Estate,
    Taunton.

STAFFORDSHIRE

INDUSTRIAL SITES

BRITISH COAL
        See Section Four for full details.

BRITISH STEEL PLC, TEESIDE DIVISION, SHELTON WORKS, ETRURIA, STOKE
Gauge : 4'8½".   (SJ 864478)

No.6	4044/23	0-6-0DE	YE	2708	1959	
No.1	2444/16	0-6-0DE	YE	2753	1959	
	JANUS	0-6-0DE	YE	2772	1960	
	WEASEL	0-4-0DE	YE	2783	1960	
	ATLAS	0-6-0DE	YE	2787	1961	
	LUDSTONE	0-6-0DE	YE	2868	1962	
	-	0-4-0DE	YE	2869	1962	Dsm

FAIRCLOUGH CIVIL ENGINEERING LTD, TUNNELLING DIVISION,
PLANT DEPOT, COLD MEECE, SWYNNERTON, STONE
Gauge : 2'0".   (SJ 850325)

S151	193.022	4wBE		CE	5955A	1972
S177	263020	4wBE		CE	B0111A	1973
S178	263043	4wBE		CE	B0111B	1973
S179	193040	4wBE		CE	B0111C	1973
S191	263.023	4wBE		CE	B0131A	1973
S200	263017	4wBE		CE	B0152A	1973
S204	263018	4wBE		CE	B0152/2A	1973
S205	263019	4wBE		CE	B0152/2B	1973
S206	263006	4wBE		CE	B0152	1973
S207	263007	4wBE		CE	B0152-1	1973
S208	263044	4wBE		CE	B0152/4	1973
S213	263024	4wBE		CE	B0183A	1974
S232	263025	4wBE		CE	B0459A	1975
S237	263008	4wBE		CE	B0471A	1975
S238	263009	4wBE		CE	B0471B	1975
S241	263.026	4wBE		CE	B0471E	1975
S242	263.027	4wBE		CE	B0471F	1975
		Rebuilt		CE	B3480/1A	1988
S260	263021	4wBE		CE	B0465	1974
S261	263.028	4wBE		CE	B0941A	1976
S263	263.029	4wBE		CE	B0941C	1976
S264	263.030	4wBE		CE	B0948.1	1976
S265	263.031	4wBE		CE	B0948.2	1976
S271	263.032	4wBE		CE	B0952.2	1976
S275	263.033	4wBE		CE	B0957B	1976
		Rebuilt		CE	B3480/1B	1988
S278	263.048	4wBE		CE	B0958.1	1976
S279	263049	4wBE		CE	B0958A	1976
S287	263050	4wBE	FLP	CE	B1551A	1977
S289	263.042	4wBE	FLP	CE	B1551B	1977
S290	193012	4wBE	FLP	CE	B1551C	1977
S291	263011	4wBE	FLP	CE	B1552	1977
	263051	4wBE		CE	B0119B	1973
	263052	4wBE		CE	B0142B	1973

Gauge : 1'6".

S104	263.013	4wBE	CE	5792B	1970
S106	263.014	4wBE	CE	5792D	1970
S128	263001	4wBE	CE	5882A	1971
S138	263002	4wBE	CE	5911B	1972
S141	263004	4wBE	CE	5926A	1972
S142	193005	4wBE	CE	5926C	1972
S143	263047	4wBE	CE	5926D	1972

S146	263015	4wBE	CE	5926/A	1972	
S147	263.016	4wBE	CE	5926/2	1972	

Locos present in yard between contracts.

G.E.C. ALSTHOM LTD, MAIN WORKS, STAFFORD
Gauge : 4'8½".   (SJ 932221)

11044	TONKA	0-4-0DE	RH	424841	1960

HEPWORTH MINERALS & CHEMICALS LTD, MONEYSTONE QUARRY, OAKAMOOR
Gauge : 4'8½".   (SK 047451)   RTC.

BRIGHTSIDE	0-4-0DH	YE	2672	1959	OOU
CAMMELL	0-4-0DH	YE	2805	1960	OOU

MARCROFT ENGINEERING, STOKE WORKS, WHIELDON ROAD, FENTON, STOKE-ON-TRENT
(A Division of CAIB U.K. Ltd)
Gauge : 4'8½".   (SJ 881439)

4	45	BASIL	4wDM	FH	3909	1959
		SARA	4wDM	FH	3951	1961
RS 206			4wDM	FH	3952	1960 OOU

MARSTONS BREWERY LTD, BOTTLE YARD, SHOBNALL, BURTON-ON-TRENT
Gauge : 4'8½".   (SK 232233)

-	0-4-0DM	Bg	3410	1955 Pvd

NATIONAL POWER, MEAFORD 'B' POWER STATION
(A Division of C.E.G.B.)
Gauge : 4'8½".   (SJ 888368)

No.1	0-4-0DH	AB	440	1958

STAFFORD BOROUGH COUNCIL, LAMMASCOTE ROAD DEPOT, STAFFORD
Gauge : 2'0".   (SJ   )

ISABEL	0-4-0ST	OC	WB	1491	1898 Pvd

Loco currently under renovation elsewhere.

SYNTHETIC CHEMICALS LTD, FOUR ASHES WORKS
Gauge : 4'8½".   (SJ 917085)

MP1	0-4-0F	OC	AB	1944	1927

PRESERVATION SITES

ALTON TOWERS RAILWAY, ALTON TOWERS, near LEEK
Gauge : 2'0".　(SK 075437)

ALTONIA	0-4-0DM	S/O	Bg		1769	1929
-	0-4-0DM	S/O	Bg		2085	1934
-	0-6-0DM	S/O	Bg		3014	1938

BASS MUSEUM, HORNINGLOW STREET, BURTON-ON-TRENT
Gauge : 4'8½".　(SK 248234)

No.9	0-4-0ST	OC	NR	5907	1901
20	4wDM		KC		1926

CHATTERLEY WHITFIELD MINING MUSEUM, TUNSTALL
Gauge : 4'8½".　(SJ 883531)

No.6		0-4-0ST	OC	R.Heath		c1886
	THE WELSHMAN	0-6-0ST	IC	MW	1207	1890
No.2		0-6-2T	IC	Stoke		1923
9		0-6-0ST	OC	YE	2521	1952
	-	0-4-0DH		NB	27876	1959
No.13D	63/000/336	6wDH		TH	181V	1967
	-	0-6-0DM		WB	3119	1956
L052		0-6-0DE		YE	2745	1960

Gauge : 2'6".

3	TOM	4wBE		Bg	3555	1961
2	JERRY	4wBE		Bg	3578	1961
No.2		4wDM		RH	480679	1961
63 000 309		4wDM		RH	7002/0767/6	1967

Gauge : 2'4".

-	4wDM		RH	375347	1954

Gauge : 2'0".

-	4wDM		RH	441424	1960

	-	0-4-0ST	OC	AB	1964	1929	
No.1		0-4-0F	OC	AB	1984	1930	
	LITTLE BARFORD	0-4-0ST	OC	AB	2069	1939	
AVONSIDE No.3		0-6-0ST	OC	AE	1919	1924	
	WHISTON	0-6-0ST	IC	HE	3694	1950	
7	WIMBLEBURY	0-6-0ST	IC	HE	3839	1956	
	MOSS BAY	0-4-0ST	OC	KS	4167	1920	
	HENRY CORT	0-4-0ST	OC	P	933	1903	
1	"IRONBRIDGE"	0-4-0ST	OC	P	1803	1933	
11		0-4-0ST	OC	P	2081	1947	
No.15	ROKER	0-4-0CT	OC	RSH	7006	1940	
No.6	LEWISHAM	0-6-0ST	OC	WB	2221	1927	
	HAWARDEN	0-4-0ST	OC	WB	2623	1940	
MEAFORD LOCOMOTIVE No.4		0-6-0DH		AB	486	1964	
	-	4wBE		EE	788	1930	
No.2		4wBE/WE		EE	1130	1939	
	-	6wDM		KS	4421	1929	
	HELEN	4wDM		MR	2262	1923	
	CORONATION	0-4-0DH		NB	27097	1953	
242915	"HERCULES"	4wDM		RH	242915	1946	+
	-	0-4-0DM		RH	395305	1956	
	GAS OIL	4wDM		RH	408496	1957	
	-	4wDH		TH/S	103C	1960	
N.C.B. 63.000.366		0-6-0DM		WB	3150	1959	
	-	4wDH		WB	3207	1961	
(900332)		2w-2PMR		Wkm	497	1932	
(900379) PT2		2w-2PMR		Wkm	580	1932	DsmT
(PWM 2204)		2w-2PMR		Wkm	4121	1946	Dsm
PWM 2807 (B170W)		2w-2PMR		Wkm	4985	1949	
TR 3		2w-2PMR		Wkm	6884	1954	Dsm
TR 5		2w-2PMR		Wkm	6900	1954	DsmT
	F.L.R.	2w-2PMR		Wkm	7139	1955	
(A 34 W)		2w-2PMR		Wkm	8501	1960	

+ Rebuilt from 2' gauge.

-	4wDM	L	39419	1953

44422		0-6-0	IC	Derby		1927
80136		2-6-4T	OC	Bton		1956
52	JOSIAH WEDGWOOD	0-6-0ST	IC	HE	3777	1952
(D3420) 08350		0-6-0DE		Crewe		1957

J. STRIKE & N. CURTIS, TAMWORTH
Gauge : 2'0".   (   )

87033	746		4wDM	MR	40SD501	1975

SUFFOLK

INDUSTRIAL SITES

DOWER WOOD & CO LTD, GRAIN FORWARDING MERCHANTS, GREEN ROAD, NEWMARKET
Gauge : 4'8½".   (TL 647630)   (Closed)

(D3174)	08108	0-6-0DE	Derby	1955 OOU

FELIXSTOWE DOCK & RAILWAY CO LTD, FELIXSTOWE
Gauge : 4'8½".   (TM 285331)

D3489	COLONEL TOMLINE	0-6-0DE	Dar	1958

JOHN APPLETON ENGINEERING,
13A MASTERLORD INDUSTRIAL PARK, STATION ROAD, LEISTON
Gauge : 2'0".   (TM 440628)

-		4wDM	MR	22212	1964 Dsm

A works with locos occasionally present for restoration.

S.B.S. SPARES LTD, PIT VALLEY FARM, off HADLEIGH ROAD, SPROUGHTON, IPSWICH
Gauge : 4'8½".   (TM 116433)

-		0-4-0DM	HC	D697	1950 OOU

PRESERVATION SITES

EAST ANGLIA TRANSPORT MUSEUM SOCIETY, EAST SUFFOLK LIGHT RAILWAY,
CHAPEL ROAD, CARLTON COLVILLE, LOWESTOFT
Gauge : 2'0".   (TM 505903)

No.1	4wDM	MR	5902	1932 Dsm
No.2	4wDM	MR	5912	1934 OOU
No.4	4wDM	RH	177604	1936

R. FINBOW, "CAITHNESS", BACTON
Gauge : 4'8½".   (TM 064682)

	(FRY)	4wVBT	VCG S		7492	1928

PLEASUREWOOD HILLS AMERICAN THEME PARK, CORTON ROAD, LOWESTOFT
Gauge : 4'8½".   (TM 545965)

	-	4wDM	RH	305315	1952

Gauge : 2'0".

97	C.P.HUNTINGTON	4w-2-4wPH	S/O Chance 73 5097-24	1973
173	C.P.HUNTINGTON	4w-2-4wPH	S/O Chance 79-50173-24	1979
167	ANNIE OAKLEY	4w-2-4wPH	S/O Chance 79.50167.24	1979

SURREY

INDUSTRIAL SITES

HEPWORTH MINERALS & CHEMICALS LTD, HOLMETHORPE, REDHILL
Gauge : 4'8½".   (TQ 288517)

	-	4wDM	FH	3832	1957 OOU
	-	4wDH	FH	4006	1963 OOU

R. MARNER, HORSEHILL FARM, NORWOOD HILL, HORLEY
(   ) Locos under renovation here occasionally.

PRESERVATION SITES

J.L. BUTLER, 5 HEATH RISE, GROVE HEATH, RIPLEY
Gauge : 60cm.   (TQ 046557)

3		0-4-0DM	Dtz	19531	c1941
12	ARCHER	4wDM	MR	4709	1936
11	BARGEE	4wDM	MR	8540	1940

CHESSINGTON WORLD OF ADVENTURES,
CHESSINGTON RAILROAD, LEATHERHEAD ROAD, CHESSINGTON
Gauge : 2'0".   (TQ 172625)

	C.P.HUNTINGTON	4w-2-4wDH	S/O Chance 64 5030 24	1964 OOU
141	C.P.HUNTINGTON	4w-2-4wDH	S/O Chance 76 50141 24	1976
166	C.P.HUNTINGTON	4w-2-4wDH	S/O Chance 79 50166 24	1979

J. CROSSKEY, SURREY LIGHT RAILWAY
Gauge : 2'0".   (   )

No.1		0-4-0ST	OC	HE	1429	1922	
	-	4wDM		HE	3621	1947	OOU
	-	4wDM		MR	20073	1950	
	-	4wDM		OK	3685	1931	
2		4wDM		RH	174535	1936	OOU
4		4wDM		RH	177642	1936	
22		4wDM		RH	226302	1944	
69		4wDM		RH	264252	1952	
24		4wDM		RH	382820	1955	
	-	2w-2PM		Rhiwbach			Dsm
3		2-4wBE		WR	887	1935	Dsm
6	AMENE	4wBE		WR	D6912	1964	

Locos are kept at a private location.

M. HAYTER, No 1 HEATHER VIEW COTTAGES, SHORTFIELD COMMON, FRENSHAM, FARNHAM
Gauge : 2'0".   (SU 843423)

| - | 2w-2PM | Wkm | 2981 | 1941 |
| - | 2w-2PM | Wkm | 3032 | 1941 |

JACKMANS GARDEN CENTRE, MAYFORD, near WOKING
Gauge : 1'3".   (SU 996564)

| - | 4wDM | S/O | L | 10498 | 1938 |

J.B. LATHAM, "CHANNINGS", KETTLEWELL HILL, WOKING
Gauge : 1'11½". (TQ 003598)

| TRIASSIC | 0-6-0ST | OC | P | 1270 | 1911 |

MOLE VALLEY DISTRICT COUNCIL, LEISURE CENTRE, WATER PARK, LEATHERHEAD
Gauge : 4'8½".   (TQ 164558)

| BIWATER EXPRESS | 0-6-0ST | OC | HL | 3837 | 1934 |

P.D. NICHOLSON, 17 CROSSLANDS ROAD, WEST EWELL
Gauge : 2'0".   (TQ 207636)

| No.1830 | PLUTO | 4wPM | FH | 1830 | 1933 |

OLD KILN LIGHT RAILWAY, THE OLD KILN AGRICULTURAL MUSEUM,
THE REEDS, REEDS ROAD, TILFORD, near FARNHAM
Gauge : 2'0".   (SU 858434)

	PAMELA		0-4-0ST	OC	HE	920	1906
M.N.No.1	ELOUISE		0-6-0WT	OC	OK	9998	1922
6			4wDM		FH	2528	1941
4	LR2832		4wPM		MR	1111	1918
5297			4wPM		MR	5297	1931
	-		4wDM		MR	5713	1936
	-		4wPM		MR	6035	1937
	LOD/758039		4wDM		MR	8887	1944
	-		4wDM		MR	8981	1946
	-		4wDM		MR	22236	1965
No.4 L12	RTT/767093	PETTER	4wDM		Wkm	3031	1941
	-		2w-2PMR		Wkm	3287	1943
	DX 68061 (TR 26) PWM 2214	2w-2DMR		Wkm	4131	1947

P. RAMPTON & FRIENDS, BURGATE FARM, VANN LANE, HAMBLEDON
Gauge : 60cm.   (TQ 001381)

	RENISHAW 4		0-4-4-0T	VCG	AE	2057	1931
7	SOTILLOS		0-6-2T	OC	Borsig	6022	1906
1	SABERO		0-6-0T	OC	Couillet	1140	1895
2	SAMELICES		0-6-0T	OC	Couillet	1209	1898
3	OLLEROS		0-6-0T	OC	Couillet	1318	1900
	-		2-6-2+2-6-2T	4C	Hano	10634	1928
101			0-4-2T	OC	Hen	16073	1918
102			0-4-0T	OC	Hen	16043	1918
103			0-4-0T	OC	Hen	16045	1918
6	LA HERRERA		0-6-0T	OC	Sabero		c1937
	RENISHAW 5		0-4-4-0T	OC	WB	2545	1936
	-		0-4-2T		WB	2895	1948
No.18			4wDE		DK		c1918

Some of the locos are stored at another, unknown, location.

SURREY & HAMPSHIRE CANAL SOCIETY, (HANTS) (Closed)
Gauge : 2'0".   (   )

1		4wDM	HE	1944	1939

I. SUTCLIFFE, SURREY INDUSTRIAL TRAMWAY
Gauge : 2'0".   (   )

	-	4wDM	Clay Cross		+
14006	"HUMMY"	4wDM	L	35811	1950
	-	4wDM	L	37911	1952

+ Constructed from parts supplied by Listers in 1961 or 1973.

THORPE PLEASURE PARK, STAINES ROAD, CHERTSEY
(A member of the R.M.C. Group)
Treasure Island Railway
Gauge : 2'0".   (TQ 027685)

TIR 002		4wDH	S/O	AK	11	1984
TIR 001		4wDH	S/O	AK	12	1984

Canada Creek Railway
Gauge : 2'0".   (TQ 035681)

89		4-4-0DH	S/O	SL	139/1.2.89	1989
89		4-4-0DH	S/O	SL	139/2.1.89	1989

WOKING MINIATURE RAILWAY, MIZENS FARM, CHERTSEY ROAD, WOKING
Gauge : 3'2¼".   (   )

No.5	WILLIAM FINLAY	0-4-0T	OC	FJ	173L	1880

EAST SUSSEX

INDUSTRIAL SITES

BRITISH GYPSUM MOUNTFIELD ROADSTONE LTD, MOUNTFIELD
Gauge : 4'8½".   (TQ 730199)

1	4wDH	TH	183V	1967
2	4wDH	TH	184V	1967

PRESERVATION SITES

BLUEBELL RAILWAY CO LTD
Gauge : 4'8½".   Locos are kept at :-

Horsted Keynes, West Sussex   (TQ 372293)
Sheffield Park   (TQ 403238)

3217	(9017)						
	EARL OF BERKELEY	4-4-0	IC	Sdn			1938
30064		0-6-0T	OC	VIW		4432	1943
(30096)	96	NORMANDY	0-4-0T	OC	9E	396	1893
(30541)	541	0-6-0	IC	Elh			1939
(30583)	488	4-4-2T	OC	N		3209	1885
30830		4-6-0	OC	Elh			1930
(30847)	847	4-6-0	OC	Elh			1936
(30928)	No.928 STOWE	4-4-0	3C	Elh			1934
(31027)	No.27 PRIMROSE	0-6-0T	IC	Afd			1910
31178		0-6-0T	IC	Afd			1910
(31263)	263	0-4-4T	IC	Afd			1905

(31323)	323	BLUEBELL	0-6-0T	IC	Afd		1910	
(31592)	No.592		0-6-0	IC	Longhedge		1901	
(31618)	1618		2-6-0	OC	Bton		1928	
31638			2-6-0	OC	Afd		1931	
(32473)	473	BIRCH GROVE	0-6-2T	IC	Bton		1898	
32636		FENCHURCH	0-6-0T	IC	Bton		1872	
(32655)	55	STEPNEY	0-6-0T	IC	Bton		1875	
33001			0-6-0	IC	Bton		1942	
34023	BLACKMOOR VALE		4-6-2	3C	Bton		1946	
34059	SIR ARCHIBALD SINCLAIR							
			4-6-2	3C	Bton		1947	
34072	257 SQUADRON		4-6-2	3C	Bton		1948	
35027	PORT LINE		4-6-2	3C	Elh		1948	
58850			0-6-0T	OC	Bow	181	1880	
73082	CAMELOT		4-6-0	OC	Derby		1955	
75027			4-6-0	OC	Sdn		1954	
78059			2-6-0	OC	Dar		1956	
80064			2-6-4T	OC	Bton		1953	
80100			2-6-4T	OC	Bton		1955	
92240			2-10-0	OC	Crewe		1958	
		THE BLUE CIRCLE	2-2-0WT	G	AP	9449	1926	
24		STAMFORD	0-6-0ST	OC	AE	1972	1927	
No.3		BAXTER	0-4-0T	OC	FJ	158	1877	
No.4		SHARPTHORN	0-6-0ST	IC	MW	641	1877	
		BRITANNIA	4wPM		H	957	1926	a
	-		2w-2PMR		Syl	14384		
	-		2w-2PMR		Wkm		1932	+
TR 16	(PWM 3951) B25		2w-2PMR		Wkm	6936	1955	
6944	PWM 3959		2w-2PMR		Wkm	6944	1955	
	(PWM 3962)		2w-2PMR		Wkm	6947	1955	DsmT
1			2w-2PMR		Wkm	6952	1955	
(900855)			2w-2PMR		Wkm	6967	1954	b
	-		2w-2PMR		Wkm	7445	1956	DsmT
(TR 39 B45W DB 965564) PWM 4306			2w-2PMR		Wkm	7509	1956	
(TR 20 B49 DB 965563) PWM 4310			2w-2PMR		Wkm	7513	1956	
	-		2w-2PMR		Wkm	7581	1956	DsmT

```
+  Carries plate Wkm 7581/1956.
a  Runs on propane gas.
b  Carries plate Wkm 7445/1956.
```

BRIGHTON RAILWAY MUSEUM, PRESTON PARK, BRIGHTON
Gauge : 4'8½".   (TQ 302061)

3845			2-8-0	OC	Sdn	1942	
34046	BRAUNTON		4-6-2	3C	Bton	1946	
34073	249 SQUADRON		4-6-2	3C	Bton	1948	+
35009	SHAW SAVILL		4-6-2	3C	Elh	1942	
35011	GENERAL STEAM NAVIGATION						
			4-6-2	3C	Elh	1944	
D3000			0-6-0DE		Derby	1952	
D3255			0-6-0DE		Derby	1956	
D3261	13261		0-6-0DE		Derby	1956	
3053	(CAR No.92)		4w-4wRER		MC	1932	
3053	(CAR No.93)		4w-4wRER		MC	1932	
3142	11161		4w-4wRER			1937	
3142	11201		4w-4wRER			1937	

-		0-4-0DM		RH	260754	1950
BESSIE		0-4-0DM		RH	260755	1950

+ Currently stored elsewhere.

CO-OP SUPERSTORE, NEVILL ROAD, HOVE
Gauge : 1'6½".   (TQ 284064)

No.1		2-2-2	IC	RSC		1860

DRUSILLA'S ZOO PARK, BERWICK, near EASTBOURNE
Gauge : 2'0".   (TQ 524050)

BILL		4wDM		MR	9409	1948
EMILY		4wDM	S/O	RH	226294	1943

THE GREAT BUSH RAILWAY, TINKERS PARK, HADLOW DOWN, near UCKFIELD
Gauge : 2'0".   (TQ 538241)

2	SEZELA No.2	0-4-0T	OC	AE	1720	1915
6	SEZELA No.6	0-4-0T	OC	AE	1928	1923
1	AMINAL	4wDM		MR		c1931
		Rebuilt		Ludlay Brick		
4	MILD	4wDM		MR	8687	1941
5	ALPHA	4wDM		RH	183744	1937
12		4wDM		FH	2163	1938
14	ALBANY	4wDM		RH	213840	1941
20		0-4-0BE		WR	4634	1951
21		4wBE		WR	5035	1954
No.22	LAMA	4wBE		WR	5033	1953
23		0-4-0BE		WR	M7534	1972
No.24	TITCH	0-4-0BE		WR	M7535	1972
	-	4wDM		MR	7469	1940

HASTINGS DIESEL PRESERVATION GROUP, ST LEONARDS MAINTENANCE DEPOT
Gauge : 4'8½".   (   )

S 60000	4w-4wDER	E1h	1957
S 60001	4w-4wDER	E1h	1957
S 60019	4w-4wDER	E1h	1957

D. MILHAM, ISFIELD STATION, near UCKFIELD
Gauge : 4'8½".   (TQ 452171)

No.945	ANNIE	0-4-0ST	OC	AB	945	1904 +
68012		0-6-0ST	IC	HE	3193	1944
		Rebuilt		HE	3887	1964
15224		0-6-0DE		Afd		1949

+ Carries plate AB 1987/1930.
a Owned by Tunbridge Wells and Eridge Railway Preservation Soc.

NATIONAL RAILWAY MUSEUM, c/o BRITISH RAIL, BRIGHTON EMU DEPOT, BRIGHTON
Gauge : 4'8½".   (TQ 307056)

    2090   (S 10656)   4w-4wRER   Lancing/Elh   1937

WEST SUSSEX

<u>INDUSTRIAL SITES</u>

MIDHURST WHITES LTD, WEDGLEN INDUSTRIAL ESTATE, BEPTON ROAD, MIDHURST
Gauge : 2'6".   (SU 877212)   RTC.

         -                     4wDM   MR   22235   1965 OOU

TUNNEQUIP LTD, NOWHURST LANE, BROADBRIDGE HEATH, HORSHAM
Gauge : 1'6".   (TQ 133325)
         -                     4wBH   Tunnequip   1980
         -                     4wBH   Tunnequip   1980
         -                     4wBH   Tunnequip   1980

<u>PRESERVATION SITES</u>

H. FRAMPTON JONES, near HORSHAM
Gauge : 2'0".   (   )

         -                     4wDM   RH   487963 1963

HOLLYCOMBE STEAM & WOODLAND GARDEN SOCIETY, HOLLYCOMBE STEAM COLLECTION,
IRON HILL, LIPHOOK
Gauge : 4'8½".   (SU 852295)

         COMMANDER B   0-4-0ST   OC   HL   2450 1899

Gauge : 3'0".

         "EXCELSIOR"   2-2-0WT   G   AP   1607 1880

Gauge : 2'0".

    70   CALEDONIA   0-4-0WT   OC   AB   1995 1931
    38   JERRY M.   0-4-0ST   OC   HE   638 1895
         -                     4wDM   RH   1941 +

         +   Either 203016 or 203019.

J. LEMON-BURTON, "PAYNESFIELD", ALBOURNE GREEN
Gauge : 1'3".   (TQ 243179)

-		0-6-0	OC	J.Lemon-Burton		c1960
-		2-6-2	OC	J.Lemon-Burton		1967 +
-		0-4-0	OC	R.H.Morse	82	1939
-		4wDM		L	51721	1960

+ Unfinished loco - only partly built.

SOUTHERN INDUSTRIAL HISTORY CENTRE TRUST,
CHALK PITS MUSEUM, HOUGHTON BRIDGE, AMBERLEY, ARUNDEL
Gauge : 3'2¼"   (TQ 031122)

1		TOWNSEND HOOK	0-4-0T	OC	FJ	172L	1880
		MONTY	4wDM		OK	7269	1936

Gauge : 3'0".

		SCALDWELL	0-6-0ST	OC	P	1316	1913

Gauge : 2'11".

		-	4wDM		MR	10161	1950

Gauge : 2'0".

2	16	LION	4-6-0	OC	BLW	44656	1917
		BARBOUILLEUR	0-4-0T	OC	Decauville	1126	1950
4		POLAR BEAR	2-4-0T	OC	WB	1781	1905
5		PETER	0-4-0ST	OC	WB	2067	1918
		-	4wDM		FH	1980	1936
		-	0-4-0DM		HC	DM686	1948
THAKEHAM TILES No.3			4wDM		HE	2208	1941
		-	4wDM		HE	3097	1944
THAKEHAM TILES No.4			4wDM		HE	3653	1948
74			4wDM		HE		
		PELDON	4wDM		JF	21295	1936
		-	4wPM		L	35421	1949
		-	4wPM		MR	872	1918
			Rebuilt		MR	3720	1925
3101			4wPM		MR	1381	1918
27			4wDM		MR	5863	1934
		IBSTOCK	4wDM		MR	11001	1956
6193		REDLAND	4wDM		OK	6193	1937
7741		THE MAJOR	4wDM		OK	7741	1937
		-	4wDM		RH	166024	1933
18			4wDM		RH	187081	1937
12			4wDM		R&R	80	1937
		-	4wBE		WR	4998	1953
2	50		4wBE		WR	5031	1953
		-	4wBE		WR	(5034	1953?)
		-	0-4-0BE		WR	T8033	1979
		-	2w-2PMR		Wkm	3403	1943

Gauge : 1'10".

23		0-4-0T	IC	Spence		1920

Gauge : 1'8".

		4wDM	L	33937	1949 Dsm
-					

TYNE & WEAR

<u>INDUSTRIAL SITES</u>

<u>BRITISH COAL</u>
    See Section Four for full details.

<u>CASTLE CEMENT (RIBBLESDALE) LTD, NEWCASTLE DEPOT, RAILWAY STREET,</u>
<u>NEWCASTLE-UPON-TYNE</u>
Gauge : 4'8½".   (NZ 244635)

No.6		0-6-0DH	JF	4240010	1960

<u>PORT OF SUNDERLAND AUTHORITY, SOUTH DOCKS, SUNDERLAND</u>
Gauge : 4'8½".   (NZ 410573, 411578)

-		0-4-0DE	RH	395294	1956
P.S.A. No.22		0-4-0DE	RH	416210	1957

<u>PORT OF TYNE AUTHORITY, TYNE DOCK, SOUTH SHIELDS</u>
Gauge : 4'8½".   (NZ 350658, 355652)   RTC.

58		0-4-0DE	RH	381755	1955 OOU
No.35		0-4-0DE	RH	418600	1958 OOU

<u>SHEPHERDS SCRAP METALS (NEWCASTLE) LTD, SCRAP DEALERS,</u>
<u>ST. PETERS STATION, NEWCASTLE</u>   RTC.
Gauge : 4'8½".   (NZ 275637)

25		4wDM	RH	275884	1949 OOU
5898	BMA 25605	4wDM	RH	441933	1960 OOU
-		4wDH	TH/S	112C	1961 OOU

TYNE WEAR METRO, NEWCASTLE-UPON-TYNE
Gauge : 4'8½".   Locos are kept at :-

               South Gosforth Car Sheds   (NZ 250686)

WL 1		0-6-0DE		BT	801	1978 +
WL 2		0-6-0DE		BT	802	1978 +
WL 3		0-6-0DE		BT	803	1978 + OOU
WL 4		0-6-0DE		BT	804	1978 +
WL 5		0-6-0DE		BT	805	1979 + OOU
	-	4wBE/WE		HE	9174	1989
	-	4wBE/WE		HE	9175	1989
	-	4wBE/WE		HE	9176	1989
PLO 17		4wDH		Permaquip		1979 OOU
D640 JWW		4wDM	R/R	Unimog		1987
F171 DUA		4wDM	R/R	Unimog	1188	1988

        +   Carry plates dated 1977.

PRESERVATION SITES

BOWES RAILWAY CO LTD, SPRINGWELL
Gauge : 4'8½".   (NZ 285589)

No.22	No.85	0-4-0ST	OC	AB	2274	1949
	W.S.T.	0-4-0ST	OC	AB	2361	1954
No.77	NORWOOD	0-6-0ST	OC	RSH	7412	1948
613	20/110/709	0-6-0DH		AB	613	1977
		Rebuilt		AB		1986
101		4wDM		FH	3922	1959
	-	0-4-0DH		HE	6263	1964

METRORAIL EXPRESS, METROLAND, METRO CENTRE, DUNSTON, near GATESHEAD
Gauge : 1'6".   (NZ 211628)

| - | 4w-4+4w-4wRE | S/O | SSt | | 1988 |

NORTH TYNESIDE METROPOLITAN BOROUGH COUNCIL, LOCKEY PARK, WIDEOPEN
Gauge : 4'8½".   (NZ 242727)

| - | 4wDM | RH | 263000 | 1949 |

TANFIELD RAILWAY PRESERVATION SOCIETY, MARLEY HILL
Gauge : 4'8½".   (NZ 207573)

	-	0-6-0ST	OC	AB	1015	1904
No.6		0-4-2ST	OC	AB	1193	1910
No.17		0-6-0T	OC	AB	1338	1913
32		0-4-0ST	OC	AB	1659	1920
	WELLINGTON	0-4-0ST	OC	BH	266	1873
No.3		0-4-0WT	OC	EB	37	1898
	IRWELL	0-4-0ST	OC	HC	1672	1937
38		0-6-0T	OC	HC	1823	1949
112	CYCLOPS	0-4-0ST	OC	HL	2711	1907
		Rebuilt		DL		1956
No.2		0-4-0ST	OC	HL	2859	1911
	STAGSHAW	0-6-0ST	OC	HL	3513	1927
COAL PRODUCTS No.3		0-6-0ST	OC	HL	3575	1923
HU 12	HUNCOAT No.3	0-6-0F	OC	HL	3746	1929
	-	0-4-0CT	OC	RSHN	7007	1940
	LYSAGHT'S	0-6-0ST	OC	RSH	7035	1940
49	9103/49	0-6-0ST	IC	RSH	7098	1943
	PROGRESS	0-6-0ST	IC	RSH	7298	1946
SIR CECIL A. COCHRANE		0-4-0ST	OC	RSH	7409	1948
No.44	9103/44	0-6-0ST	OC	RSH	7760	1953
38		0-6-0ST	OC	RSH	7763	1954
21		0-4-0ST	OC	RSHN	7796	1954
	-	0-6-0ST	OC	RSH	7800	1954
No.16		0-6-0ST	OC	RSH	7944	1957
No.3		0-4-0ST	OC	RWH	2009	1884
No.4		4wVBT	VCG	S	9559	1953
2502/7		0-6-0ST	IC	WB	2779	1945
	F.G.F.	0-4-0DH		AB	552	1968
9		Bo-BoWE		AEG	1565	1913
No.2		0-4-0DE		AW	D22	1933
	-	2w-2DHR		Bg	3565	1962
	-	0-4-0DM		RSHN	6980	1940
3		Bo-BoWE		RSHN	7078	1944
1		0-4-0DM		RSHN	7901	1958

Gauge : 60cm.

No.11	ESCUCHA	0-4-0T	OC	BH	748	1883

TYNE & WEAR JOINT MUSEUM SERVICE
Stephenson Railway Museum, Middle Engine Lane, near Chirton
Gauge : 4'8½".   (NZ 323693)

(60019)	2509 SILVER LINK	4-6-2	3C	Don	1866	1937
(65894)	2392	0-6-0	IC	Don		1923
A.No.5		0-6-0PT	IC	K	2509	1883
	BILLY	0-4-0	VC	RS	A4	1826
401	VULCAN	0-6-0ST	OC	WB	2994	1950
03078	(D2078)	0-6-0DM		Don		1959
12098		0-6-0DE		Derby		1952
(3267)	DE 900730	4w-4wRER				1904
10		0-6-0DM		Consett	No.10	1958
4		Bo-BoWE		Siemens		1909

Washington "F" Pit Museum, Washington New Town
Gauge : 2'0".   (NZ 303574)

-	0-4-0DM	RH	392157	1956

WARWICKSHIRE

INDUSTRIAL SITES

BLUE CIRCLE INDUSTRIES LTD, HARBURY CEMENT WORKS
Gauge : 4'8½".   (SP 394587)

No.4	4wDH	S	10007	1959

BRITISH COAL
        See Section Four for full details.

JOHNSTON CONSTRUCTION LTD, PLANT DEPOT,
WATLING STREET, SHAWELL, near LUTTERWORTH
Gauge : 1'6".   (SP 538793)

-	2w-2BE	Iso	EV 558	1973	OOU
-	2w-2BE	Iso	EV 686	1972	OOU

LILLEY PLANT LTD, PLANT DEPOT,
HAUNCHWOOD COLLIERY SITE, GALLEY COMMON, near NUNEATON
Gauge : 2'0".   (SP 313917)

EL 1		0-4-0BE	WR	G6304	1960
EL 3		4wBE	WR	C6575	1963
EL 5		4wBE	WR	E6808	1965
EL 6		4wBE	WR	E6809	1965
EL 11		4wBE	CE	5852B	1970
EL 12		4wBE	CE	5852/1	1970
EL 18W		0-4-0BE	WR	D6879	1964
EL 20W	66/4/1	0-4-0BE	WR	4475	1950
EL 21W		0-4-0BE	WR	4476	1950
EL 24W		0-4-0BE	WR	3219	1945
EL 25W		4wBE	WR	4213	1949
EL 26W		4wBE	WR	4355	1950
EL 27W		4wBE	WR	4653	1951
EL 28W		4wBE	WR	4654	1951
EL 29W		4wBE	WR	4352	1950
EL 30W		4wBE	WR	4897	1952
EL 31W		4wBE	WR	4898	1952
EL 32W		4wBE	WR	5005	1952
EL 33W		4wBE	WR	5007	1952
EL 34W		4wBE	WR		
EL 35W		4wBE	WR		

EL 42W	66/4/7	0-4-0BE	WR	D6878	1964
	–	4wBE	FLP CE	B1547A	1977
	–	4wBE	CE	B3214A	1985
	–	4wBE	CE	B3214B	1985

Gauge : 1'6".

EL 4		0-4-0BE	WR	G7216	1967
EL 7		4wBE	CE	5373/1	1967
EL 8		4wBE	CE	5740	1970
EL 9		4wBE	CE	5740/2	1970
EL 10		4wBE	CE	5464	1968
EL 11		4wBE	CE	5373/2	1967
EL 14		4wBE	CE	5953A	1972
EL 15		4wBE	CE	5953B	1972
EL 16		4wBE	CE	B0110A	1973
EL 17		4wBE	CE	B0110B	1973
EL 19W	66/4/4	0-4-0BE	WR	4580	1950
EL 22W	66/4/6	0-4-0BE	WR	3788	1948
EL 23W	66/4/3	0-4-0BE	WR	4579	1950
EL 36W	PN66/8/1	4wBE	CE	5920	1972
EL 37W	66/6/1	4wBE	CE	5827	1970

Locos present in yard between contracts.
See also entry under Strathclyde.

MILLER CONSTRUCTION LTD, PLANT DEPOT, WATLING STREET, RUGBY
Gauge : 2'0".   (SP 533788)

L 18	432/29	4wBE	CE	5446	1968
L 19		4wBE	CE	5481/3	1968
L 21		4wBE	CE	B0987.1	1976
L 20		4wBE	CE	B0987.2	1976
L 22		4wBE	Omam		1990
L 23		4wBE	Omam		1990
L 24		4wBE	Omam		1990
L 25		4wBE	Omam		1990
L 26		4wBE	Omam		1990
L 27		4wBE	Omam		1990

Gauge : 1'6".

L 10		4wBE	CE	5827	1970
L 11		4wBE	CE	5920	1972
L 12		4wBE	CE	5965A	1973
L 13		4wBE	CE	5965B	1973
L 14		4wBE	CE	5965C	1973
L 15		4wBE	CE	B0109A	1973
L 16		4wBE	CE	B0109B	1973
L 17		4wBE	CE	5431	1968

Locos present in yard between contracts.

MINISTRY OF DEFENCE, ARMY RAILWAY ORGANISATION
Kineton Depot
        See Section Five for full details.

<u>Long Marston Depot</u>
        See Section Five for full details.

<u>PRISON SERVICE COLLEGE, NEWBOLD REVEL, STRETTON UNDER FOSSE, near RUGBY</u>
Gauge : 2'0".   (SP 455808)

	-	4wDM	L		+

    + Currently in store at M.O.D. Army Vehicle Store,
      Branston, Staffs.

<u>SEVERN LAMB LTD, WESTERN ROAD, STRATFORD-UPON-AVON</u>
(SP 197554) New miniature locos under construction & locos in for repair
            usually present.

<u>SHEPPARD (GROUP) LTD, GEORGE COHEN, MIDLANDS DIVISION,</u>
<u>METAL MERCHANTS AND PROCESSORS, KINGSBURY</u>
Gauge : 4'8½".   (SP 219969)

9/7354	E.S.C. No.42	0-4-0DH		NB	27939	1959	OOU
	-	4wDH		S	10059	1961	

<u>THE RUGBY GROUP PLC, NEW BILTON WORKS, near RUGBY</u>
Gauge : 4'8½".   (SP 487756)

	-	4wDH		TH	173V	1966

<u>WHITE WAGTAIL LTD, SCRAPYARD, GUN RANGE FARM,</u>
<u>SHILTON LANE, SHILTON, near COVENTRY</u>
Gauge : 4'8½".   (SP 389838)

(D2141)	03141	0-6-0DM		Sdn		1960	OOU
(D2145)	03145	0-6-0DM		Sdn		1961	OOU
	-	0-4-0DH		NB	27814	1958	OOU
	-	0-4-0DH		NB	27940	1959	OOU

PRESERVATION SITES

<u>THE 1857 SOCIETY, COVENTRY STEAM RAILWAY CENTRE,</u>
<u>ROWLEY ROAD, BAGINTON, COVENTRY</u>
Gauge : 4'8½".   (SP 354751)

	-	0-4-0F	OC	AB	1772	1922
NORTH GAWBER No.6 AREA		0-6-0T	OC	HC	1857	1952
	-	0-4-0DM		HC	D604	1936
No.9510		0-4-0DE		RH	268881	1949

DOUGLAS KEMPTON, 73 THE HAMLET, LEEK WOOTTON, near KENILWORTH
Gauge : 2'0".   (   )

2	LADY LUXBOROUGH	0-4-0ST	OC	WB	2088	1919

NORTH WARWICKSHIRE RAILWAY SOCIETY, NUNEATON
Gauge : 2'0".   (   )

	-	4wDM	MR	5881	1935 +
T2		4wDM	MR	8739	1942
	-	4wDM	MR	9411	1948

+ Currently in store at Leamington Spa, Warwickshire.

WARWICK DISTRICT COUNCIL
Newbold Comyn Park, Leamington Spa
Gauge : 4'8½".   (SP 333657)

75170		0-6-0ST	IC	WB	2758	1944

Victoria Park, Leamington
Gauge : 4'8½".   (SP 311655)

D2182		0-6-0DM	Sdn	1962

P. WESTMACOTT, 13 ALCESTER ROAD, STUDLEY
Gauge : 1'11½". (SP 074639)

-		0-6-0T	OC	AB	1578	1918

WEST MIDLANDS

INDUSTRIAL SITES

ALBRIGHT & WILSON LTD, CHEMICAL MANUFACTURERS, OLDBURY
Gauge : 4'8½".   (SO 994883)

1		4wDM	FH	3658	1953
2		4wDM	FH	3686	1954

ALLEN ROWLAND & CO LTD, STATION WORKS, WARWICK ROAD, TYSELEY
Gauge : 4'8½".   (SP 109840)

No.6	872/51	0-4-0DM	JF	4210116	1956	OOU
No.7	872/52	0-4-0DM	JF	4210125	1957	OOU
T 7760		4wDM	RH	224346	1945	OOU
	-	4wDM	RH	279597	1949	
	IVOR	4wDM	RH	495994	1963	

AUSTIN-ROVER GROUP LTD
Cofton Hackett Factory
Gauge : 4'8½". (SP 011764)

LONGBRIDGE	4wDH	TH	276V	1977

Longbridge Works, Birmingham
Gauge : 4'8½". (SP 009775, 011774)

RACHAEL	0-6-0DH	HE	7003	1971
LAURA	0-6-0DH	HE	8805	1978
EMMA	0-6-0DH	HE	8902	1978
LICKEY	0-6-0DH	RR	10221	1965
FRANKLEY	4wDH	TH	283V	1978

BRITISH COAL
        See Section Four for full details.

BRITISH STEEL CORPORATION, TUBES DIVISION, GENERAL TUBES WORKS GROUP,
BROMFORD WORKS, WHEELWRIGHT ROAD, ERDINGTON, BIRMINGHAM
Gauge : 4'8½". (SP 113899)

	WILLIAM A.	0-4-0DH	EEV	D1279	1969 OOU
	BARABEL	0-4-0DH	RR	10202	1964
59	-	0-4-0DH	S	10098	1962
	-	0-4-0DH	S	10099	1962 Dsm
	-	2w-2wDE	YE	2598	1956 Dsm

CHAS. B. PUGH (WALSALL) LTD, BRIQUETTING STEEL SWARF & CAST IRON BORINGS,
STAFFORD STREET, WEDNESBURY   RTC.
Gauge : 4'8½". (SO 985947)

-	4wDM	Bg/DC	2107	1937 OOU
-	4wDM	RH	235515	1945 OOU
-	4wDM	RH	349038	1954 OOU

HODSONS (BLOXWICH) LTD, BLOXWICH ROAD, BIRCHILLS, WALSALL
Gauge : 4'8½". (SK 006006)

PAUL	0-4-0DH	JF	4220011	1961 OOU
-	0-4-0DH	JF	4220034	1965 OOU

J.J. GALLAGHER & CO LTD, PLANT DEPOT,
ARMOURY CLOSE, LITTLE GREEN LANE, BIRMINGHAM
Gauge : 1'6". (SP 097864)

-	4wBE	CE	5956	1972
-	4wBE	CE	B0922A	1975
-	4wBE	CE	B0922B	1975

Locos present in yard between contracts.

J.P.M. PARRY & ASSOCIATES LTD, CORNGREAVES TRADING ESTATE, OVEREND ROAD, CRADLEY HEATH
Gauge : 2'0".   (SO 948853)

-		2w-2FER	Parry		1989
-		2w-2BER	J.Peat		

L.C.P. FUEL CO, PENSNETT TRADING ESTATE, SHUT END
Gauge : 4'8½".   (SO 903898)

(D2853)	02003	PETER	0-4-0DH	YE	2812	1960 OOU
(D2868)	SAM		0-4-0DH	YE	2851	1961 OOU

L.C.P. HOLDINGS LTD, PENSNETT TRADING ESTATE, SHUT END
Gauge : 4'8½".   (SO 900894, 901897)

2025	WINSTON CHURCHILL	0-6-0ST	IC	MW	2025	1923 Pvd
	-	4wDM		RH	215755	1942 Pvd

METROPOLITAN-CAMMELL CARRIAGE & WAGON CO LTD, WASHWOOD HEATH
Gauge : 4'8½".   (SP 103889)

-	0-4-0DH	HE	7259	1971
PETER	0-4-0DH	HE	7424	1978
57	0-6-0DH	S	10053	1961

Also new MC railcars under construction usually present.

ROUND OAK RAIL LTD, BRIERLEY HILL
Gauge : 4'8½".   (SO 925879)

No.1		0-4-0DE	YE	2593	1955 OOU
No.2	EXCALIBUR	0-4-0DE	YE	2614	1957
No.3		0-4-0DE	YE	2662	1957 OOU
No.4		0-4-0DE	YE	2774	1959 OOU
No.5		0-4-0DE	YE	2784	1960
No.8		0-4-0DE	YE	2881	1962 OOU
No.9		0-4-0DE	YE	2882	1962 OOU
No.10	MERLIN	0-4-0DE	YE	2883	1963
No.11		0-4-0DE	YE	2532	1953 OOU
	-	0-4-0DE	YE	2821	1961

SEVERN TRENT WATER AUTHORITY, TAME DIVISION
Gauge : 2'0".   Locos are kept at :-

Lagoon Works, Water Orton	(SP 159913)	
Minworth Main Depot & Workshops	(SP 156917)	

PLANT No.34			4wDM	MR	40SD502	1975 OOU
	87035	748	4wDM	MR	40SD503	1975 OOU
PLANT No.27	8740B		4wDM	SMH	40SD515	1979 OOU
		739	4wDM	SMH	40SD516	1979 OOU

TARMAC CONSTRUCTION LTD, PLANT DEPOT,
WARD STREET, ETTINGSHALL, WOLVERHAMPTON
Gauge : 2'0".   (   )

| 8 | 428006 | 4wBE | CE | B0428 | 1974 + |
| | 428007 | 4wBE | CE | B0428 | 1974 + |

    + To Egypt c1985 for use on Cairo Wastewater Scheme.
    Locos present in yard between contracts.

T.M.A. ENGINEERING LTD., TYBURN ROAD, ERDINGTON, BIRMINGHAM
New locos under construction, and locos for repair, occasionally present.

UNITED ENGINEERING STEELS LTD, (BRYMBO DIVISION),
CABLE STREET MILLS, CABLE STREET, WOLVERHAMPTON
Gauge : 2'0".   (SO 925977)

| No.2 | CM 9075 | 4wBE | WR | C6716 | 1963 |
| No.1 | CM 9076 | 4wBE | WR | C6717 | 1963 |

WELLMAN, SMITH, OWEN ENGINEERING CORP LTD, DARLASTON
(SO   )   New WSO locos under construction occasionally present.

W.T. HUNT, c/o HUNT BROS (OLDBURY) LTD, WEST BROMWICH LANE, OLDBURY
Gauge : 1'3".   (SO 989899)

No.1	SUTTON BELLE	4-4-2	OC	Cannon		1933
		Rebuilt				1953
No.2	SUTTON FLYER	4-4-2	OC	Cannon		
No.3	PRINCE OF WALES	4-4-2	OC	BL	11	1914
	-	4w-4wPM		Guest		1946

    Locos in store.

PRESERVATION SITES

BIRMINGHAM CORPORATION,
MUSEUM OF SCIENCE & INDUSTRY, NEWHALL STREET, BIRMINGHAM
Gauge : 4'8½".   (SP 064874)

| 46235 | CITY OF BIRMINGHAM | 4-6-2 | 4C | Crewe | | 1939 |

Gauge : 2'8".

| | SECUNDUS | 0-6-0WT | OC | B&S | | 1874 |

Gauge : 2'0".

| No.56 | (LORNA DOONE) | 0-4-0ST | OC | KS | 4250 | 1922 |
| 1 | LEONARD | 0-4-0ST | OC | WB | 2087 | 1919 |

CHASEWATER LIGHT RAILWAY AND MUSEUM, CHASEWATER PLEASURE PARK, BROWNHILLS
Gauge : 4'8½".   (SK 034070)

```
    3                              0-4-0ST    OC  AB     1223  1911
No.8        INVICTA                0-4-0ST    OC  AB     2220  1946
            -                      0-6-0ST    OC  HC      431  1895
   10       WHIT No.4              0-6-0T     OC  HC     1822  1949
    4       ASBESTOS               0-4-0ST    OC  HL     2780  1909
No.11       ALFRED PAGET           0-4-0ST    OC  N      2937  1882
    6                              0-4-0ST    OC  P       917  1902
No.2        THE COLONEL            0-4-0ST    OC  P      1351  1914
   59632                           4wVBT      VCG S      9632  1957
   37       TOAD                   0-4-0DH        JF  4220015  1962
   21                              4wDM           KC     1612  1929
    1                              4wPM           MR     1947  1919
   97457                           0-4-0DE        RH   458641  1963
(E50416 DB 975005) C 102 R 2-2w-2w-2DMR         Wkm     7346  1957
```


DUDLEY ZOO MINIATURE RAILWAY, DUDLEY
Gauge : 1'3".   (SO 947911)

```
            -              4-4w-4-4w-4DER        RRS           1983
```


ROBIN GORE, COVENTRY
Gauge : 90cm.   (   )

```
            -                      0-4-0WT    OC  OK     5102  1912
```


STANDARD GAUGE STEAM TRUST, BIRMINGHAM RAILWAY MUSEUM,
THE STEAM DEPOT, WARWICK ROAD, TYSELEY, BIRMINGHAM
Gauge : 4'8½".   (SP 105841)

```
 2807                              2-8-0      OC  Sdn          1905
 2873                              2-8-0      OC  Sdn          1918
 3803                              2-8-0      OC  Sdn          1939
 4983       ALBERT HALL            4-6-0      OC  Sdn          1931
 5043       EARL OF MOUNT EDGCUMBE
                                   4-6-0      4C  Sdn          1936
 7029       CLUN CASTLE            4-6-0      4C  Sdn          1950
 7760                              0-6-0PT    IC  NB    24048  1930
 9600                              0-6-0PT    IC  Sdn          1945
(46115)     6115
            THE SCOTS GUARDSMAN 4-6-0        3C  NBQ   23610  1927
No.1                               0-4-0T     OC  AE     1977  1925
            HENRY                  0-4-0ST    OC  HL     2491  1901
29781                              0-6-0ST    IC  MW     2015  1921
            -                      0-4-0ST    OC  P      1722  1926
    1                              0-4-0ST    OC  P      2004  1941
No.670                             2-2-2      IC  Tyseley       1989
(D318    40118)   97408            1Co-Co1DE     (EE    2853  1961
                                                -(RSH   8148  1961
(D3029   08021)   13029            0-6-0DE        Derby        1953
D5410    (27059)                   Bo-BoDE        BRCW DEL 253 1962
            -                      0-4-0PM        Bg      800  1920
```

Gauge : 1'3".

<table>
<tr><td>COUNT LOUIS</td><td>4-4-2</td><td>OC BL</td><td>32 1923</td></tr>
</table>

WILTSHIRE

AUSTIN-ROVER GROUP LTD, MANUFACTURING DIVISION,
SWINDON BODY PLANT, BRIDGE END ROAD, STRATTON ST MARGARET, SWINDON
Gauge : 4'8½".   (SU 167864, 168868)

SBL 3		0-4-0DH	JF	4220009	1960 OOU
SBL 4		0-4-0DH	JF	4220017	1961
SBL 5		0-4-0DH	JF	4220018	1961
1		0-4-0DH	JF	4220032	1965

BATH & PORTLAND STONE LTD, MONKS PARK MINE, CORSHAM
Gauge : 2'6".   (ST 878682)

8		4wBE	GB	2920	1958
	-	4wDM	RH	398101	1956

Locos work underground.

BLUE CIRCLE INDUSTRIES LTD, WESTBURY WORKS
Gauge : 4'8½".   (ST 885527)

BC 1		0-4-0DH	EEV	8449	1965
	-	0-6-0DH	TH	278V	1978

COOPERS (METALS) LTD, BRIDGE HOUSE, GIPSY LANE WORKS, GIPSY LANE, SWINDON
Gauge : 4'8½".   (SU 165860)

	-	0-4-0DM	JF	4210082	1953 OOU

DEAN AND CHAPTER OF SALISBURY CATHEDRAL, c/o M.O.D., A.F.D. CHILMARK DEPOT
Gauge : 2'0".   (ST 976312)

	-	4wDM	MR	9932	1972
		Rebuilt	AK		1988

E.C.C. CALCIUM CARBONATES LTD, BROADLANDS QUARRY, QUIDHAMPTON, SALISBURY
Gauge : 4'8½".   (SU 114314)

		0-4-0DH	RSHD/WB	8367	1962

MINISTRY OF DEFENCE, AIR FORCE DEPARTMENT
Chilmark Depot
Gauge : 4'8½".   (ST 982302)

Also uses M.O.D., A.R.O. locos.
See Section Five for full details.

Gauge : 2'0".   (ST 976312)

NG 23	4wBE	BD	3702	1973
	Rebuilt	AB		1987
NG 24	4wBE	BD	3703	1973
	Rebuilt	AB		1986
NG 25	4wBE	BD	3704	1973
	Rebuilt	AB	6526	1987
NG 51	4wDH	AB	720	1987
NG 52	4wDH	AB	721	1987

Dinton Depot
Gauge : 4'8½".   (SU 008308)

Uses M.O.D.,A.R.O. locos. See Section 5 for full details.

Gauge : 2'0".

NG 50	4wDH	AB	719	1987
NG 53	4wDH	AB	764	1988
NG 54	4wDH	AB	765	1988

MINISTRY OF DEFENCE, ARMY RAILWAY ORGANISATION, TIDWORTH DEPOT
See Section Five for full details.

MRS. WHITES GARDEN LTD, NURSERY at ?
Gauge : Monorail. (   )

		2a-2DH	AK	M001	1988

PRESERVATION SITES

N. JEARY, DYKES FARM, BLACKLANDS, near CALNE
Gauge : 4'8½".   (SU 014686)

No.4	0-4-0ST	OC	AB	2047	1937

LITTLECOTE HOUSE, near HUNGERFORD
Gauge : 1'3".   (SU 307703)

362	SYDNEY		2-4-2	OC	Guest	18	1963
	CROMWELL		4wDH		AK	13R	1984
		Rebuilt from	4wDM		RH	452280	1960

LONGLEAT LIGHT RAILWAY, LONGLEAT, WARMINSTER
Gauge : 1'3".   (ST 808432)

3	DOUGAL		0-6-2T	OC	SL		1970
4			4-4wDHR		Longleat		1984
	-		2-8-2DH	S/O	Longleat		1989
	CEAWLIN		2-8-2DH	S/O	SL	75 356	1975 Dsm

SCIENCE MUSEUM, ANNEXE, WROUGHTON, near SWINDON
Gauge : 2'0".   (SU 131790)

807		2w-2-2-2wRE	EE	807	1931

Gauge : 1'11½".

	-		0-4-0DM	HE	4369	1951

Loco not on public display.

SWINDON & CRICKLADE RAILWAY SOCIETY, BLUNSDON ROAD STATION, near SWINDON
Gauge : 4'8½".   (SU 110897)

5637			0-6-2T	IC	Sdn		1925
7903	FOREMARKE HALL		4-6-0	OC	Sdn		1949
1	RICHARD TREVITHICK		0-4-0ST	OC	AB	2354	1954
1371	MERLIN/MYRDDIN		0-4-0ST	OC	P	1967	1939
2022	(03022)		0-6-0DM		Sdn		1958
(D3180	08114)	1506	0-6-0DE		Derby		1955
5222	(25072)		Bo-BoDE		Derby		1963
	-		0-4-0DM		JF	21442	1936
	-		0-4-0DM		JF	4210137	1958
	-		2w-2PMR		Wkm	8089	1958

SWINDON WORKS HERITAGE CENTRE, SWINDON
Port Line Group
Gauge : 4'8½".   (   )

30053		0-4-4T	IC	9E	1905
80104		2-6-4T	OC	Bton	1955

Swindon Railway Workshops Ltd.
Gauge : 4'8½". (   )

4121		2-6-2T	OC	Sdn		1937
4141		2-6-2T	OC	Sdn		1946
4612		0-6-0PT	IC	Sdn		1942
5521		2-6-2T	OC	Sdn		1927
5526		2-6-2T	OC	Sdn		1928
80072		2-6-4T	OC	Bton		1953
D249		0-4-0WT	OC	KS	3063	1918
D1015	WESTERN CHAMPION	C-CDH		Sdn		1962
(D2152)	03152	0-6-0DM		Sdn		1960
(D3308)	08238	0-6-0DE		Dar		1956

THAMESDOWN BOROUGH COUNCIL, ARTS & RECREATION GROUP, MUSEUMS DIVISION, GREAT WESTERN RAILWAY MUSEUM, FARINGDON ROAD, SWINDON
Gauge : 7'0¼".   (SU 145846)

NORTH STAR		2-2-2	IC	Sdn		1925 +

+ Replica, incorporating parts of the original, RS 150/1837.

Gauge : 4'8½".

2516		0-6-0	IC	Sdn	1557	1897
4003	LODE STAR	4-6-0	4C	Sdn	2231	1907
9400		0-6-0PT	IC	Sdn		1947
(W4W)	No.4	4w-2w+2DMR		AEC		1934
A38W		2w-2PMR		Wkm	8505	c1960

THE COLLEGE, SWINDON, DEPARTMENT OF ENGINEERING, NORTH STAR SITE, NORTH STAR AVENUE, SWINDON
Gauge : 1'3".   (SU 148855)

NORTH STAR		2-2-2	IC	Sdn C.		c1972

UNDERGROUND QUARRY MUSEUM, PICKWICK QUARRY, CORSHAM
Gauge : 2'0".   (   )

-		4wDM	MR	60S318	1964 a +
A.M.W. No.189		4wDM	RH	200512	1940 a
-		4wDM	RH	359169	1953

+ Brake column is stamped 11164 in error.
a Property of Tapegrey Ltd, in store.

P.S. WEAVER, NEW FARM, LACOCK, near CORSHAM
Gauge : 1'9".   (ST 899691)

-		0-4-0VBT	VCG	P.Weaver		1978

NORTH YORKSHIRE

INDUSTRIAL SITES

B.O.C.M. SILCOCK LTD, OLYMPIA WORKS, BARLBY ROAD, SELBY
Gauge : 4'8½".   (SE 624326, 625327)

-	0-4-0DM	JF	4200003	1946
-	4wDM	R/R S&H	7501	1966

BRITISH RAIL ENGINEERING (1988) LTD, YORK WORKS, HOLGATE ROAD, YORK
Gauge : 4'8½".   (SE 585518)

D3236	(08168)	0-6-0DE	Dar	1956

BRITISH SUGAR CORPORATION LTD, POPPLETON FACTORY, YORK
Gauge : 4'8½".   (SE 576531)

-	0-4-0DM	RH	327964	1953
-	0-4-0DM	RH	395304	1956

COSTAIN LTD, PLANT DEPOT, THORPE ARCH INDUSTRIAL ESTATE, WETHERBY
Gauge : 1'6"/2'0". (SE 443467)

045	4wBE	CE	5882C	1971
046	4wBE	CE	5911A	1972
-	4wBE	CE	5940A	1972
-	4wBE	CE	5940B	1972
-	4wBE	CE	5961B	1972
-	4wBE	CE	5961C	1972
-	4wBE	CE	5961D	1972
-	4wBE	CE	B0107A	1973
-	4wBE	CE	B0107B	1973
-	4wBE	CE	B0107C	1973
-	4wBE	CE	B0113A	1973
-	4wBE	CE	B0113B	1973
-	4wBE	CE	B0119A	1973
-	4wBE	CE	B0119C	1973
-	4wBE	CE	B0142C	1973
-	4wBE	CE	B0166	1974

Locos present in yard between contracts.

MINISTRY OF DEFENCE, ARMY RAILWAY ORGANISATION, HESSAY DEPOT
See Section Five for full details.

THE POPPLETON NURSERY LIGHT RAILWAY, (G. WARNER), YORK
Gauge : 2'0".   (   )

		4wDM	MR	7494	1940
-					

SELBY STORAGE & FREIGHT LTD, (MEMBER OF POTTER GROUP), RAIL DISTRIBUTION CENTRE, SELBY
Gauge : 4'8½".   (SE 629322)

-	4wDM	RH		275881	1949
HERCULES	0-4-0DM	RH		281271	1950
-	0-6-0DH	RR		10220	1965

TILCON LTD, SWINDEN LIMEWORKS, GRASSINGTON, near SKIPTON
Gauge : 4'8½".   (SD 983614)

(D3067)	08054	201277 M 414	0-6-0DE	Dar	1953
12083		201276 M 413	0-6-0DE	Derby	1950

YORK HANDMADE BRICK CO, ALNE, near EASINGWOLD
Gauge : 2'0".   (SE 522663)   RTC

		4wDM	MR	8694	1943	00U
-						

PRESERVATION SITES

NEIL CLAYTON, c/o CLAYTON EXPRESS, FISHER GREEN, RIPON
Gauge : 3'0".   (SE 319708)

		4wDM	MR	40S280	1968
-					

Gauge : 1'11½".

No.8		4wDM	L	50191	1957

FLAMINGOLAND, FLAMINGO PARK ZOO, KIRBY MISPERTON, near PICKERING
Gauge : 1'3".   (SE 778800)

278		2-8-0PH	S/O SL	R9	1976
278	7	2-8-0DH	S/O SL	1/84	1984

GREAT YORKSHIRE RAILWAY PRESERVATION SOCIETY, Murton, near York
Gauge : 4'8½".   (SE 651537)

CHURCHILL	0-4-0DM	JF	4100005	1947

<u>Starbeck Railway Centre, Prospect Road, Starbeck, Harrogate</u>
Gauge : 4'8½". (SE 331556)

	JOYCE	0-4-0ST	OC	P	2103	1950
	ORMSBY	0-4-0DM		JF	22077	1938
ED6		0-4-0DM		JF	4200022	1948
DS 48	RYAN	4wDM		RH	305306	1952
DS 48	JIM	4wDM		RH	417892	1959
DS 88	OCTAVIOUS ATKINSON	4wDM		RH	466630	1962

<u>LIGHTWATER VALLEY LEISURE LTD, LIGHTWATER VALLEY FARM, RIPON</u>
Gauge : 1'3". (SE 285756)

	LITTLE GIANT	4-4-2	OC	BL	10	1905
	KING GEORGE	4-4-2	OC	BL	21	1912
6100	ROYAL SCOT	4-6-0	OC	Carland		
111	YVETTE	4-4-0	OC	E.A.Craven		1946
1326	BLACOLVESLEY	4-4-4PM	S/O	BL		1909
278	7	2-8-0DH	S/O	SL	17.6.79	1979

<u>NATIONAL RAILWAY MUSEUM, LEEMAN ROAD, YORK</u> (SE 594519)
Gauge : 7'0¼".

	IRON DUKE	4-2-2	IC	Resco		1985 +

        + Incorporates parts of RSH 7135/1944.

Gauge : 4'8½".

	ROCKET	0-2-2	OC	RS	4089	1934 +
No.1		4-2-2	OC	Don	50	1870
No.3		0-4-0	IC	Bury		1846
66	AEROLITE	2-2-4T	IC	KTH	281	1851
82	BOXHILL	0-6-0T	IC	Bton		1880
214	GLADSTONE	0-4-2	IC	Bton		1882
251		4-4-2	OC	Don	991	1902
563		4-4-0	OC	9E	380	1893
673		4-2-2	IC	Derby		1897
790	HARDWICKE	2-4-0	IC	Crewe	3286	1892
990	HENRY OAKLEY	4-4-2	OC	Don	769	1898
No.1275		0-6-0	IC	D	708	1874
1621		4-4-0	IC	Ghd		1893
1868		2-2-2	OC	Crewe	20	1845
2818		2-8-0	OC	Sdn	2122	1905
(3717)	3440 CITY OF TRURO	4-4-0	IC	Sdn	2000	1903
(30245)	245	0-4-4T	IC	9E	501	1897
(30925)	925 CHELTENHAM	4-4-0	3C	Elh		1934
(31737)	737	4-4-0	IC	Afd		1901
34051	WINSTON CHURCHILL	4-6-2	3C	Bton		1946
35029	ELLERMAN LINES	4-6-2	3C	Elh		1949
(41000)	1000	4-4-0	3C	Derby		1902
(42700)	2700	2-6-0	OC	Hor		1926
46229	DUCHESS OF HAMILTON	4-6-2	4C	Crewe		1938
(50621)	1008	2-4-2T	IC	Hor	1	1889
(60022)	4468 MALLARD	4-6-2	3C	Don	1870	1938
(60800)	4771 GREEN ARROW	2-6-2	3C	Don	1837	1936
(63601)	102	2-8-0	OC	Gorton		1912

(65567)	1217	0-6-0	IC	Str		1905
(68633)	87	0-6-0T	IC	Str	1249	1904
(68846)	1247	0-6-0ST	IC	SS	4492	1899
92220	EVENING STAR	2-10-0	OC	Sdn		1960
IMPERIAL No.1		0-4-0F	OC	AB	2373	1956
1439		0-4-0ST	IC	Crewe	842	1865
	HODBARROW	0-4-0ST	OC	HE	299	1882
	-	0-6-0ST	IC	HE	3696	1950
	ROCKET	0-2-2	OC	Loco Ent	No.2	1979
No.15	EUSTACE FORTH	0-4-0ST	OC	RSHN	7063	1942
5	FRANK GALBRAITH	4wVBT	VCG	S	9629	1957
607		4-8-4	OC	VF	4674	1935
D200	(40122)	1Co-Co1DE		(EE	2367	1958
				(VF	D395	1958
(D2090)	03090	0-6-0DM		Don		1960
D2860		0-4-0DH		YE	2843	1960
(D3079)	08064	0-6-0DE		Dar		1954
D8000	(20050)	Bo-BoDE		(EE	2347	1957
				(VF	D375	1957
(D9002)	55002	Co-CoDE		(EE	2907	1960
THE KINGS OWN YORKSHIRE LIGHT INFANTRY				(VF	D559	1960
E5001	(71001)	Bo-BoRE		Don		1958
26020	(76020)	Bo-BoWE		Gorton	1027	1951
(26500)	No.1	Bo-BoWE/RE		BE		1905
27000	ELECTRA	Co-CoWE		Gorton	1065	1954
(E3036)	84001	Bo-BoWE		NB	27793	1960
	41001	Bo-BoDE		Crewe		1972
APT-E	PC1/PC2	4w-4w-4wArtic	GTE	Derby		1972
(DS 75)	75S	4wRE		Siemens	6	1898
No.1	(BEL 2)	4wBE		Stoke		1917
8143	1293	4w-4RER		MV/MC		1925
11179	3131	4w-4RER		EE/Elh		1938
28249		4w-4wRER		Oerlikon M.C.		1915
RDB 975874	LEV 1	4wDMR		Leyland		1978
	-	0-4-0DE		AW	D21	1933
	-	4wPM		MR	4217	1931
	-	0-6-0DM		RSHN	7746	1954
960209		2w-2PMR		Wkm	899	1933

+ Replica of loco in original condition as built in 1829.

Gauge : 3'0".

719		0-6-0DM	HC	DM719	1950

Gauge : 2'0".

809		2w-2-2-2wRE	EE	809	1931 +
	-	4wDM	RH	187105	1937
	-	4w Atmospheric Car			c1865

+ Built 1931 but originally carried plates dated 1930.

Gauge : 1'11½".

No.1	K	0-4-0+0-4-0T	4C	BP	5292	1909
No.3	LIVINGSTON THOMPSON					
		0-4-4-0T	4C	Boston Lodge		1885
No.2		4wPM		MR	1377	1918

Gauge : 1'6".

WREN		0-4-0STT	OC	BP	2825	1887

Some locomotives are usually under renovation, or stored,
in the Museum Annexe at Leeman Road Goods Station and at
the N.C.L. Depot, York.

Certain locos will be used on 'Special Runs' and also
exhibited at other sites.

NORTH BAY RAILWAY, NORTHSTEAD MANOR GARDENS, NORTH BAY, SCARBOROUGH
Gauge : 1'8".   (TA 035898)

1931	NEPTUNE	4-6-2DM	S/O	HC	D565	1931
1932	TRITON	4-6-2DM	S/O	HC	D573	1932

NORTH YORKSHIRE MOORS RAILWAY PRESERVATION SOCIETY
Gauge : 4'8½".   Locos are kept at :-

Goathland	(NZ 836013)
Grosmont	(NZ 828049, 828053)
Levisham	(NZ 818909)
New Bridge	(NZ 803854)
Pickering	(NZ 797842)

3814		2-8-0	OC	Sdn		1940
6619		0-6-2T	IC	Sdn		1928
30825		4-6-0	OC	Elh		1927 Dsm
(30841)	No.841	4-6-0	OC	Elh		1936
30926	REPTON	4-4-0	3C	Elh		1934
34010	SIDMOUTH	4-6-2	3C	Bton		1945
44767	GEORGE STEPHENSON	4-6-0	OC	Crewe		1947 +
45428	ERIC TREACY	4-6-0	OC	AW	1483	1937
(62005)	2005	2-6-0	OC	NBQ	26609	1949
(63395)	2238	0-8-0	OC	Dar		1918
(63460)	901	0-8-0	3C	Dar		1919
(65894)	2392	0-6-0	IC	Dar		1923
75014		4-6-0	OC	Sdn		1951
80135		2-6-4T	OC	Bton		1956
No.3180	ANTWERP	0-6-0ST	IC	HE	3180	1944
No.29		0-6-2T	IC	K	4263	1904
WD 3672	DAME VERA LYNN	2-10-0	OC	NBH	25458	1944
5		0-6-2T	IC	RS	3377	1909
D821		B-BDH		Sdn		1960
D2207		0-6-0DM		⌐(VF	D208	1953
				⌐(DC	2482	1953
D5032	(24032) HELEN TURNER	Bo-BoDE		Crewe		1959
D7029		B-BDH		BPH	7923	1962
(D7541)	25191 THE DIANA	Bo-BoDE		Derby		1965
D7628	(25278) SYBILLA	Bo-BoDE		BP	8038	1965
D8568	290-068-6	Bo-BoDE		CE	4365U/69	1963
(D9009)	55009	Co-CoDE		⌐(EE	2914	1960
	ALYCIDON			⌐(VF	D566	1960

```
DEPARTMENTAL LOCOMOTIVE No.16     0-4-0DM      Bg/DC     2164    1941
   12139   REDCAR                 0-6-0DE      EE        1553    1948
No.21                             0-4-0DM      JF     4210094    1954
No.1                              4wDH         TH        129V    1963
No.2                              4wDH         TH        131V    1963
STANTON No.44                     0-4-0DE      YE        2622    1956
No.1     FRED                     2w-2PMR      Wkm        578    1932
No.2     PAULINE                  2w-2PMR      Wkm
   3     FRANK   DB 965053        2w-2PMR      Wkm       7576    1956
   4     NELSON DB 965108         2w-2PMR      Wkm       7623    1957
   5     GRAHAM                   2w-2PMR      Wkm        593    1932
   6     KEN    7                 2w-2PMR      Wkm       1724    1934
   7                              2w-2PMR      Wkm       7565    1956
         -                        2w-2PMR      Wkm        417    1931 DsmT
         -                        2w-2PMR      Wkm       1305    1933 DsmT
         -                        2w-2PMR      Wkm       1523    1934 DsmT
```

+ Currently under renovation at Morpeth.

THE EMBSAY STEAM RAILWAY, EMBSAY STATION, EMBSAY, near SKIPTON
Gauge : 4'8½".   (SE 007533)

```
No.22                             0-4-0ST   OC  AB      2320    1952
No.8                              0-6-0ST   OC  HC      1450    1922
SLOUGH ESTATES LTD No.5           0-6-0ST   OC  HC      1709    1939
No.140                            0-6-0T    OC  HC      1821    1948
AIREDALE No.3                     0-6-0ST   IC  HE      1440    1923
   S112    SPITFIRE               0-6-0ST   IC  HE      2414    1941
No.7       BEATRICE               0-6-0ST   IC  HE      2705    1945
   S 134   WHELDALE               0-6-0ST   IC  HE      3168    1944
   S 121   PRIMROSE No.2          0-6-0ST   IC  HE      3715    1952
No.69                             0-6-0ST   IC  HE      3785    1953
No.6       N.C.B.MONCKTON No.1    0-6-0ST   IC  HE      3788    1953
No.9                              0-4-0ST   OC  P       1159    1908
No.4                              0-4-0ST   OC  RSHN    7661    1950
           ANN                    4wVBT     VCG S       7232    1927
No.3                              0-4-0ST   OC  YE      2474    1949
   (D2203)                        0-6-0DM      —(VF     D145    1952
                                                (DC     2400    1952
   D9513   N.C.B. 38              0-6-0DH      Sdn              1964
   MDE 15                         4wDM         Bg/DC    2136    1938 Dsm
No.36                             0-6-0DM      HC       D1037   1958
           H.W. ROBINSON          0-4-0DM      JF     4100003   1946
           -                      4wDM         RH      294263   1950
887                               4wDM         RH      394009   1955
DB 965095                         2w-2PMR      Wkm       7610   1957
```

Gauge : 2'0".

```
           -                      4wPM         L         9993   1938 +
           -                      4wPM         L        10225   1938
P 1215                            4wDM         MR        5213   1930
           -                      4wDM         MR        8979   1946
           -                      4wDM         RH      175418   1936
Y.W.A. L2                         4wDM         RH
```

+ Currently under renovation at Barnoldswick, near Colne.

SOUTH YORKSHIRE

AMALGAMATED CONSTRUCTION CO LTD, MINING & CIVIL ENGINEERS,
WHALEY ROAD, BARUGH, BARNSLEY
Gauge : 2'0".   (SE 320085)

2	-	4wDM	HE		1947 +
1		4wDM	RH	249565	1947
		4wDM	RH	481552	1962

+ Either 3510 or 3512.

BARNSLEY METROPOLITAN BOROUGH COUNCIL,
ELSECAR & CORTONWOOD PROJECT GROUP, ELSECAR WORKSHOPS, WOMBWELL
Gauge : 2'2".   (SE 390003)

-	4wDM	RH	382808	1955

BOOTH ROE METALS LTD, SCRAP MERCHANTS,
CLARENCE METAL WORKS, ARMER STREET, ROTHERHAM
Gauge : 4'8½".   (SK 421924)

	-	0-4-0DH	AB	478	1963	
2303/63		0-6-0DH	AB	491	1964	
DL 13		0-4-0DH	HC	D1279	1963	
DL16		0-4-0DH	HC	D1387	1967	
D 24	PLANT No. 72241	4wDH	RR	10241	1966	OOU
D10		4wDH	TH	170V	1966	

Yard with locos for scrap usually present.

BRITISH COAL
        See Section Four for full details.

BRITISH STEEL PLC,
B.S.C. (CHEMICALS) LTD, CARBONISATION & ELECTRODE COATING WORKS GROUP,
ORGREAVE WORKS, ORGREAVE, SHEFFIELD
Gauge : 4'8½".   (SK 426874)

1		0-4-0WE	WSO	4424/1	1946	
	-	0-4-0WE	WSO	4571	1947	OOU
	-	0-4-0WE	WSO	8266	1962	
No.2	2444/17	0-6-0DE	YE	2754	1960	OOU
No.3	2444/18	0-6-0DE	YE	2866	1962	

CEMENTATION MINING LTD, PLANT DEPOT, BENTLEY WORKS, DONCASTER
(Part of TH Group Services)
Gauge : 2'0".   (SE 563056)

1	70214	4wBE	CE	B0145A	1973
2	17001	4wBE	CE	B0167	1974
3	70217	4wBE	CE	B0145D	1973
4	70213	4wBE	CE	B0132B	1973
5	70216	4wBE	CE	B0145C	1973
	70215	4wBE	CE	B0132A	1973

Locos present in yard between contracts.

CENTRAL ELECTRICITY GENERATING BOARD, WOODHEAD TUNNEL & DUNFORD BRIDGE
Gauge : 2'0".   (SK 114998, 156022)

-	4wBE	CE	5843	1971
-	4wDM	RH	444208	1961

COALITE FUELS & CHEMICALS LTD, GRIMETHORPE WORKS, BARNSLEY
Gauge : 4'8½".   (SE 403084)

1		4wDH	RR	10258	1966
2		4wDH	RR	10259	1966
3		4wDH	S	10164	1963
No.4	2	4wDH	S	10179	1964

COOPERS (METALS) LTD, (Incorporating Marple & Gillot),
EAST COAST ROAD, ATTERCLIFFE, SHEFFIELD
Gauge : 4'8½".   (SK 373888)

CLAUDE THOMPSON	0-4-0DM	JF	4210142	1958	
-	4wDH	TH	140V	1964	OOU

DAVY McKEE (SHEFFIELD) LTD, PRINCE OF WALES ROAD, DARNALL, SHEFFIELD
Gauge : 4'8½".   (SK 395875)

7600	4wDH	TH	189C	1967

FISONS PLC, HORTICULTURE DIVISION,
HATFIELD PEAT WORKS, STAINFORTH MOOR, THORNE, DONCASTER
Gauge : 3'0".   (SE 713084)

113	4wDM	HE	7367	1974	OOU
02-04	4wDM	MR	40S378	1971	
-	4wDM	SMH	40SD527	1983	

HARTWOOD EXPORTS (MACHINERY) LTD.,
HANGMANSTONE DEPOT, SHEFFIELD ROAD, BIRDWELL, near BARNSLEY
Gauge : 4'8½".   (SE 349006)

237	B11	SA	0-4-0DM	_(VF	4863	1942	Dsm
				(DC	2171	1942	

Yard with locos for scrap or resale occasionally present.

ROTHERHAM ENGINEERING STEELS
(A Division of United Engineering Steels Ltd)
Aldwarke Works, Rotherham
Gauge : 4'8½".   (SK 447951, 449953, 451954, 456957)

		-	0-4-0ST	OC	RSH	7020	1941	Dsm a
	31	624/83	0-6-0DE		YE	2904	1964	
	32	624/84	0-6-0DE		YE	2935	1964	
	34	624/86	0-6-0DE		YE	2947	1965	
	36		0-6-0DH		HE	7001	1971	OOU
B.S.C.	37	PLANT NO.2032/0630	0-6-0DH		HE	7002	1971	
B.S.C.	40		0-6-0DH		HE	7357	1973	OOU
	84		0-4-0DE		_(BP	7877	1960	OOU +
					(BT	330	1960	
	88		0-4-0DE		_(BP	7941	1960	OOU +
					(BT	334	1960	
	93		0-6-0DE		YE	2889	1962	
	95		0-6-0DE		YE	2891	1963	
	97		0-6-0DE		YE	2906	1963	
B.S.C.	98		0-6-0DE		YE	2907	1963	OOU

+ Locos can work in tandem.
a Frame used for weighbridge testing.

Templeborough Works, Rotherham
Gauge : 4'8½".   (SK 413917, 417920, 419912)

20	624/67	0-4-0DE	YE	2688	1959	
25	624/72	0-4-0DE	YE	2861	1962	Dsm
94		0-6-0DE	YE	2890	1962	
96		0-6-0DE	YE	2905	1963	

R.F.S. ENGINEERING LTD,
Doncaster Works, Hexthorpe Road, Doncaster
Gauge : 4'8½".   (SE 569031)

(D3110)	08085		0-6-0DE	Derby	1955	
(D3232)	08164)	002 PRUDENCE	0-6-0DE	Dar	1956	
(D3342)	08272)		0-6-0DE	Derby	1957	Dsm
(D3401)	08331)	001 TERENCE	0-6-0DE	Derby	1957	
(D3932)	08764	003 FLORENCE	0-6-0DE	Hor	1961	

Vanguard Works, Hooton Road, Kilnhurst
Gauge : 4'8½".   (SK 465975)

(D3238)	08170		0-6-0DE	Dar		1956
(D3769)	08602	CLARENCE	0-6-0DE	Derby		1959
-			0-4-0DH	EEV	D1122	1966
10212	BADDESLEY		0-6-0DH	RR	10212	1964
-			0-6-0DH	RR	10213	1964
			Rebuilt	TH		1988 +
-			0-6-0DH	RR	10216	1965
P.B.A. 40			0-6-0DH	RR	10219	1969
378			0-6-0DH	RR	10288	1969
-			4wDH	S	10037	1960
-			0-6-0DH	TH/S	124C	1963 Dsm
53			0-6-0DH	TH	261V	1976
-			2w-2DH	R/R TH	V331	1989
-			4wDM	R/R S&H	7502	1966

   + Carries plate TH V334/1988 in error.

Gauge : 3'0".

-		4wBERC	TH	SE109	1980

      Locos under construction and repair usually present;
      Locos are also hired out to various concerns.

S. HARRISON & SONS (TRANSPORT) LTD, 310 SHEFFIELD ROAD, TINSLEY, SHEFFIELD
Gauge : 4'8½".   (SK 399910)

-		0-4-0ST	OC	AB	2217	1947 Pvd
No.3		0-4-0ST	OC	AB	2360	1954 Pvd

STOCKSBRIDGE ENGINEERING STEELS,
STOCKSBRIDGE WORKS, STOCKSBRIDGE, SHEFFIELD
(A Division of United Engineering Steels Ltd)
Gauge : 4'8½".   (SK 260990, 267992)

No.1	714/37	4wDM	Robel	21.12 RK3	1969
No.27		0-6-0DE	YE	2720	1958 OOU
No.28		0-6-0DE	YE	2721	1958
No.30		0-6-0DE	YE	2750	1959
No.31		0-6-0DE	YE	2751	1959 OOU
No.33		0-6-0DE	YE	2740	1959
34		0-6-0DE	YE	2594	1956
35		0-6-0DE	YE	2635	1957
36		0-6-0DE	YE	2798	1961
37		0-6-0DE	YE	2736	1959
STOCKSBRIDGE RAILWAY CO		0-6-0DE	YE	2739	1959

TICKHILL PLANT LTD, PLANT DEPOT, APY HILL LANE, TICKHILL, near DONCASTER
Gauge : 2'0".   (SK 583931)

MBS 002	4wBE	WR	4817	1951
MBS 008	0-4-0BE	WR	5157	1953
MBS 010	0-4-0BE	WR	5244	1954
MBS 323	4wBE	WR	5115	1953
MBS 324	4wBE	WR	5316	1955
MBS 347	4wBE	WR	H7067	1968
MBS 348	0-4-0BE	WR	H7049	1968
MBS 521	4wBE	WR	1199	1938

Gauge : 1'6".

MBS 213	0-4-0BE	WR	6703	1962
MBS 432	0-4-0BE	WR	H7185	1968
MBS 433	0-4-0BE	WR	4320	1950
MBS 492	0-4-0BE	WR	2063	1941
MBS 493	0-4-0BE	WR	6131	1959
MBS 494	0-4-0BE	WR		
MBS 520	0-4-0BE	WR	7068	1968

Locos present in yard between hirings.

TINSLEY WIRE (SHEFFIELD) LTD, ATTERCLIFFE COMMON, SHEFFIELD
Gauge : 4'8½".   (SK 393902)   RTC.

-	4wDM	FH	3677	1954	OOU

TRACKWORK ASSOCIATES LTD, THOM LANE, LONG SANDALL, DONCASTER
Gauge : 4'8½".   (SE 606069)

CHARLES	4wDM	RH	417889	1958

PRESERVATION SITES

P. BRIDDON, SHEFFIELD
Gauge : 2'0".   (   )

-	4wDM	OK	(3444	1930?)

KELHAM ISLAND INDUSTRIAL MUSEUM, off ALMA STREET, SHEFFIELD
Gauge : 4'8½".   (SK 352882)

1	0-4-0DE	YE	2481	1950

7			0-4-0ST	OC	HC	1689	1937
S102	CATHRYN		0-6-0T	OC	HC	1884	1955
8	TL60 WR28		0-6-0ST	IC	HE	3183	1944
	-		0-6-0ST	IC	HE	3192	1944
			Rebuilt		HE	3888	1964
	GEORGE		4wVBT	VCG	S	9596	1955
(D213)	40013 ANDANIA		1Co-Co1DE		(EE	2669	1959
					(VF	D430	1959
(D2070)	17		0-6-0DM		Don		1959
(D2199)	ROCKINGHAM COLLIERY 1						
			0-6-0DM		Sdn		1961
D2229	No.5		0-6-0DM		(VF	D278	1955
					(DC	2552	1955
D2284			0-6-0DM		(RSH	8102	1960
					(DC	2661	1960
(D2334)	No.33		0-6-0DM		(RSH	8193	1961
					(DC	2715	1961
D2337	No.3 DOROTHY		0-6-0DM		(RSH	8196	1961
					(DC	2718	1961
D2854			0-4-0DH		YE	2813	1960
(D2420	06003) 97804		0-4-0DM		AB	435	1959
D2953			0-4-0DM		AB	395	1955
(D2985	07001)		0-6-0DE		RH	480686	1962
D3019			0-6-0DE		Derby		1953
D3476	73603		0-6-0DE		Dar		1957
D4092			0-6-0DE		Dar		1962
12074			0-6-0DE		Derby		1950
12088			0-6-0DE		Derby		1951
	SPEEDY		0-4-0DM		AB	361	1942
	HOTWHEELS		0-6-0DM		AB	422	1958
	-		4wDM		FH	3817	1956
No.21			0-6-0DM		HC	D707	1950
	ENTERPRISE		0-6-0DM		HC	D810	1953
No.44			0-6-0DH		HE	6684	1968
No.48			0-6-0DH		HE	7279	1972
406			0-4-0DH		NB	27427	1955
No.2	TRX 10		4wDM		RH	432479	1954
No.67	KEN		0-6-0DH		S	10180	1964
S 116-1	BIGGA		0-4-0DH		TH	102C	1960
		Rebuild of	0-4-0DM		JF	4200019	1947
No.47			0-6-0DH		TH	249V	1974
No.2	ROTHERHAM		0-4-0DE		YE	2480	1950
	-		2w-2DM				
	-		4wDMR		Wkm	9688	1965

WEST YORKSHIRE

BARGATE MOTOR SPARES, MANCHESTER ROAD, LINTHWAITE, HUDDERSFIELD
Gauge : 4'8½".   (SE 097145)

YD No.94		0-4-0DM	HE	2641	1941	OOU

BRITISH COAL
        See Section Four for full details.

CENTRAL ELECTRICITY GENERATING BOARD,
THORNHILL POWER STATION, DEWSBURY
Gauge : 4'8½".   (SE 229201)

-		0-4-0DH	JF	4210002	1949	OOU

CROSSLEY EVANS LTD, SHIPLEY STATION
Gauge : 4'8½".   (SE 148372)

42	PRINCE OF WALES	0-4-0DM	HE	7159	1969	
0040		4wDM	RH	284838	1950	OOU
9	BETH	4wDM	RH	425483	1958	OOU

HUNSLET G.M.T. LTD, JACK LANE, LEEDS
Gauge : 4'8½".   (SE 305321)

        Locos under construction & repair usually present.

MONCKTON COKE & CHEMICAL CO LTD, ROYSTON, near BARNSLEY
Gauge : 4'8½".   (SE 376120)

-		4wWE	GB	420452-2	1979

PLASMOR LTD, CONCRETE BLOCK MANUFACTURERS, WOMERSLEY ROAD, KNOTTINGLEY
Gauge : 4'8½".   (SE 503228)

-		4wDH	Plasmor		c1972

POWER GEN, ELLAND POWER STATION
(A Division of C.E.G.B.)
Gauge : 4'8½".   (SE 119219)

   ELLAND No.2   0-4-0DH   JF   4220010 1960 OOU

ROCKWARE GLASS LTD, HEADLANDS GLASS WORKS, KNOTTINGLEY
Gauge : 4'8½".   (SE 496232)

            -            4wDM   R/R Unimog   12/993 1982

THE OLTON HALL PRESERVATION SOCIETY, c/o PROCOR (U.K.) LTD,
HORBURY JUNCTION WORKS, WAKEFIELD
Gauge : 4'8½".   (SE 306176)

  5972   OLTON HALL   4-6-0   OC Sdn   1937 Pvd

THYSSEN (GB) LTD, PLANT DEPOT,
LANGTHWAITE GRANGE INDUSTRIAL ESTATE, SOUTH KIRBY, PONTEFRACT
Gauge : 2'0".   (   )

            -            4wBE FLP _(BGB   18/4/002 1987
                                    (WR   544201 1987

VICKERS DEFENCE SYSTEMS LTD, ROYAL ORDNANCE FACTORY, CROSSGATES, LEEDS
Gauge : 4'8½".   (SE 371346)   RTC.

            -            4wDM   FH   3918 1959 OOU

YORKSHIRE WATER AUTHORITY, WESTERN DIVISION,
NORTH BIERLEY PURIFICATION WORKS, OAKENSHAW
Gauge : 2'0".   (SE 179277)

  L8   4wDM   HE   7195 1974

PRESERVATION SITES

CITY OF BRADFORD METROPOLITAN COUNCIL ART GALLERIES & MUSEUMS,
BRADFORD INDUSTRIAL MUSEUM, MOORSIDE MILLS, MOORSIDE ROAD, BRADFORD
Gauge : 4'8½".   (SE 184353)

       NELLIE   0-4-0ST   OC HC   1435 1922

Gauge : 2'0".

            -            4wDM   (FH   3627 1953 ?)

HARTON COLLIERY No.2		4wWE		Siemens		1908

KEIGHLEY & WORTH VALLEY LIGHT RAILWAY LTD
Gauge : 4'8½". Locos are kept at :-

	Haworth	(SE 034371)
	Ingrow	(SE 058399)
	Oakworth	(SE 052389)
	Oxenhope	(SE 032355)

5775		0-6-0PT	IC	Sdn		1929
30072		0-6-0T	OC	VIW	4446	1943
34092	CITY OF WELLS	4-6-2	3C	Bton		1949
41241		2-6-2T	OC	Crewe		1949
42765		2-6-0	OC	Crewe	5757	1927 a
43924		0-6-0	IC	Derby		1920
45212		4-6-0	OC	AW	1253	1935
45596	BAHAMAS	4-6-0	3C	NBQ	24154	1935
		Rebuilt		HE	5596	1968
47279		0-6-0T	IC	VF	3736	1924
48431		2-8-0	OC	Sdn		1944 +
51218		0-4-0ST	OC	Hor	811	1901
52044		0-6-0	IC	BP	2840	1887
68077		0-6-0ST	IC	AB	2215	1947
75078		4-6-0	OC	Sdn		1956
78022		2-6-0	OC	Dar		1954
80002		2-6-4T	OC	Derby		1952
2226		0-4-0ST	OC	AB	2226	1946
No.2258	TINY	0-4-0ST	OC	AB	2258	1949
752		0-6-0ST	IC	BP	1989	1881
402	LORD MAYOR	0-4-0ST	OC	HC	402	1893
31	HAMBURG	0-6-0T	IC	HC	679	1903
67		0-6-0T	IC	HC	1369	1919
118	BRUSSELLS	0-6-0ST	IC	HC	1782	1945
	BELLEROPHON	0-6-0WT	OC	HF	C	1874
19		0-4-0ST	OC	Hor	1097	1910
5820		2-8-0	OC	Lima	8758	1945
1210	SIR BERKELEY	0-6-0ST	IC	MW	1210	1891
52		0-6-2T	IC	NR	5408	1899
	FRED	0-6-0ST	IC	RSH	7289	1945
No.57	SAMSON	0-6-0ST	IC	RSHN	7668	1950
63	CORBY	0-6-0ST	IC	RSHN	7761	1954
1931	THE BRONTES OF HAWORTH					
		2-8-0	OC	VF	5200	1945
D2511		0-6-0DM		HC	D1202	1961
D3336	(08266)	0-6-0DE		Dar		1957
D5209	(25059)	Bo-BoDE		Derby		1963
E79962		4wDMR		WMD	1267	1958
(E79964)		4wDMR		WMD	1269	1958
D 226	VULCAN	0-6-0DE		(EE	2345	1956
				(VF	D226	1956
23	MERLIN	0-6-0DM		HC	D761	1951
32	HUSKISSON	0-6-0DM		HE	2699	1944
	-	0-4-0DM		P	5003	1961
	-	2w-2PMR		Wkm		DsmT

+ Currently under renovation elsewhere.
a Currently under renovation at Ian Riley Engineering,
   Liverpool.

KIRKLEES LIGHT RAILWAY CO LTD, CLAYTON WEST, near HUDDERSFIELD
Gauge : 1'3".   (SE 258112)

		2-6-2T	OC	B.Taylor		1989
20		2-6-2DH	S/O	G&S	20	1964

LEEDS CITY COUNCIL, DEPARTMENT OF LEISURE SERVICES,
LEEDS INDUSTRIAL MUSEUM, ARMLEY MILLS, CANAL ROAD, LEEDS
Gauge : 4'8½".   (SE 275342)

(GWR 252)		0-6-0	IC	EW		1855 Dsm
	ELIZABETH	0-4-0ST	OC	HC	1888	1958
111	ALDWYTH	0-6-0ST	IC	MW	865	1882
	-	4wBE		GB	1210	1930
	-	0-4-0WE		GB	2543	1955
	SOUTHAM 2	0-4-0DM		HC	D625	1942
B16	ND 3066	0-4-0DM		HE	2390	1941
	-	0-4-0DM		JF	22060	1937
	-	0-4-0DM		JF	22893	1940 +

+ Currently stored at Kirkstall Brewery site.

Gauge : 3'6".

	PIONEER	0-6-0DM	HC	DM634	1946
No.3		0-6-0DM	HC	DM733	1950

Gauge : 3'0".

	LORD GRANBY	0-4-0ST	OC	HC	633	1902
4057		0-6-0DM		HE	4057	1953
	-	2-4-0DM		JF	20685	1935

Gauge : 2'11".

	-	4wDM	HC	D571	1932

Gauge : 2'8".

H881		0-4-0DM	HE	3200	1945

Gauge : 2'6".

	JUNIN	2-6-2DM	HC	D557	1930 +
20018		0-4-0DM	HE	3411	1947
	-	4wBE	HT	9728	1985

+ Currently stored at Kirkstall Brewery site.

Gauge : 2'1½".

No.5		0-4-0DMF		HE	4019	1949

Gauge : 2'0".

	CHEETAL	0-6-0WT	OC	JF	15991	1923
	BARBER	0-6-2ST	OC	TG	441	1908 +
	-	0-4-0DM	FLP	HC	DM1164	1959
1368		4wDM	FLP	HC	DM1368	1965
1		0-4-0DM		HE	2008	1939
2959		4wDM		HE	2959	1944
8		4wDM	FLP	HE	4756	1954
12		0-4-0DM		HE	5340	1957
2		0-4-0DM		HE	6048	1961
36863		4wDM		HU	36863	1929
21294	LAYER	4wDM		JF	21294	1936
LEEDS CORPORATION No.1		4wPM		MR	1369	1918

+ Currently under renovation at Bradford Industrial Museum.

Gauge : 1'6".

	JACK	0-4-0WT	OC	HE	684	1898
5A	4	4wBE		GB	1325	1933
	-	4wBE		GB	1326	1933 Dsm

P.N. LOWE, ABBEY LIGHT RAILWAY, BRIDGE ROAD, KIRKSTALL, LEEDS
Gauge : 2'0".   (SE 262357)

No.1	LOWECO	4wDM	L	20449	1942
No.2		4wDM	HE	2463	1944
No.3		4wDM	MR	5859	1934
No.4	DRUID	4wDM	MR	8644	1941
No.4	A.M.W. No.197 VULCAN	4wDM	RH	198287	1940
No.10	79/190	4wDM	OK	5926	1935
	-	4wPM	HU	39924	1924
	-	4wPM	MH	A110	1925
	-	4wDM	RH	235654	1946

MIDDLETON RAILWAY TRUST, TUNSTALL ROAD, HUNSLET, LEEDS
Gauge : 4'8½".   (SE 305310)

(68153)						
DEPARTMENTAL LOCO No.54		4wVBT	VCG	S	8837	1933
1823	HARRY	0-4-0ST	OC	AB	1823	1924
53	WINDLE	0-4-0WT	OC	EB	53	1909
1310		0-4-0T	IC	Ghd	38	1891
Nr.385		0-4-0WT	OC	Hartmann	2110	1895
	HENRY DE LACY II	0-4-0ST	OC	HC	1309	1917
	MIRVALE	0-4-0ST	OC	HC	1882	1955
No.6	PERCY	0-4-0ST	OC	HL	3860	1935
17	ARTHUR	0-6-0ST	IC	MW	1601	1903
	JOHN BLENKINSOP	0-4-0ST	OC	P	2003	1941
	-	0-4-0ST	OC	WB	2702	1943
	-	0-4-0DE		(BT	91	1958
				(BP	7856	1958

	MARY	0-4-0DM	HC	D577	1932
	CARROLL	0-4-0DM	HC	D631	1946
L.M.S. 7401 JOHN ALCOCK		0-6-0DM	HE	1697	1932
	"COURAGE"	4wDM	HE	1786	1935
	-	0-4-0DM	JF	3900002	1945
	-	0-4-0DH	JF	4220038	1966
No.3		4wDM	RH	441934	1960
	-	4wDH	TH/S	138C	1964
(DE 320467 DB 965049)		2w-2PMR	Wkm	7564	1956
	-	2w-2PMR	Wkm		DsmT

T. STANHOPE, ARTHINGTON STATION, near LEEDS
Gauge : 4'8½". (SJ 933218)

PWM 2779	2w-2PMR	Wkm	6878	1954

Gauge : 1'3".

-	4wVBT	VCG T.Stanhope		1987
-	4wDM	T.Stanhope		c1977

WEST YORKSHIRE TRANSPORT MUSEUM TRUST,
LUDLAM STREET DEPOT, MILL LANE, BRADFORD
Gauge : 4'8½". (SE 164322)

No.1	4wBE/WE	EE	905	1935
E10	4wWE	Siemens	862	1913

YORKSHIRE MINING MUSEUM, CAPHOUSE COLLIERY, NEW ROAD, OVERTON, WAKEFIELD
Gauge : 4'8½". (SE 253164)

40	0-6-0DH	HE	7307	1973

Gauge : 2'6".

STEPHANIE	4w-4wDHF	GMT	0593	1981
KIRSTEN	4w-4wDH	GMT		1981
746	0-4-0DM	HC	DM746	1951
-	0-4-0DM	HC	DM748	1951
1356	0-4-0DM	HC	DM1356	1965

Gauge : 2'4".

-	4wBE	FLP Atlas	2463	1944

Gauge : 2'3".

-	4wDH	HE	8831	1978
CAPHOUSE FLYER	4wDH	HE	8832	1978

Gauge : 2'1½".

-	4wDM	RH	379659	1955

Gauge : 2'0".

No.2	0-4-0DM	HC	DM655	1949

SECTION 2 SCOTLAND

+   No known locomotives exist.

BORDERS

INDUSTRIAL SITES

PENICUIK PEAT COMPANY LTD, WHIM MOSS, LAMANCHA STATION,
COWDENBURN, near LEADBURN
Gauge : 2'0".   (NT 206531)

	-	4wDH	AK	No.8	1982
	-	4wDM	MR	22253	1965
120	81A138	4wDM	RH	462365	1960

Locos move frequently between this site and
Springfield Moss, Lothian.

R.A. WATSON, WHITE MOSS PEAT WORKS, BOGSBANK, WEST LINTON
Gauge : 2'0".   (NT 150497)

	-	2wPM	R.A. Watson	OOU

PRESERVATION SITES

DR. R.P. JACK, THE STATION, EDDLESTON
Gauge : 2'0".   (NT 242471)

	-	0-4-0T	OC	AB	1871	1925
	-	4wDM		HE	2927	1944

CENTRAL

INDUSTRIAL SITES

ALCAN (U.K.) LTD, DAVIDS LOAN, FALKIRK
Gauge : 4'8½". (NS 892818) RTC.

	-	4wDH	EEV	D908	1964	OOU
No.2		0-4-0DM	JF	4100015	1948	OOU

B.P. OIL GRANGEMOUTH REFINERY LTD, GRANGEMOUTH REFINERY
Gauge : 4'8½". (NS 942817, 952822)

10		0-6-0DH	AB	600	1976
11		0-6-0DH	AB	649	1980
No.2	144-2	0-6-0DH	EEV	D917	1965
		Rebuilt	AB		1981
		Rebuilt	HAB	6694	1990
No.3	144-3	0-6-0DH	EEV	D1232	1968
		Rebuilt	AB		1980
9		0-6-0DH	HE	7304	1972
12		0-6-0DH	TH	290V	1980

CALEDONIAN PEAT PRODUCTS LTD, GARDRUM MOSS, SHIELDHILL, near FALKIRK
Gauge : 3'0".   (NS 885757)

	-	4wDM	MR	60S362	1968
	-	4wDM	MR	60S382	1969
LO 1		4wDM	RH	394022	1956
2		4wDM	RH		

COAL CONTRACTORS LTD, ROUGHCASTLE OPENCAST SITE,
TAMFOURHILL ROAD, CAMELON, FALKIRK
Gauge : 4'8½".   (NS 845796)

	-	4wDM	FH	3716	1955

FYNEGOLD EXPLORATION LTD, CONONISH FARM, near TYNDRUM
Gauge : 2'0".   (NN 292286)

5		4wBE	CE	B2905	1981 +
		Rebuilt	CE	B3550A	1988
6		4wBE	CE	B2983	1983
		Rebuilt	CE	B3550B	1988
	-	4wBE	WR		
		Rebuilt	WR	556001	1988

+   Plate reads B2903.

L. & P. PEAT PRODUCTS LTD, LETHAM MOSS WORKS, near AIRTH STATION
Gauge : 2'0".   (NS 871867)

	-	4wDM	LB	53225	1962
	-	4wDM	MR	5402	1932
	-	4wDM	MR	21505	1955
	-	4wDM	MR	40S343	1969
No.2		4wDM	RH	223698	1944
		Rebuilt	Richardsons		1976 OOU

SCOTTISH GRAIN DISTILLERS LTD, CAMBUS DISTILLERY, CAMBUS, near ALLOA
Gauge : 4'8½".   (NS 854940)

(D3558)	08443	0-6-0DE	Derby		1958

PRESERVATION SITES

FALKIRK DISTRICT MUSEUM, ABBOTSGRANGE MUSEUM STORE, GRANGEMOUTH
Gauge : 4'8½".   (   )

No.1	0429	0-4-0DM	JF	22902	1943

I. HUGHS, WHITE COTTAGE, CALLANDER
Gauge : 2'0".   (   )

		4wDM		MR	22237	1965
-						

SCOTTISH RAILWAY PRESERVATION SOCIETY
BO'NESS & KINNEIL RAILWAY, BO'NESS STATION, UNION STREET, BO'NESS
Gauge : 4'8½".   (NT 003817)

(55189)	419		0-4-4T	IC	St Rollox		1908
(62712)	No.246	MORAYSHIRE	4-4-0	3C	Dar	1391	1928
(65243)	No.673	MAUDE	0-6-0	IC	N	4392	1891
80105			2-6-4T	OC	Bton		1955
3	LADY VICTORIA		0-6-0ST	OC	AB	1458	1916
No.3	CLYDESMILL		0-4-0ST	OC	AB	1937	1928
B.A.CO.LTD No.3			0-4-0ST	OC	AB	2046	1937
THE WEMYSS COAL CO LTD No.20			0-6-0T	IC	AB	2068	1939
(No.6)			0-4-0CT	OC	AB	2127	1942
No.17			0-4-0ST	OC	AB	2296	1952
24			0-6-0T	OC	AB	2335	1953
	CITY OF ABERDEEN		0-4-0ST	OC	BH	912	1887
	(ELLESMERE)		0-4-0WT	OC	H(L)	244	1861
No.17			0-6-0ST	IC	HE	2880	1943
N.C.B. No.19			0-6-0ST	IC	HE	3818	1954
No.5			0-6-0ST	IC	HE	3837	1955
	SIR JOHN KING		0-4-0ST	OC	HL	3640	1926
1313			4-6-0	OC	Motala	586	1917
No.13	KELTON FELL		0-4-0ST	OC	N	2203	1876
(No.1)	(LORD ROBERTS)		0-6-0T	IC	NR	5710	1902
	RANALD		4wVBT	VCG	S	9627	1957
	DENIS		4wVBT	VCG	S	9631	1958
No.7			0-6-0ST	IC	WB	2777	1945
D5351	(27005)		Bo-BoDE		BRCW	DEL 194	1961
(D7585)	25235		Bo-BoDE		Dar		1964
(D9524)	B.P.No. 8		0-6-0DH		Sdn		1964
DS1			0-6-0DM		AB	343	1941
	-		2w-2DM		Arrols		c1966
	-		4wWE		BTH		1908
F.82			4wBE		EE	1131	1940
	TEXACO		0-4-0DM		JF	4210140	1958
	KILBAGIE		4wDM		RH	262998	1949
DS 3			4wDM		RH	275883	1949
DS 4	P 6687		0-4-0DE		RH	312984	1951
DS 6			0-4-0DE		RH	421439	1958
	-		0-4-0DE		RH	423658	1958
	-		0-4-0DE		RH	423662	1958
D 88/003			4wDM		RH	506500	1965
970213			2w-2PMR		Wkm	6049	
	-		2w-2PMR		Wkm	10482	1970

Gauge : 3'0".

	BORROWSTOUNNESS	0-4-0T	OC	AB	840	1899
	-	4wDH		MR	110U082	1970

DUMFRIES & GALLOWAY

<u>INDUSTRIAL SITES</u>

<u>IMPERIAL CHEMICAL INDUSTRIES PLC, PETROCHEMICALS AND PLASTICS DIVISION,</u>
<u>DUMFRIES WORKS, DRUNGANS, DUMFRIES</u>
Gauge : 4'8½".   (NX 947751)

DUMFRIES No.1		0-6-0DM	AB	385	1952
No.2		4wDM	RH	398613	1956 +

+   Carries plate 398163 in error.

<u>J. WALKER, SCRAP MERCHANT, SHAWHILL STATION YARD, ANNAN</u>
Gauge : 4'8½".   (NY 201664)

244		0-4-0DH	TH	130C	1963 OOU
	Rebuild of	0-4-0DM	JF	22971	1942

<u>L. & P. PEAT LTD, NUTBERRY WORKS, EASTRIGGS</u>
Gauge : 2'0".   (NY 249671)

-	4wDM	AK	7	1982	
-	4wDM	FH	3756	1955	
-	4wDM	MR	40S383	1971	
-	4wDM	RH	174532	1936 Dsm	
-	4wDM	RH	211641	1941	

<u>MINISTRY OF DEFENCE, ROYAL ORDNANCE FACTORY, EASTRIGGS</u>
See Section Five for full details.

<u>NOBELS EXPLOSIVES CO LTD, POWFOOT, ANNAN.</u>
(Subsidiary of I.C.I. Ltd)
Gauge : 4'8½".   (NY 163663)   RTC.

No.1		0-4-0DM	AB	347	1941 OOU
220		0-4-0DM	AB	359	1941 OOU

Gauge : 2'6".   (NY 163661)

7329		4wDM	HE	7329	1973
7330		4wDM	HE	7330	1973
1		4wDM	RH	422569	1959

SCOTTISH NUCLEAR LTD, CHAPELCROSS WORKS, ANNAN
(A Division of C.E.G.B.)
Gauge : 5'4". (NY 216695)

No.1	4301/G/0001	4wDM	RH	411320	1958
No.2	4301/G/0002	4wDM	RH	411321	1958

FIFE

INDUSTRIAL SITES

BRITISH COAL
See Section Four for full details.

MINISTRY OF DEFENCE, AIR FORCE DEPARTMENT, LEUCHARS
Gauge : 4'8½". (NO 453212, 456209)

Uses M.O.D., A.R.O. locos, see Section Five for full details.

MINISTRY OF DEFENCE, NAVY DEPARTMENT
Royal Naval Armament Depot, Crombie
Gauge : 4'8½". (NT 04x84x)

YARD No.1100		4wDH	CE	B1844	1979	

Gauge : 2'6". (NT 044845)

YARD No. 873		2w-2BE		GB	3537	
YARD No. 874		2w-2BE		GB	3538	
YARD No. 875	PM 11	2w-2BE		GB	3539	
YARD No.1066	B 2	4wBE	FLP	CE	B0483	1976
YARD No.1067		4wBE	FLP	CE	B0483	1976
YARD No.1068		4wBE	FLP	CE	B0483	1976
YARD No.1075		4wDM		HE	7447	1976
YARD No.1076		4wDM		HE	7448	1976
YARD No.1073		4wDM		HE	7450	1976
YARD No.1074		4wDM		HE	7451	1976

Royal Naval Dockyard, Rosyth
Gauge : 4'8½". (NT 108821)

	-	4wDM	FH	3739	1955	OOU
YARD No.3080	Y.S.M.	4wDM	FH	3777	1956	
YARD No.3081	YSM JINGLING GEORDIE	4wDM	FH	3778	1956	
YARD No.3082		4wDM	FH	3779	1956	
262		4wDH	TH	302V	1982	

R. & M. SUPPLIES (INVERKEITHING) LTD, THE BAY, INVERKEITHING
(Leased from Forth Ports Authority)
Gauge : 4'8½".   (NT 127823)

-		0-4-0DM	JF	4210138	1958
-		0-4-0DH	JF	4220003	1959 00U

SCOTTISH GRAIN DISTILLERS LTD, CAMERON BRIDGE DISTILLERY, WINDYGATES, LEVEN
Gauge : 4'8½".   (NO 348001)

-		0-6-0DH	EEV	D1193	1967
-		4wDM	RH	321733	1952

SCOTTISH POWER,
Kincardine Power Station (Closed)
Gauge : 4'8½".   (NS 923887)   RTC.

	-		0-4-0DH	AB	517	1966 00U
8010	H/P 54		0-4-0DE	RH	402801	1956 00U
8010	H/P 95	9	0-4-0DE	RH	431764	1960 00U
8010			0-4-0DE	RH	449753	1961 00U

Methil Power Station
Gauge : 4'8½".   (NO 382001)

No.2	HP 515	0-4-0DH	AB	515	1966
No.1	HP 299	0-4-0DH	AB	516	1966

THOS MUIR HAULAGE & METALS LTD, KIRKCALDY
Gauge : 4'8½".   (NT 282926)

No.12	H 662	0-4-0DH	NB	27732	1957 00U

T. MUIR, SCRAP MERCHANTS, EASTER BALBEGGIE, near THORNTON
Gauge : 4'8½".   (NT 291962)

No.3		0-4-0ST	OC	AB	946	1902 00U
	-	0-4-0ST	OC	AB	1069	1906 00U
No.10		0-6-0T	OC	AB	1245	1911 00U
	-	0-4-0ST	OC	AB	1807	1923 00U
No.47		0-4-0ST	OC	AB	2157	1943 00U
No.15		0-6-0ST	IC	AB	2183	1945 00U
No.6		0-4-0ST	OC	AB	2261	1949 00U
No.7		0-4-0ST	OC	AB	2262	1949 00U
	-	0-4-0ST	OC	GR	272	1894 00U
	-	4wDM		RH	235521	1945 00U

Gauge : 2'8".

-		4wDM	RH	506415	1964 00U

Gauge : 2'6".

		0-4-0DH	RH	476133	1967 OOU

Gauge : 2'0".

554	H 643	0-4-0DM	HE	3286	1945 Dsm

TULLIS RUSSELL, AUCHTERMUCHTY MILL, GLENROTHES
Gauge : 4'8½".   (NO 276017)

-		4wDM	R/R Unilok	2195	c1988

PRESERVATION SITES

J. CAMERON
Gauge : 4'8½".   (   )

60009	OSPREY	4-6-2	3C	Don	1853	1937

CARNEGIE DUNFERMLINE TRUST, PITTENCRIEFF PARK, DUNFERMLINE
Gauge : 4'8½".   (NT 086872)

No.29		0-4-0ST	OC	AB	1996	1934

FIFE REGIONAL COUNCIL, LOCHORE MEADOWS COUNTRY PARK, LOCHORE
Gauge : 4'8½".   (NT 172963)

No.30		0-4-0ST	OC	AB	2259	1949

LOCHTY PRIVATE RAILWAY CO, near ANSTRUTHER
Gauge : 4'8½".   (NO 522080)

10	FORTH	0-4-0ST	OC	AB	1890	1926
No.21		0-4-0ST	OC	AB	2292	1951
	-	0-4-0ST	OC	P	1376	1915
No.16		0-6-0ST	IC	WB	2759	1944
400	RIVER EDEN	0-4-0DH		NB	27421	1955
405	RIVER TAY	0-4-0DH		NB	27426	1955
N.C.B. No.10		0-6-0DH		NB	27591	1957
No.4	NORTH BRITISH	4wDM		RH	421415	1958
	-	0-4-0DH		RH	506399	1964

NORTH EAST FIFE DISTRICT COUNCIL, DEPARTMENT OF RECREATION,
RIO-GRANDE MINIATURE RAILWAY, CRAIGTOUN COUNTRY PARK, near ST. ANDREWS
Gauge : 1'3".   (NO 482141)

278	RIO GRANDE	2-8-0PH	S/O	SL	R8	1976

GRAMPIAN

INDUSTRIAL SITES

CAPTAIN J. HAY OF HAYFIELD, DELGATIE CASTLE, TURRIFF, ABERDEEN
Gauge : 2'0".   (NJ 754507)

		4wDM	LB	52610	1961	

GEORGE WATSON & SONS, PEAT WORKS, MIDDLEMUIR WORKS, NEW PITSLIGO
Gauge : 2'0".   (NJ 901573)   RTC.

		4wDM	LB	53162	1962	OOU
		4wDM	LB	54781	1965	OOU
2563		4wPM	OK	2563	1929	OOU

PRESERVATION SITES

ABERDEEN CITY COUNCIL, SEATON PARK, ABERDEEN
Gauge : 4'8½".   (NJ 943092)

		0-4-0ST	OC AB	2239	1947

ALFORD VALLEY RAILWAY CO LTD, ALFORD
Gauge : 2'0".   Locos are kept at :-

Alford Station   (NJ 579159)
Haughton House   (NJ 584165)

	"SACCHARINE"	0-4-2T	OC JF	13355	1912	
A.V.R. No.1	HAMEWITH	4wDM	L	3198	c1930	
	-	4wDM	MR	5342	1931	Dsm
	-	4wDM	MR	9215	1946	Dsm
	THE BRA'LASS	4wDM	S/O MR	9381	1948	
	-	4wDM	S/O MR	22129	1962	
87022		4wDM	MR	22221	1964	

BANFF & BUCHAN DISTRICT COUNCIL LEISURE & RECREATION DEPARTMENT
FRASERBURGH MINI-RAILWAY
Gauge : 2'0".   (NK 001659)

677	KESSOCK KNIGHT	4wDM	S/O LB	53541	1963

HIGHLAND

INDUSTRIAL SITES

BRITISH ALCAN HIGHLAND SMELTERS LTD, PRIMARY ALUMINIUM DIVISION,
LOCHABER WORKS, FORT WILLIAM
Gauge : 4'8½".   (NN 126751)

-		4wDM R/R Unimog 424121-10-072555 c1980		

KEY & KRAMER LTD, COATING DIVISION, SALTBURN, near INVERGORDON   (Closed)
Gauge : 900mm.   (NH 724702)

-	4wDH	Schöma	3848	1973	OOU
-	4wDH	Schöma	3849	1973	OOU
-	4wDH	Schöma	3850	1973	OOU

Locos stored at James Jacks (Evanton) Ltd,
Evanton Industrial Estate, Evanton (NH 629685).

WIGGINS TEAPE (UK) PLC, FORT WILLIAM MILL, ANNAT POINT,
CORPACH, FORT WILLIAM
Gauge : 4'8½".   (NN 083766)

(D3102	08077)	0-6-0DE	Derby		1955
	DANBYDALE	0-6-0DE	YE	2714	1958

PRESERVATION SITES

THE DUKE OF SUTHERLAND, DUNROBIN STATION, near BRORA
Gauge : 2'0".   (   )

	BRORA	0-4-0PM	S/O Bg	1797	1930
		Rebuilt	Bg	2083	1934

STRATHSPEY RAILWAY CO LTD
Gauge : 4'8½".   Locos are kept at :-

	Aviemore		(NH 898131)			
	Boat of Garten		(NH 943189)			
(45025)	5025	4-6-0	OC	VF	4570	1934
46512		2-6-0	OC	Sdn		1952
(57566)	828	0-6-0	IC	St Rollox		1899
No.20		0-6-0ST	OC	AB	1833	1924
No.17		0-6-0T	IC	AB	2017	1935
	-	0-4-0ST	OC	AB	2020	1936
No.1	DAILUAINE	0-4-0ST	OC	AB	2073	1939
48	9103/48	0-6-0ST	IC	HE	2864	1943
No.60		0-6-0ST	IC	HE	3686	1948

No.9	9114/9	CAIRNGORM	0-6-0ST	IC	RSH		7097	1943	
2996	VICTOR		0-6-0ST	OC	WB		2996	1951	
(D3605)	08490		0-6-0DE		Hor			1958	
D5394	(27050)		Bo-BoDE		BRCW	DEL 237		1962	
Sc 79979			4wDMR		A.C.Cars			1958	Dsm
	INTER VILLAGE 12-5		4wDM		MR		5763	1957	
3			0-4-0DH		NB		27549	1956	
010	CLAN		0-4-0DM		RH		260756	1950	
	QUEEN ANNE		4wDM		RH		265618	1948	
No.39	MADALENE		0-4-0DM		RH		304471	1951	
813	SHARA		2w-2PMR		Wkm		1288	1933	

LOTHIAN

INDUSTRIAL SITES

BLUE CIRCLE INDUSTRIES LTD, OXWELLMAINS CEMENT WORKS, DUNBAR
Gauge : 4'8½".   (NT 708768)

No.1	ADAM	4wDH	S	10022	1959	
No.2		4wDH	RR	10266	1967	
No.3		4wDH	S	10006	1959	
1		4wDH	RR	10247	1966	OOU
2		4wDH	RR	10248	1966	
3		4wDH	RR	10249	1966	
3		4wDH	S	10021	1959	
	-	4wDH	S	10033	1960	

BRITISH COAL
      See Section Four for full details.

PENICUIK PEAT COMPANY LTD, SPRINGFIELD MOSS, SPRINGFIELD ROAD,
WELLINGTON REFORMATORY, near LEADBURN
Gauge : 2'0".   (NT 233567)

| - | 4wDM | MR | 8738 | 1942 |

Locos move frequently between this site and Whim Moss, Borders.

SCOTTISH AGRICULTURAL INDUSTRIES PLC, LEITH FERTILISER WORKS,
LEITH DOCKS, EDINBURGH
Gauge : 4'8½".   (NN 278772)

| 3-161 | 4wDH | R/R NNM | 83505 | 1984 |

PRESERVATION SITES

LOTHIAN REGIONAL COUNCIL, RAIL & MINING HERITAGE CENTRE, LADY VICTORIA COLLIERY, NEWTONGRANGE
Gauge : 4'8½".   (NT 332638)

No.21		0-4-0ST	OC AB	2284	1949

Gauge : 3'6".

TRAINING LOCO No.2		0-6-0DM	HE	4074	1955

Gauge : 2'6".

-		4wBE	CE	5871A	1971

MIDLOTHIAN DISTRICT COUNCIL, DANDERHALL CHILDRENS RECREATION PARK, EDMONSTONE ROAD, DANDERHALL
Gauge : 4'8½".   (NT 308698)

N.C.B. No.29		0-4-0ST	OC AB	1142	1908

NATIONAL MUSEUM OF SCOTLAND, CHAMBERS STREET, EDINBURGH
Gauge : 5'0".   (NT 258734)

"WYLAM DILLY"	4w	VCG Wm Hedley	1827-1832 +

+ Incorporates parts of loco of same name built c1814.

Gauge : 1'7".

WYLAM DILLY	4w	VCG RSM	1885

POLKEMMET COUNTRY PARK, WHITBURN
Gauge : 4'8½".   (NS 924649)

-		0-6-0ST	OC AB	1175	1909 +

+ Carries plate AB 1296/1912 in error.

PRESTONGRANGE MINING MUSEUM, PRESTONGRANGE
Gauge : 4'8½".   (NT 374737)

No.6		0-4-0ST	OC AB	2043	1937
17		0-4-0ST	OC AB	2219	1946
No.7	PRESTONGRANGE	0-4-2ST	OC GR	536	1914
No.1	TOMATIN	4wDM	MR	9925	1963
No.33		4wDM	RH	221647	1943
	-	4wDM	RH	458960	1962

Gauge : 2'0".

N.C.B. No.10		4wDM	HE	4440	1952

D. RITCHIE & FRIENDS, EDINBURGH
Gauge : 3'0".   (   )

		4wDM	RH	466591	1961

Gauge : 2'6".

		4wDM	RH	189992	1938
	-	4wDM	RH	242916	1946
	-	4wDM	RH	273843	1949
10553		0-4-0DM	RH	338429	1955
P 9303	YARD No.1018	4wBE	VE	7667	

Gauge : 2'0".

	-	4wDM	HE	2654	1942
	TERRAS	4wDM	MR	7189	1937
CCC 51		4wDM	MR	7330	1938
	-	4wDM	MR	9982	1954
	-	4wDM	RH	179005	1936
	-	4wDM	RH	249530	1947

ROYAL MUSEUM OF SCOTLAND, NEWBATTLE ABBEY STORE, near DALKEITH
Gauge : 2'6".   (NT 333663)

OIL COMPANY No.2	4wWE	BLW	20587	1902

ORKNEY

PRESERVATION SITES

LYNESS INTERPRETATION CENTRE, HOY
Gauge : Metre.   (   )

ND 3646	4wDM	RH	210961	1941

Gauge : 2'6".

ND 3305 B40	4wBE	WR	3805	1948

Gauge : 60cm.

AD 27	4wDM	RH	229633	1944

STRATHCLYDE

INDUSTRIAL SITES

ANDREW DICK & SON (ENGINEERS) LTD, NORTHBURN ROAD, COATBRIDGE
Gauge : 4'8½".   (NS 740662)

-	4wVBT	VCG	S	9561	1953	Pvd

BRITISH STEEL PLC, STRIP PRODUCTS DIVISION, RAVENSCRAIG WORKS, MOTHERWELL
Gauge : 6'9½".   (NS 778562)   Used in Strip Mill.

-	4wBE	GB	6064	1962	

Gauge : 4'8½".

No.1	0-6-0DH	S	10151	1964	c
2	0-6-0DH	RR	10290	1970	b
3	0-4-0WE	RSH	8207	1961	
4	0-4-0WE	RSH	8298	1962	
32	0-6-0DH	RR	10286	1969	
33	0-6-0DH	RR	10262	1967	a
39	0-6-0DH	RR	10289	1970	
40	6wDH	RR	10265	1967	
41	6wDH	RR	10277	1968	
42	6wDH	TH	V316	1987	e
43	6wDH	TH	V317	1987	
244	0-6-0DH	TH	244V	1972	Dsm
246	0-6-0DH	TH	255V	1975	
248	0-6-0DH	TH	254V	1975	
249	0-6-0DH	TH	256V	1975	
250	0-6-0DH	TH	252V	1974	
292	0-6-0DH	RR	10292	1971	
602	0-6-0DH	HE	7190	1970	
	Rebuilt	AB	6808	1981	
7804	0-4-0WE	RSH	7804	1954	
DH 7	0-4-0DH	AB	605	1976	
DH 8	0-4-0DH	HE	7049	1971	
DH 9	0-4-0DH	HE	7423	1976	
DH 10	0-4-0DH	AB	591	1974	OOU
DH 11	0-4-0DH	AB	602	1975	
DH 16	4wDH	TH	266V	1976	Dsm
DH 403	0-4-0DH	HE	7322	1972	
-	DH				Dsm +
-	0-4-0DH	AB	590	1974	Dsm +
-	0-4-0DH	AB	595	1975	Dsm +
-	0-4-0DH	AB	610	1976	Dsm +
-	0-4-0WE	EEV	5360	1971	
	Rebuilt	AB	6082	1983	
-	0-4-0WE	EEV	5361	1971	
	Rebuilt	AB	6083	1983	
-	0-4-0WE	_(GECT	5370	1973	
		‾(BD	3684	1973	
-	0-4-0WE	GECT	5574	1979	
-	0-4-0DH	HE	7262	1972	Dsm +
-	0-4-0DH	HE	7263	1972	Dsm +

-	0-6-0DH	HE	7306	1972 Dsm
-	4wWE	RFSK	334V	1990
-	4wWE˙	RFSK	335V	1990
-	0-6-0DH	S	10032	1960 Dsm
-	4wDH	TH	267V	1976 Dsm

+ Frames in use as Barrier Wagons.
a Carries plate RR 10287.
b Carries plate RR 10216/1965.
c Frame of S 10151 rebuilt by TH with parts of RR 10213.
d Currently under overhaul at RFS Engineering Ltd.
e Plate reads P214 1987.

BRITISH STEEL PLC, TUBES DIVISION,
Imperial Works, Martyn Street, Airdrie
Gauge : 4'8½".   (NS 752648)   RTC.

-	0-4-0DH	EEV	3991	1970

Clydesdale Works, Bellshill
Gauge : 4'8½".   (NS 753592)

43	0-4-0DH	GECT	5575	1979
44	0-4-0DH	GECT	5576	1979
45	0-4-0DH	GECT	5577	1979

B.S.C. DIVERSIFIED ACTIVITIES, FULLWOOD FOUNDRIES,
NEW STEVENSTON WORKS, NEW STEVENSTON, MOTHERWELL
Gauge : 4'8½".   (NS 758600)

-	4wBE	CE	B1571	1978

CALEDONIAN PAPER PLC, LONG DRIVE, SHEWALTON, IRVINE
Gauge : 4'8½".   (NS 335354)

-	0-6-0DH	HE	9092	1988

CASTLE CEMENT (CLYDE) LTD, GARTSHERRIE WORKS, HOLLANDHURST ROAD, COATBRIDGE
Gauge : 4'8½".   (NS 730661)

-	0-6-0DH	EEV	3998	1970
-	0-6-0DH	RR	10217	1965

COSTAIN DOW MAC, COLTNESS FACTORY, NEWMAINS
Gauge : 4'8½".   (NS 823553)

-	4wDH	HE	7430	1977
-	4wDM	RH	326065	1952
COSTAIN 1	4wDM	RH	408494	1957 OOU

DEANSIDE TRANSIT LTD, DEANSIDE ROAD, HILLINGTON, GLASGOW
Gauge : 4'8½".   (NS 52x65x)

(D3362)	08292	0-6-0DE	Derby		1957
(D3415)	08345	0-6-0DE	Derby		1958
(D3896)	08728	0-6-0DE	Crewe		1960
(D3904)	08736	0-6-0DE	Crewe		1960

DELTA COAL AND ENERGY CORPORATION, MOSSEND COAL DEPOT, BELLSHILL
Gauge : 4'8½".   (NS 749605)

| C2/29 | | 0-4-0DM | JF | 4210126 | 1957 |

DRUMS & ORMSARY ESTATES, (SIR WILLIAM LITHGOW),
HARDRIDGE RAILWAY, KILMALCOLM & NETHERTON, LANGBANK
Gauge : 2'0".   (NS 313675, 393722)

	-	4wPM	MR	2097	1922	
	-	4wPM	MR	2171	1922	
No.2		4wDM	MR	8700	1941	OOU

EDMUND NUTTALL LTD, CIVIL ENGINEERS, PLANT DEPOT, KILSYTH
Gauge : 2'6".   (NS 702776)

| No.3 | EN 78 | 4wBE | SIG | 706-716 | 1976 |
| No.2 | EN 79 | 4wBE | SIG | 706-717 | 1976 |

Gauge : 2'0".

| EN 56 | | 4wBE | CE | 5590/5 | 1969 |

> Locos present in yard between contracts.

ESSO PETROLEUM CO LTD, BOWLING TERMINAL, DUNGLASS, BOWLING, near DUMBARTON
Gauge : 4'8½".   (NS 436738)

| ESSO | | 4wDH | RR | 10199 | 1964 |

HEWDEN & STEWART LTD, MOODIESBURN PEAT WORKS, CHRYSTON
Gauge : 2'0".   (NS 705708)

| - | | 4wDM | MR | 9846 | 1952 |

HUNSLET-BARCLAY LTD, LOCOMOTIVE BUILDERS, CALEDONIA WORKS, WEST LONGLANDS STREET, KILMARNOCK
Gauge : 4'8½".   (NS 425382)

42	-	0-4-0DH	AB	482	1963	
		0-6-0DH	AB	592	1974	
	-	0-6-0DH	AB	646	1979	
		Rebuilt	HAB	6767	1990	
	-	0-6-0DH	AB	659	1982	
		Rebuilt	HAB	6768	1990	
	-	0-6-0DH	AB	660	1982	
		Rebuilt	HAB	6769	1990	
63 000 329	HOLDITCH No.2	0-6-0DH	HE	6663	1969	
63 000 314	HOLDITCH No.1	0-6-0DH	HE	7018	1971	
	-	0-6-0DH	S	10093	1962	OOU
	-	0-4-0DH	S	10096	1962	OOU
	-	4wDH	TH	214V	1969	OOU
		Rebuilt	HAB	6651		
No.4		0-6-0DH	TH	285V	1979	

Locos under construction & repair usually present.

I.C.I. EXPLOSIVES, NOBELS EXPLOSIVES COMPANY LTD, ARDEER WORKS
Gauge : 4'8½".   (NS 278414)

-	0-4-0DH	AB	551	1968	
-	4wDH	S	10052	1961	

IRELAND FERROUS LTD, SCRAP DEALERS, LANGLOAN WORKS, COATBRIDGE
Gauge : 4'8½".   (NS 725643)   RTC.

-	0-4-0DM	HE	3125	1946	Dsm

JOHN WOODROW (BUILDERS) LTD, PLANT DEPOT, MAIN STREET, BRIDGE OF WEIR
Gauge : 2'0".   (NS 390656)

-	4wDM	MR	22012	1958	OOU

LILLEY PLANT LTD, PLANT DEPOT, CHARLES STREET, SPRINGBURN, GLASGOW
Yard (NS 609667) with locos present between contracts.

See entry under Warwickshire for loco fleet details.

L. & P. PEAT, FANNYSIDE WORKS, LONGRIGGEND
Gauge : 2'6".   (NS 801739)   (Closed)   RTC.

-	4wDM	LB	55870	1968	Dsm

M.C. METAL PROCESSING LTD, SPRINGBURN WORKS, SPRINGBURN ROAD, GLASGOW
Gauge : 4'8½".   (NS 608668)

| 227 | | 0-4-0DM | ―(VF | 5262 | 1945 |
| | | | (DC | 2181 | 1945 |

MINISTRY OF DEFENCE, ARMY RAILWAY ORGANISATION, INCHTERF GUN RANGE, MILTON OF CAMPSIE, near KIRKINTILLOCH
See Section Five for full details.

MINISTRY OF DEFENCE, NAVY DEPARTMENT, ROYAL NAVAL ARMAMENT DEPOT, GIFFEN, BEITH
Gauge : 4'8½".   (NS 345510)

| 00 NZ 66 | | 4wDH | BD | 3732 | 1977 |

MOTHERWELL BRIDGE & ENGINEERING CO LTD, MOTHERWELL
Gauge : 4'8½".   (NS 747575)

| - | | 0-4-0DH | RH | 457299 | 1962 OOU |

M. TURNER & SONS, DEMOLITION CONTRACTORS, 12 MUIRKIRK ROAD, LUGAR, CUMNOCK
Gauge : 3'6".   (NS 593213)

| TRAINING LOCO No.1 | | 0-6-0DM | HE | 4075 | 1955 OOU |

N.A.T.O. MOORING & SUPPORT DEPOT, FAIRLIE
Gauge : 4'8½".   (NS 209558)

| YARD No.5198 | | 4wDM | FH | 3747 | 1955 |

NOBELS EXPLOSIVES CO LTD, ARDEER WORKS
(Subsidiary of I.C.I. Ltd)
Gauge : 2'6".   (NS 290401, 290405)

21		4wDH	AB	554	1970
22		4wDH	AB	555	1970
23		4wDH	AB	556	1970
24		4wDH	AB	557	1970
26		4wDH	AB	560	1971
	05/579	4wDH	AB	561	1971
25		4wDH	AB	562	1971
32	-	4wDH	AK	18	1985
31	-	4wDH	AK	19	1985
	05/582	4wDH	AK	21	1987
	05/583	4wDH	AK	22	1987

R.B. TENNENT, CAST ROLLMAKERS, WHIFFLET FOUNDRY, COATBRIDGE
(Member of Sheffield Forgemasters Ltd)
Gauge : 4'8½".   (NS 738643)

| PLANT No.0911255 SENTINEL No.1 | 4wDH | S | 10002 | 1959 | OOU |
| PLANT No.0911563 SENTINEL No.2 | 4wDH | S | 10026 | 1960 | + |

+   Carries plate S 10043.

ROCHE PRODUCTS LTD, DALRY
Gauge : 4'8½".   (NS 295503)

| - | 4wDH | TH | 185V | 1967 |
| | Rebuilt | AB | 6941 | 1982 |

ROYAL ORDNANCE PLC, ROYAL ORDNANCE FACTORY, BISHOPTON
Gauge : 4'8½".   (NS 438703)

868		0-4-0DM	AB	338	1939	OOU
-		4wDM	FH	3894	1958	
		Rebuilt	AB		1988	
-		0-4-0DH	RSHD/WB	8364	1962	
LOCO No.9		4wDH	TH	277V	1977	

Gauge : 2'6".

1		4wBE		BV	562	1970	
2		4wBE		BV	563	1970	
3		4wBE		BV	307	1968	
4		4wBE		BV	308	1968	
5		4wBE		BV	690	1974	
6		4wBE		BV	694	1974	
7		4wBE		BV	611	1972	
8		4wBE		BV	695	1974	
9		4wBE		BV	696	1974	
(9)		4wBE		WR	2334	1942	Dsm +
10		4wBE		BV	697	1974	
11		4wBE		BV	564	1970	
12		4wBE		BV	565	1970	
13		4wBE		BV	691	1974	
15		4wBE		BV	309	1968	
16	38/540	4wBE		BV	1142	1976	
17		4wBE		BV	614	1972	
18		4wBE		BV	1143	1976	
19		4wBE		BV	613	1972	
20		4wBE		BV	612	1972	
25		4wBE		BV	692	1974	OOU
25		4wBE		WR	6214	1959	
26		4wBE		BV	693	1974	
28		4wBE		BV	306	1968	
29		4wBE		WR	1614	1940	OOU
35		4wBE		GB	898	1939	OOU
36		4wBE		GB	899	1939	Dsm
(36)	36/1335	4wBE		WR	6215	1959	OOU
37		4wBE		GB	988	1940	Dsm
38	38/518	4wBE		BV	698	1974	
	MAINTENANCE DEPOT 38	4wBE	FLP	WR	5766	1957	OOU

39	38/471	4wBE	BV	608	1971	
39		4wBE	GB	990	1940	Dsm
40		4wBE	BV	699	1974	
41		4wBE	GB	992	1940	Dsm
42	38/514	4wBE	BV	700	1974	
43		4wBE	BV	701	1974	
(44)		4wBE	GB			Dsm
45		4wBE	BV	703	1974	
46	38/472	4wBE	BV	609	1971	
46		4wBE	GB	3166		Dsm
47	38/516	4wBE	BV	702	1974	
48	38/473	4wBE	BV	610	1971	
(48)		4wBE	GB	3168		Dsm
53		4wBE	GB	3815		OOU
54		4wBE	GB	3816		OOU
55		4wBE	GB	3825		
3583		4wBE FLP	GB	1698	1940	
1		4wDH	HE	7513	1976	
2		4wDH	HE	7512	1976	
3		4wDH	HE	8827	1979	
4		4wDH	HE	8828	1979	
5		4wDH	HE	8829	1979	
6		4wDH	HE	8964	1979	Dsm
7		4wDH	HE	8965	1979	
8		4wDH	HE	8830	1979	
9		4wDH	HE	8967	1980	
10		4wDH	HE	8966	1980	
11		4wDH	HE	8968	1980	
12		4wDH	HE	8969	1980	
13		4wDH	HE	9082	1984	
14		4wDH	HE	9081	1984	
15		4wDH	HE	9080	1984	
16		4wDH	HE	9079	1984	
-		4wDM	SMH	40SPF522	1981	

+ Frame used as a battery stand.

SHEPPARD GROUP LTD, SCRAP METAL MERCHANTS & PROCESSORS,
INSHAW WORKS, MOTHERWELL
Gauge : 4'8½".   (NS 776549)   RTC.

-	0-4-0DH	HE	7043	1971	Dsm
-	4wDM	RH	321731	1952	Dsm

Also other locos for scrap occasionally present.

STRATHCLYDE PASSENGER TRANSPORT EXECUTIVE, GOVAN WORKSHOPS, GOVAN ROAD
Gauge : 4'0".   (NS 555655)

L2	4wBE	CE	B0965B	1977	
L3	4wBE	CE	B0965A	1977	
L4	4wBE	CE	B0186	1974	
	Rebuilt	CE	B3542	1988	
L5	4wBE	CE	B0186	1974	Dsm
-	4wBE R/R	NNM	78101E	1979	

PRESERVATION SITES

AYRSHIRE RAILWAY PRESERVATION GROUP, SCOTTISH INDUSTRIAL RAILWAY CENTRE, MINNIVEY COLLIERY & DUNASKIN SHED, WATERSIDE RAILWAY, DALMELLINGTON
Gauge : 4'8½".   (NS 443083, 475073)

No.16		0-4-0ST	OC	AB	1116	1910	
No.8		0-6-0T	OC	AB	1296	1912	
No.19		0-4-0ST	OC	AB	1614	1918	
3		0-4-0ST	OC	AB	1889	1926	
-		0-4-0F	OC	AB	1952	1928	
No.10		0-4-0ST	OC	AB	2244	1947	
No.25		0-6-0ST	OC	AB	2358	1954	
No.1		0-4-0ST	OC	AB	2368	1955	
(12052)	MP 228	0-6-0DE		Derby		1949	
(12093)	MP 229	0-6-0DE		Derby		1951	
YARD No.AC 118 M3571		0-4-0DM		AB	366	1943	
No.7	144-7	0-4-0DM		AB	399	1956	
-		4wDMR		Donelli Spa	163	1979	
YARD No. 107		0-4-0DM		HE	3132	1944	
-		0-4-0DM		JF	22888	1939	
-		0-4-0DM		JF	4200028	1948	Dsm
YARD No. BE 1116	DY322	4wDM		RH	224352	1943	
M/C 324	BLINKIN' BESS	4wDM		RH	284839	1950	
	JOHNNIE WALKER	4wDM		RH	417890	1959	
-		0-4-0DM		RH	421697	1959	
-		4wDH		S	10012	1959	
DX 68003	DB 965331	2w-2PMR		Wkm	10179	1968	
68002	DB 965330	2w-2PMR		Wkm	10180	1968	

Gauge : 3'0".

-	4wDH	HE	8816	1981
-	4wDM	RH	256273	1948

Gauge : 2'6".

No.1	4wDM	RH	211681	1942
2	4wDM	RH	183749	1937
No.3	4wDM	RH	210959	1941

CITY OF GLASGOW DISTRICT COUNCIL, MUSEUM OF TRANSPORT, KELVIN HALL, BUNHOUSE ROAD, GLASGOW
Gauge : 4'8½".   (NS 565663)

103		4-6-0	OC	SS	4022	1894
123		4-2-2	IC	N	3553	1886
(62277)	49					
	GORDON HIGHLANDER	4-4-0	IC	NB	22563	1920
(62469)	256 GLEN DOUGLAS	4-4-0	IC	Cowlairs		1913
-		0-4-0VBT	VCG	Chaplin	2368	1885
No.1		0-6-0F	OC	AB	1571	1917
9		0-6-0T	OC	NB	21521	1917

Gauge : 4'0".

-	4wBE		(JF	16559	1925
			(WR	583	1927
SUBWAY CAR No.1	4w-4wRER		Oldbury		1896

LOWTHERS RAILWAY SOCIETY, LEADHILLS
Gauge : 2'6".   (NS 888145)

1072		4wBE	CE		B0943	1976

Gauge : 2'0".

No.1			4wDM	FH	2631	1943	
1	THE GULLIVER		4wPM	JF	18892	1931	
2 L114	LBC KDL7	ELVAN	4wDM	MR	9792	1955	
4	QE 102-95	NITH	4wDM	RH	7002/0467/2	1966	
5	QE 102-94	SNAR	4wDM	RH	7002/0467/6	1966	
8	THE PERIL		4wDH	HE	6347	1975	
8	8564		4wDM	MR	8564	1940	Dsm +
	K 11129		4wDM	MR	8863	1944	Dsm
	K 11139		4wDM	MR	8884	1944	Dsm
	-		4wDM	FH	2586	1941	

+ Converted into a brake van.

MONKLANDS DISTRICT COUNCIL, SUMMERLEE HERITAGE PARK, WEST CANAL STREET, COATBRIDGE CENTRAL
Gauge : 4'8½".   (NS 728655)

No.11		0-4-0ST	OC	GH		1898
No.9		0-6-0T	IC	HC	895	1909
	ROBIN	4wVBT	VCG	S	9628	1957
No.5	2589	0-4-0DH		AB	472	1961
		Rebuilt from		AB		1966

ROYAL MUSEUM OF SCOTLAND, BIGGAR GASWORKS MUSEUM, GASWORKS ROAD, BIGGAR
Gauge : 2'0".   (NT 039376)

5		0-4-0T	OC	AB	988	1903

UNIVERSITY OF STRATHCLYDE, ENGINEERING APPLICATIONS CENTRE, 141 ST JAMES ROAD, GLASGOW
Gauge : 4'8½".   (   )

N.C.B. No.23		0-4-0ST	OC	AB	2260	1949

TAYSIDE

<u>INDUSTRIAL SITES</u>

JOHN DEWAR & SONS LTD, WHISKY BLENDERS, INVERALMOND, PERTH
Gauge : 4'8½".   (NO 097258)

-		4wDM	RH	458957	1961

PRESERVATION SITES

<u>CALEDONIAN RAILWAY (BRECHIN) LTD, BRECHIN, near MONTROSE, & BRIDGE OF DUN</u>
Gauge : 4'8½".   (NO 603603, 663587)

46464		2-6-0	OC	Crewe		1950
-		0-4-0ST	OC	AB	807	1897
-		0-4-0ST	OC	AB	1863	1926
-		0-6-0ST	IC	HE	2879	1943
N.C.B. 6		0-6-0ST	IC	WB	2749	1944
(D2866)		0-4-0DH		YE	2849	1961
D3059 (08046) BRECHIN CITY		0-6-0DE		Derby		1954
(D5347) 27001		Bo-BoDE		BRCW	DEL 190	1961
(Sc 51017) 126 413		2-2w-2w-2DMR		Sdn		1958
(Sc 51043)		2-2w-2w-2DMR		Sdn		1958
		4wDM		FH	3743	1955
-		4wDM		RH	275880	1949
-		0-4-0DM		RH	421700	1959
-		0-4-0DE		RH	449747	1960

<u>DALMUNZIE RAILWAY, DALMUNZIE HOTEL, DALMUNZIE</u>
Gauge : 2'0".   (NO 091713)

DALMUNZIE	4wPM	MR		2014	1920

<u>IAN N. FRASER, "PALACE GATES", VIEWFIELD ROAD, ARBROATH</u>
Gauge : 1'3".   (NO 629408)

THE FLOWER OF THE FOREST	2w-2VBT	VCG Ravenglass	5	1985

WESTERN ISLES

INDUSTRIAL SITES

<u>STORNOWAY WATERWORKS, ISLE OF LEWIS</u>
Gauge : 2'0".   (NB 410375)

-	4wPM	(MR?)		Dsm

SECTION 3 WALES

ADRAN 3 CYMRU

CLWYD

ALYN & DEESIDE DISTRICT COUNCIL, QUEENSFERRY LEISURE CENTRE,
QUEENSFERRY, DEESIDE
Gauge : 4'8½".   (SJ 315684)

41		0-4-0DH	HC	D1020	1956	Pvd

BRITISH STEEL PLC, STRIP PRODUCTS, SHOTTON WORKS, DEESIDE
Gauge : 4'8½".   (SJ 305719, 306703, 311695, 314699)

12	0-6-0DH	EEV	5352	1971
16	0-6-0DH	GECT	5391	1973
17	0-6-0DH	GECT	5392	1973
24	0-6-0DH	GECT	5402	1975

Gauge : 2'6".   (SJ 302704, 305705)   Cold Strip Mill.

No.1	A	4wBE	GB	420330/1	1972	OOU
No.2		4wBE	GB	420330/2	1972	OOU
No.3		4wBE	WR	Q7628	1976	Dsm
4		4wBE	WR	Q7807	1976	
No.5		4wBE	WR	Q7808	1976	
6		4wBE	WR	Q7809	1976	
No.1	.	4wDH	HE		c1985	
No.2		4wDH	HE		c1985	
48		0-6-0DM	HB	D1417	1971	
49		0-6-0DM	HB	D1418	1971	
50		0-6-0DM	HB	D1419	1971	
51		0-6-0DM	(HE	8847	1981	
			(HC	DM1447	1981	

CASTLE CEMENT (PADESWOOD) LTD, PADESWOOD HALL CEMENT WORKS, BUCKLEY
Gauge : 4'8½".   (SJ 290622)

6	4wDH	RR	10235	1965
7	4wDH	RR	10276	1967
-	0-4-0DE	YE	2854	1961

C.C. CRUMP & CO, DENTITH WAGON REPAIR WORKS, CONNAH'S QUAY
Gauge : 4'8½".   (SJ 296697)

MARIE	0-4-0DH	JF	22882	1939	
M.O.P. No.7	0-4-0DM	JF	4210144	1958	Dsm
-	0-4-0DE	YE	2732	1959	

CROXDEN HORTICULTURAL PRODUCTS LTD,
MANOR HOUSE WORKS, WHIXALL MOSS, near BETTISFIELD
Gauge : 2'0".   (SJ 478367, 503368)   RTC.

-	4wDM	RH	191679	1938	OOU

DEESIDE TITANIUM LTD, WEIGHBRIDGE ROAD, SEALAND, DEESIDE
Gauge : 4'8½".   (SJ 313716)

		4wDM	R/R Unilok	2186	1982
-					

MOSTYN DOCKS LTD, MOSTYN DOCK, HOLYWELL
(Subsidiary of Faber Prest Plc)
Gauge : 4'8½".   (SJ 156811)

1	0-4-0DE	YE	2627	1956
2	0-4-0DE	YE	2819	1960

SHOTTON PAPER CO PLC, WEIGHBRIDGE ROAD, SHOTTON, DEESIDE
Gauge : 4'8½".   (SJ 303717)

D672 NWX	4wDM	R/R Unimog	1145/86	1986

UNITED ENGINEERING STEELS LTD, BRYMBO STEELWORKS
Gauge : 4'8½".   (SJ 296536)

		0-4-0DE	RR	10254	1967	
9111/85		0-4-0DE	YE	2623	1956	OOU
	SPENCER	0-6-0DE	YE	2659	1957	Dsm
No.29		0-6-0DE	YE	2722	1958	
No.32		0-6-0DE	YE	2752	1959	
	ESMOND	0-6-0DE	YE	2792	1961	
	WILLIAM	0-6-0DE	YE	2800	1962	
	AUSTIN	0-4-0DE	YE	2855	1961	
	CHARLES	0-4-0DE	YE	2858	1961	
	EMRYS	0-6-0DE	YE	2867	1962	OOU
	NEVILLE	0-6-0DE	YE	2884	1962	
	WINSTON	0-6-0DE	YE	2942	1965	OOU

PRESERVATION SITES

R. IRELAND, "PLATFORM THREE", COLWYN BAY RAILWAY STATION
Gauge : 4'8½".   (SH 852793)

1	FIREFLY	0-6-0T	OC	HC	1864	1952

I.B. JOLLY, 1 LLEWELLYN DRIVE, BRYN-Y-BAAL, near MOLD
Gauge : 4'8½".   (SJ 261647)

		4wDM	MR	1944	1919 +	
TR 27		2w-2PMR	Wkm	4132	1947	Dsm +
DB 965051		2w-2PMR	Wkm	(7574?)	1956	Dsm +
3	(9036)	2w-2PMR	Wkm	8196	1958 +	

Gauge : 2'7".

4		4wDM	MR	5025	1929	Dsm +

Gauge : 60cm.

264		4wPM	MR		c1916
LR 2718		4wPM	MR	997	1918 Dsm +
-		4wDM	MR	4803	1934 +
	A.H. WORTH	4wDM	MR	5852	1933 Dsm
-		4wDM	MR	7201	1937 +
-		4wDM	MR	8723	1941 Dsm +
No.9		4wDM	MR	9547	1950 +
-		0-4-0BE	WR	N7661	1974 +
-		2w-2PM	Wkm	3030	1941 +

Gauge : 1'11½".

-	4wDM	L	30233	1946
-	4wPM	MR	6013	1931
-	4wDM	MR	20558	1955 +

+   These locos are elsewhere for renovation.

LESLEY LEISURE LTD, RHYL   (Closed)
Gauge : 1'3".   (SJ 003805)

2	JOAN	4-4-2	OC	A.Barnes	101	1920 +
1	LOUISE	4-4-2	OC	A.Barnes	102	1922 +
No.104		4-4-2	OC	A.Barnes	104	1928 +
	(MICHEAL)	4-4-2	OC	A.Barnes	105	1930
	CLARA	0-4-2DM	S/O	G&S		1961
	-	4w-4wPH				1964 +

+   Locos in storage.

LLANGOLLEN STEAM RAILWAY SOCIETY
Gauge : 4'8½".   Locos are kept at :

Llangollen Shed	(SJ 212422)
Llangollen Station	(SJ 215422)
Berwyn Loop	(SJ 195432)
Canal Sidings	(SJ 209432)

2859		2-8-0	OC	Sdn		1918
5199		2-6-2T	OC	Sdn		1934
5532		2-6-2T	OC	Sdn		1928
5952	COGAN HALL	4-6-0	OC	Sdn		1935
7754		0-6-0PT	IC	NB	24042	1930
7821	DITCHEAT MANOR	4-6-0	OC	Sdn		1950
7822	FOXCOTE MANOR	4-6-0	OC	Sdn		1950
7828	ODNEY MANOR	4-6-0	OC	Sdn		1950
1243	RICHBORO	0-6-0T	OC	HC	1243	1917
DARFIELD No. 1		0-6-0ST	IC	HE	3783	1953
5459 1	BURTONWOOD BREWER	0-6-0ST	IC	K	5459	1932
3		0-6-0ST	IC	P	1567	1920
D2162	(03162)					
BIRKENHEAD SOUTH 1879-1985		0-6-0DM		Sdn		1960
D3265	(08195) MARK	0-6-0DE		Derby		1956
D7629	(25279)	Bo-BoDE		BP	8039	1965
(D7663)	25313	Bo-BoDE		Derby		1966
D9502	9312/97	0-6-0DH		Sdn		1964
M 51618		2-2w-2w-2DHR		Derby C&W		1959

D 4	-	4wDM	FH		3953	1961
D 2	ELISEG	0-4-0DM	HC		D1012	1956
D 3		0-4-0DM	JF		22753	1939
DR 90005	NEPTUNE	0-4-0DM	JF		4000007	1947
D 1		4wDMR	Matisa		PV6 627	1967
		0-4-0DH	NB		27734	1958
	-	0-4-0DE	S/O RH		416207	1957
	WINNINGTON	0-4-0DE	RH		416213	1957
	RICHARD BORRETT	0-6-0DE	YE		2669	1958
	PILKINGTON	0-4-0DE	YE		2782	1960

RHYL TOWN COUNCIL // CYNGOR TREF Y RHYL, RHYL STATION, RHYL
Gauge : 1'3".   (SJ 009812)

106	BILLY	4-4-2	OC A.Barnes	106	1934

SLATE MINE MUSEUM, // CHWAREL LLECHI WYNNE, WYNNE QUARRY, GLYN CEIRIOG
Gauge : 2'0".   (SJ 199379)

26		4wDM	MR	8720	1941
	BEAR	4wDM	RH	339209	1952

DYFED

INDUSTRIAL SITES

ALFRED McALPINE QUARRIES LTD, MINERALS DIVISION,
CILYRYCHEN QUARRY, LLANDYBIE
Gauge : 2'11".   (SN 616168)   RTC.

	-	4wDM	RH	404976	1957 OOU

A.M.G. RESOURCES LTD, NEVILL'S DOCK, LLANELLI
Gauge : 4'8½".   (SS 505990)

D2		0-4-0DH	EEV	D1204	1967
	-	0-4-0DE	RH	544998	1969

AMOCO (U.K.) LTD, HERBRANDSTON, MILFORD HAVEN
Gauge : 4'8½".   (SM 888085)

	-	4wDH	TH	286V	1979
	-	0-4-0DH	YE	2808	1960 OOU

BRITISH COAL
     See Section Four for full details.

BRITISH STEEL PLC, STRIP PRODUCTS DIVISION, TROSTRE WORKS, LLANELLI
Gauge : 4'8½".   (SS 531994)

No.1		0-4-0DE	BT/WB	3096	1956
3		0-6-0DH	NB	28044	1961 OOU
6		0-6-0DE	BT/WB	3021	1951

GULF OIL REFINING LTD, WATERSTON, near MILFORD HAVEN
Gauge : 4'8½". (SM 935055)

(D2046)	2	0-6-0DM	Don		1958
		Rebuilt	HE	6644	1967
(D2113	03113)	0-6-0DM	Don		1960 Dsm
219		4wDH	TH	193V	1968
245		0-4-0DH	TH	239V	1971

MILFORD DOCKS CO LTD, MILFORD HAVEN
Gauge : 4'8½". (SM 899061)   RTC.

MARGARET BRISTOWE	0-4-0DM	HC	D850	1954 OOU
CHARLES NEWBON	0-4-0DM	JF	4200015	1947 Dsm
F.B. ROBJENT	0-4-0DM	JF	4200016	1947 OOU

MINISTRY OF DEFENCE, NAVY DEPARTMENT, ROYAL NAVAL ARMAMENT DEPOT,
Milford Haven
Gauge : 4'8½". (SM 917051)

No.4	YD No.26656	4wDH	BD	3733	1977
	YARD No.8690	4wDM	FH	3857	1957
ND 10022	YARD No.9677	4wDM	FH	3968	1961

Gauge : Metre.

ND 10551	4wDH	BD	3775	1983
ND 10552	4wDH	BD	3776	1983

Trecwn
Gauge : 4'8½". (SM 970324)

ND 10491	ARMY 412 C2SA	0-4-0DH	NB	27647	1959
ND 10490	ARMY 413 C2SA	0-4-0DH	NB	27648	1959

Gauge : 2'6". (SM 970325)

ND 10261	A 1	4wDM	BD	3751	1980
ND 10262	A 2	4wDM	BD	3752	1980
ND 10260	T 0003	4wDM	BD	3753	1980
ND 10392	T 0004	4wDM	BD	3755	1982
ND 10393	A 5	4wDM	BD	3756	1982
ND 10493	T 0006	4wDM	BD	3764	1983
ND 10554	A 7	4wDH	BD	3779	1983
ND 10555	T 0008	4wDH	BD	3780	1983
ND 10556	T 0009	4wDH	BD	3781	1983
ND 10557	T 0010	4wDH	BD	3782	1983
ND 10570	T 0011	4wDH	BD	3783	1984
ND 10571	T 0012	4wDH	BD	3784	1984
ND 2965	YARD No. 54 T 0235	4wDMR	FH	2196	1940
ND 3306	YARD No.B 48	2w-2BE	GB	3545	1948
ND 3307	YARD No.B 49	2w-2BE	GB	3546	1948
ND 3308	YARD No.B 50	2w-2BE	GB	3547	1948

Also uses M.O.D., A.R.O. locos.
See Section Five for full details.

THE NATIONAL TRUST, DOLAUCOTHI GOLD MINES, PUMPSAINT
Gauge : 1'10½". (SN 666404)

```
            -             0-4-0BE        WR                            +
            -             4wBE           WR              899  1935  OOU
```

     + Either WR 1080/37 or WR 5311/55.

PRESERVATION SITES

BRECON MOUNTAIN RAILWAY CO LTD,
VALE OF RHEIDOL LIGHT RAILWAY // LEIN FACH CWM RHEIDOL, ABERYSTWYTH
Gauge : 1'11½". (SN 587812)

```
    7       OWAIN GLYNDWR      2-6-2T    OC  Sdn               1923
    8       LLYWELYN           2-6-2T    OC  Sdn               1923
    9       PRINCE OF WALES    2-6-2T    OC  DM          2     1902
   10                          0-6-0DH       BMR        002    1987 +
   68804                       2w-2DMR       Permaquip  005    1985
    -                          2w-2PMR       Wkm      10943    1976
```

     + Built with parts supplied by BD.

DAVIES FARMS CAREW LTD, STEPHENS GREEN FARM, MILTON, near PEMBROKE
Gauge : 4'8½". (SN 041019)

```
            -             0-4-0ST    OC  WB            2565  1936
```


DYFED COUNTY COUNCIL, SCOLTON MANOR MUSEUM & COUNTRY PARK,
SCOLTON, near HAVERFORDWEST
Gauge : 4'8½". (SM 991222)

```
GWENDRAETH RAILWAY CO No.2
No.1378  KIDWELLY  MARGARET  0-6-0ST    IC  FW          410   1878
  A 123 W                    2w-2PMR        Wkm        3361   1942
```


GWILI RAILWAY CO LTD // Y RHEILFFORDD GWILI CYF
Gauge : 4'8½". Locos are kept at :-

```
            Bronwydd Arms              (SN 417239)
            Cynwyl Elfed Station       (SN 386264)
            Llwyfan Cerrig             (SN 405258)
            Pensarn, Carmarthen,       (SN 413193)
```

```
ROSYTH No.1                    0-4-0ST    OC  AB       1385   1914 a
No.1                           0-6-0ST    OC  HC       1885   1955
          LITTLE LADY          0-4-0ST    OC  P        1903   1936
          OLWEN                0-4-0ST    OC  RSHN     7058   1942
   71516                       0-6-0ST    IC  RSH      7170   1944
SWANSEA VALE No.1              4wVBT     VCG  S        9622   1958 a
   2784   DYLAN THOMAS         0-4-0DH        NB      27654   1956 a
          -                    4wDM           RH     183062   1937
          IDRIS                4wDM           RH     207103   1941
    3     SWANSEA JACK         4wDM           RH     393302   1955 a
          TRECATTY             0-6-0DM        RH     421702   1959
```

```
02101                          0-4-0DE      YE      2779  1960
B154W    PWM 2222              2w-2PMR      Wkm     4139  1947 +
```

         + Awaiting renovation at Brynteg School, Bridgend.
         a Owned by Y Clwb Rheil Cymru.

LLANELLI BOROUGH COUNCIL // CYNGOR BWRAEISTREF LLANELLI,
Kidwelly Industrial Museum // Amgueddfa Diwydiannol Cydweli,
Llangadog, Kidwelly
Gauge : 4'8½".   (SN 422078)

```
          -                    0-4-0ST   OC  AB      1081  1909
          -                    0-6-0ST   OC  P       2114  1951
No.2                           0-4-0DM       AB       393  1954
```

Gauge : 2'0".

```
          -                    4wDM          RH    398063  1956
                               Rebuilt       ESCA          1986
```

Pembrey Country Park // Parc Wledig Pen-Bre, Pembrey
Gauge : 2'0".   (SN 402004)

```
       MERLIN                  4wDH     S/O HU      LX1001  1968
```


LLANELLI & DISTRICT RAILWAY PRESERVATION SOCIETY,
GREAT MOUNTAIN RAILWAY, CYNHEIDRE
Gauge : 4'8½". (   )

```
   520/7                       0-6-0DH       EEV    D1199  1967
   102      2751               0-4-0DM       RH    418790  1958
```


OAKWOOD ADVENTURE AND LEISURE COMPLEX, CANASTON BRIDGE, NARBERTH
Nutty Jakes Gold Mine
Gauge : 60cm.   (SN 072124)

```
          -                    4w-4wRE  S/O  SL            1990
          -                    4w-4wRE  S/O  SL            1990
```

Gauge : 1'3".

```
          -                    4-4wDHR       Goold         1989
       LINDY-LOU               2-8-0DH  S/O  SL     7218  1972
       LENKA                   4w-4DHR       SL     7322  1973 +
```

         + 4-wheel power unit incorporates the main frames of
           4wDM L 7280/1936.

TEIFI VALLEY RAILWAY PLC // RHEILFFORDD DYFFRYN TEIFI,
HENLLAN, near LLANDYSUL
Gauge : 2'0".   (SN 357406)

```
       ALAN GEORGE             0-4-0ST   OC  HE       606  1894
       SHOLTO                  4wDM          HE      2433  1941
       SIMON                   4wDM          MR      7126  1936
          -                    4wDM          MR      7215  1938 Dsm
       SAMMY                   4wDM          MR     11111  1959
```

GWENT

ALPHA STEEL LTD, CORPORATION ROAD, NEWPORT
Gauge : 4'8½".   (ST 334844)   RTC.

-		0-4-0DM	RH	252686	1949 OOU

A.R.C./POWELL DUFFRYN QUARRIES LTD, MACHEN QUARRY, near NEWPORT
Gauge : 4'8½".   (ST 223886)

-	4wDH	FH	3890	1959
-	4wDH	RR	10222	1965

BRITISH COAL
        See Section Four for full details.

BRITISH STEEL PLC, STRIP PRODUCTS DIVISION, LLANWERN WORKS, NEWPORT
Gauge : 4'8½".   (ST 385863)

101	153/1521	0-6-0DH	EEV	D1246	1968
102	153/1522	0-6-0DH	EEV	D1247	1968 OOU
103	153/1523	0-6-0DH	EEV	D1248	1968
104	153/1524	0-6-0DH	EEV	D1249	1968
		Rebuilt	GECT		1975
105	153/1525	0-6-0DH	EEV	D1253	1968
106	153/1526	0-6-0DH	EEV	D1226	1967
107	153/1527	0-6-0DH	EEV	D1251	1968 OOU
		Rebuilt	GECT		1975
108	153/1534	0-6-0DH	EEV	D1254	1968 OOU
		Rebuilt	GECT		1975
109	153/1535	0-6-0DH	EEV	D1252	1968 OOU
		Rebuilt	GECT		1975
301	153/1528	0-6-0DH	GECT	5378	1972
302	153/1529	0-6-0DH	GECT	5379	1972
303	153/1530	0-6-0DH	GECT	5380	1972
304	153/1531	0-6-0DH	GECT	5381	1972 OOU
305	153/1532	0-6-0DH	GECT	5382	1973
306	153/1533	0-6-0DH	GECT	5383	1973
D.E.1	153/2503	6wDE	GECT	5409	1976
D.E.2	153/2504	6wDE	GECT	5410	1976
D.E.3	153/2505	6wDE	GECT	5411	1976
D.E.4	153/2506	6wDE	GECT	5412	1976
D.E.5	153/2507	6wDE	GECT	5413	1976
1		0-4-0WE	GB	2998	1961
2		0-4-0WE	Llanwern		1977
	-	0-4-0WE	GB	2999	1961
	-	0-4-0WE	GECT	5389	1973
	-	0-4-0WE	GECT	5390	1973

BRITISH STEEL, STRIP PRODUCTS, TINPLATE, EBBW VALE WORKS, EBBW VALE
(A Division of British Steel Plc)
Gauge : 4'8½".   (SO 173070, 174073, 174078)

170		0-8-0DH	HE	7063	1971
171		0-8-0DH	HE	7064	1971
172		0-8-0DH	HE	7065	1971
173		0-8-0DH	HE	7200	1972
	-	2w-2PMR	Wkm	9359	1963

GWENT COAL DISTRIBUTION CENTRE, NEWPORT
Gauge : 4'8½".   (ST 317875)

	PRIDE OF GWENT	0-6-0DM	HC	D1186	1959
	Rebuilt	HE		8526	

ISIS-LINK LTD, ORB WORKS, NEWPORT
Gauge : 4'8½".   (ST 323862)

-		4wDM	RH	312433	1951
-		4wDH	RR	10198	1965

NATIONAL POWER, USKMOUTH POWER STATION, WEST NASH, NEWPORT
(A Division of C.E.G.B.)
Gauge : 4'8½".   (ST 327837)

-	0-4-0DM	AB	445	1959
USKMOUTH 3	0-4-0DH	HE	5622	1960

Gauge : 2'0".   (ST 331840)   Pleasure Line.

-	4wDM	HE	6013	1961

ROYAL ORDNANCE AMMUNITION LTD, GLASCOED, USK
Gauge : 4'8½".   (SO 345016)

19092	R.O.F.GLASCOED No.3	0-4-0DH	RSHD/WB	8365	1962	
19091	R.O.F.GLASCOED No.2	0-4-0DH	RSHD/WB	8366	1962	OOU
31469	R.O.F. No.1	0-4-0DH	TH	132C	1963	OOU
	Rebuild of	0-4-0DM	JF	22982	1942	
31527	R.O.F. No.4	4wDH	TH	292V	1980	

TALYWAIN BUILDING SUPPLIES AND GARDEN CENTRE, TALYWAIN
Gauge : Metre.   (SO 259045)

436	OLD BUCKS LEVEL	4wDH	MR	121UA117	1974

PRESERVATION SITES

BIG PIT MINING MUSEUM, BIG PIT COLLIERY SITE, BLAENAVON
Gauge : 2'0".   (   )

3		0-4-0DM	HE	6049	1961

PONTYPOOL & BLAENAVON RAILWAY COMPANY (1983) LTD,
PONTYPOOL & BLAENAVON RAILWAY, BLAENAVON
Gauge : 4'8½".   (SO 237093)

2874		2-8-0	OC	Sdn			1918
3855		2-8-0	OC	Sdn			1942
4253		2-8-0T	OC	Sdn			1917
5668		0-6-2T	IC	Sdn			1926
5967	BICKMARSH HALL	4-6-0	OC	Sdn			1937
4	NORA	0-4-0ST	OC	AB	1680		1920
	MENELAUS	0-6-0ST	OC	P	1889		1935
No.8		0-6-0ST	IC	RSH	7139		1944
		Rebuilt		HE	3880		1961
	BROOKFIELD	0-6-0PT	OC	WB	2613		1940
RT 1		0-6-0DM		JF	22497		1938
	-	4wDH		S	10083		1961
	-	0-6-0DM		_(VF	D78		1948
				⁻(DC	2252		1948
	PWM 3962	2w-2PMR		Wkm	6947		1955

GWYNEDD

INDUSTRIAL SITES

ANGLESEY ALUMINIUM METAL LTD, PENRHOS WORKS, HOLYHEAD
Gauge : 4'8½".   (SH 264807)

56-007		0-4-0DH	HE	7183	1970

A.R.C. AGGREGATES LTD, PENMAENMAWR GRANITE QUARRIES
Gauge : 3'0".   (SH 701758)   RTC.

(PENMAEN)	0-4-0VBT	VC	DeW		1878 Dsm +

+ Loco dumped on Level Two.

ASSOCIATED OCTEL CO LTD, AMLWCH
Gauge : 4'8½".   (SH 446936)

	-	0-4-0DH	HE	7460	1977
1		4wDM	RH	321727	1952

CARNARVON MINING CO LTD, CLOGAU ST DAVIDS GOLD MINE,
BONTDDU, near DOLGELLAU
Gauge : 2'0".   (SH 667195, 672201)

-		4wBE	WR	M7556	1972
		Rebuilt	WR	10114	1984
-		0-4-0BE	WR	L1009	1981

GREAVES WELSH SLATE CO LTD
Llechwedd Slate Quarries, Blaenau Ffestiniog
Gauge : 2'0".   (SH 702468)   RTC.

-		4wDM	RH	174542	1935 OOU

Maen Offeren Slate Quarry, Blaenau Ffestiniog
Gauge : 2'0".   (SH 713467, 716465)   (Some locos work underground)

-	4wBE	CE	5688/2	1969	
-	4wDM	RH	174536	1936	00U
-	4wDM	RH	175127	1935	00U
-	4wBE	WR	918	1936	00U
-	4wBE	WR			

GWYNFYNNYDD & BEDDCOED GOLD MINE, COED-Y-BRENIN FOREST,
GANLLWYD, near DOLLGELLAU   (Closed : on care & maintenance basis)
Gauge : 1'11½". (SJ 737281)

-	4wBE	CE	5370	1967	00U
-	4wBE	CE	5885	1971	00U
-	4wBE	WR	1277	1938	00U
-	4wBE	WR	5537	1956	00U

INIGO JONES & CO LTD, TUDOR SLATE WORKS, GROESLON, near PENYGROES
(Subsidiary of Wincilate Ltd)
Gauge : 2'3".   (SH 471551)

-	4wBE	LMM		Pvd

WINCILATE LTD, ABERLLEFENNI SLATE QUARRIES, ABERLLEFENNI
Gauge : 2'3".   (SH 769102)

-	4wBE	CE	B0457	1974

PRESERVATION SITES

BALA LAKE RAILWAY LTD // RHEILFFORDD LLYN TEGID CYF, LLANUWCHLLYN
Gauge : 1'11½". (SH 881300)

No.3	HOLY WAR	0-4-0ST	OC	HE	779	1902	
	ALICE	0-4-0ST	OC	HE	780	1902	Dsm
No.4	MAID MARIAN	0-4-0ST	OC	HE	822	1903	
	-	4wDM		FH	2544	1942	
	-	4wDM		HE	1974	1939	
727/69		4wDM		HU	38384	c1930	
	-	4wDM		L	34025	1949	
	-	4wDM		MR	5821	1934	
	-	4wDM		RH	189972	1938	
No.12	CHILMARK	4wDM		RH	194771	1939	
	INDIAN RUNNER	4wDM		RH	200744	1940	
No.11	CERNYW	4wDM		RH	200748	1940	
1		4wDM		RH	209430	1942	
	-	4wDM		RH	283512	1949	
D1087	MEIRIONYDD	4w-4wDH		SL	22	1973	

BUTLINS PWLLHELI HOLIDAY WORLD, PENYCHAIN
Gauge : 2'0".   (SH 434363)

157	C.P.HUNTINGTON	4w-2-4wDM	S/O Chance 78-50157-24	1978

CONWY VALLEY RAILWAY MUSEUM // AMGUEDDFA RHEILFFORDD DYFFRYN CONWY,
BETWS-Y-COED
Gauge : 60cm.   (SH 796565)

	SGT. MURPHY	0-6-0T	OC	KS	3117	1918

Gauge : 1'3".

70000	BRITANNIA	4-6-2	OC	(Longfleet Eng	1968
				(TMA 8753	1987

CORRIS RAILWAY SOCIETY // CYMDEITHAS RHEILFFORDD CORRIS, MAESPOETH
Gauge : 2'6".   (SH 753069)

WMD 2	CORRIS No.6	4wDH	RH	518493	1966

Gauge : 2'3".

No.5	ALAN MEADEN	4wDM	MR	22258	1965

FFESTINIOG MOUNTAIN TOURIST CENTRE LTD, GLODDFA GANOL, BLANEAU FFESTINIOG
Gauge : 2'0".   (SH 693470)

87025	L201N		4wDM	MR	22238	1965
87028	L204N	28	4wDM	MR	40S308	1967
87032	L202N	32	4wDM	MR	40S412	1973

FESTINIOG RAILWAY CO // Y RHEILFFORDD FFESTINIOG
Gauge : 1'11½". Locos are kept at :-

	Boston Lodge Shed & Works	(SH 584379, 585378)
	Ffestiniog Railway Museum,	
	Porthmadog Goods Shed	(SH 571384)
	Glan-Y-Pwll Depot	(SH 693461)
	Minffordd P.W. Depot	(SH 599386)
	Porthmadog Station	(SH)

No.LR 10003	MOUNTAINEER	2-6-2T	OC	AL	57156	1916
No.10	MERDDIN EMRYS	0-4-4-0T	4C	Boston Lodge		1879
	EARL OF MERIONETH /					
	IARLL MEIRIONYDD	0-4-4-0T	4C	Boston Lodge		1979
No.2	PRINCE	0-4-0STT	OC	GE	(199?)	1863
No.1	(PRINCESS)	0-4-0STT	OC	GE	(200?)	1863
No.4	PALMERSTON	0-4-0STT	OC	GE		1864
No.5	WELSH PONY	0-4-0STT	OC	GE	(234?)	1867
	BLANCHE	2-4-0STT	OC	HE	589	1893
	LINDA	2-4-0STT	OC	HE	590	1893
1	BRITOMART	0-4-0ST	OC	HE	707	1899
9	HARLECH CASTLE	0-6-0DH		BD	3767	1983
	MOELWYN	2-4-0DM		BLW	49604	1918
	ASHOVER	4wDM		FH	3307	1948
	UPNOR CASTLE	4wDM		FH	3687	1954
YD No.1016	CONWY CASTLE /					
	CASTELL CONWY	4wDM		FH	3831	1958
	MOEL HEBOG	0-4-0DM		HE	4113	1955
No.LR 10550	MARY ANN	4wDM		MR	(590?	1917) +
	-	4wDM		MR	3694	1924
		Rebuild of MR			1895	1919 Dsm

```
      JGF 1     "JANE"                  4wDM       MR          8565  1940 OOU
    No.6        THE COLONEL             4wDM       MR          8788  1943
      JGF 2     THE LADY DIANA          4wDM       MR         21579  1957
                -                       4wDM       MR         21615  1957 Dsm
      JGF 3     SANDRA    19/01         4wDM       MR         22119  1961 Dsm a
                ANDREW                  4wDM       RH        193984  1939 Dsm
                ALISTAIR                4wDM       RH        201970  1940 OOU
    1543                                2w-2PMR    Wkm         1543  1934
```

        + Carries plate 507/1917.
        a In use as a winch wagon.

Gauge : 750mm.

```
        SANTA THEREZA        2-6-0    OC  BLW        15511  1897
```

NARROW GAUGE RAILWAY CENTRE OF NORTH WALES,
Blaenau Ffestiniog B.R. Station
Gauge : 2'0".   (SH 696461)

```
        -                    4wDM       HE          2207  1941
```

Craft Centre, Site of Braich Goch Quarry, Upper Corris
Gauge : 2'4".   (SH 751077)

```
        CORRIS   398102      4wDM       RH        398102  1956
```

Dolwyddelan B.R. Station
Gauge : 2'6".   (SH 738522)

```
    2209    DOLWYDDELAN      4wDM       HE          2209  1941
```

Gloddfa Ganol, Blaenau Ffestiniog (Not all locos are on public display).
Gauge : 4'8½".   (SH 693470)

```
WARRINGTON No.3                       0-4-0DM    JF         22898  1940
                -                     0-4-0DM    JF         22900  1941
    TR 8        4151                  2w-2PMR    Wkm         4151  1948
    TR 12   (PWM 2188)    A156W       2w-2PMR    Wkm         4165  1948
    TR 18   PWM 4301                  2w-2PMR    Wkm         7504  1956
            PWM 4302                  2w-2PMR    Wkm         7505  1956
            PWM 4311                  2w-2PMR    Wkm         7514  1956
    TR 7    A33M                      2w-2PMR    Wkm         8503  1960
```

Gauge : Metre.

```
    50823 1915        .                4wPM       RP         50823  1915
```

Gauge : 3'0".

```
        LLANFAIR              0-4-0VBT VC  DeW              1895
    3930044                   4wDM       JF       3930044  1950
    105H006                   4wDH       MR       105H006  1969
    1082                      0-4-0DM    Ruhr        1082 c1936
    C 37                      2w-2PMR    Locospoor B7281E
    (C 13)                    2w-2PMR    Wkm         2449  1938 Dsm
    6/501                     2w-2PMR    Wkm         4091  1946 Dsm
```

W6/504		2w-2PMR		Wkm	4092	1946	Dsm
C 18	4808	2w-2PMR		Wkm	4808	1948	
C 20	4810	2w-2PMR		Wkm	4810	1948	
C 23		2w-2PMR		Wkm	4813	1948	Dsm
(C 26)		2w-2PMR		Wkm	4816	1948	Dsm

Gauge : 2'6".

984	THE WEE PUG	0-4-0T	OC	AB	984	1903
45913		4wPM		HU	45913	1932
YD No.988		4wDM		RH	235729	1944

Gauge : 2'0".

14005	STEAM TRAM	4wVBT	G	L	14005	1940	
	Rebuilt from	4wPM				1969	
3114	DERW BACH	0-4-0ST	OC	KS	3114	1918	
1568	DOROTHY	0-4-0ST	OC	WB	1568	1899	
	-	4wBE		BE	16303	c1917	
No.760		0-4-0PM		BgC	760	1918	
774	OAKELEY	0-4-0PM		BgC	774	1919	
No.3205	BREDBURY	2w-2PM		Bredbury		c1954	
257081	DELTA	0-4-0DM		Dtz	257081	c1931	
No.1568		4wPM		FH	1568	1927	
	-	4wPM		FH	1747	1931	Dsm
1881		4wPM		FH	1881	1934	
	-	4wDM		FH	2025	1937	
3424		4wPM		FH	3424	1949	
	-	4wDM		FH	4008	1963	
1840		4wBE		GB	1840	1942	
	-	4wPM		H	982	1930	Dsm
D564		4wDM		HC	D564	1930	
2666		4wDM		HE	2666	1942	
6018	TWUSK	4wDM		HE	6018	1961	
6299		4wPM		L	6299	1935	
	RAIL TAXI	4-2-0PMR		R.P.Morris		1967	
	-	4wDM		MR	22128	1961	
164350		4wDM		RH	164350	1933	
	-	4wDM		RH	166028	1932	Dsm
	-	4wDM		RH	198297	1939	
235711	PEN-YR-ORSEDD	4wDM		RH	235711	1945	
ZM 32	416214	4wDM		RH	416214	1957	
	-	4wBE		Spondon		1926	
	THAKEHAM	4wPM		Thakeham		c1946	
640	WELSH PONY	4wWE		WR	640	1926	
1298	LITTLE GEORGE	0-4-0BE		WR	1298	1938	

Gauge : 60cm.

2442	-	0-6-0T	OC	KS	2442	1915
3010		0-6-0T	OC	KS	3010	1916
3014		0-6-0T	OC	KS	3014	1916
No.646		0-4-0PM		BgC	646	1918
No.736		0-4-0PM		BgC	736	1918
1835		4wDM		HE	1835	1937
4470		4wPM		OK	4470	1930
211647		4wDM		RH	211647	1941

Gauge : 1'11½".

No.3	FESTINIOG	4wDM	Festiniog Rly	1974

Gauge : 1'10¾".

	KATHLEEN	0-4-0VBT	VC	DeW		1877

Gauge : 1'6".

551		4wBE		WR	551	1924

Gauge : 1'3".

6502	WHIPPIT QUICK	4wPMR		L	6502	1935
		Rebuilt		Fairbourne		

NARROW GAUGE RAILWAY MUSEUM TRUST, WHARF STATION, TYWYN
Gauge : 2'6".   (SH 586004)

LBC L1 5 NUTTY	4wVBT	ICG	S	7701	1929

Gauge : 2'0".

No.2	0-4-0WT	OC	KS	721	1901

Gauge : 1'10¾".

GEORGE HENRY	0-4-0VBT	VC	DeW		1877
ROUGH PUP	0-4-0ST	OC	HE	541	1891
JUBILEE 1897	0-4-0ST	OC	MW	1382	1897

Gauge : 1'10".

13	0-4-0T	IC	Spence	1895

Gauge : 1'6".

DOT	0-4-0WT	OC	BP	2817	1887
PET	0-4-0ST	IC	Crewe		1865

NATIONAL MUSEUM OF WALES // AMGUEDDFA CHWARELI GOGLEDD CYMRU, WELSH SLATE MUSEUM, GILFACHDDU, LLANBERIS
Gauge : 1'11½". (SH 586603)

UNA	0-4-0ST	OC	HE	873	1905
-	4wBE		BE		1917
-	4wPMR		Oakeley		

THE NATIONAL TRUST, INDUSTRIAL RAILWAY MUSEUM, PENRHYN CASTLE // CASTELL PENRHYN, LLANDEGAI, near BANGOR
Gauge : 4'8½". (SH 603720)

	HAWARDEN	0-4-0ST	OC	HC	526	1899
	VESTA	0-6-0T	IC	HC	1223	1916
No.1		0-4-0WT	OC	N	1561	1870
	HAYDOCK	0-6-0T	IC	RS	2309	1876 +

+ Works plate reads 1879.

Gauge : 4'0".

| | FIRE QUEEN | 0-4-0 | OC | AH | | 1848 |

Gauge : 3'0".

| KETTERING FURNACES No.3 | 0-4-0ST | OC | BH | 859 | 1885 |
| (WATKIN) | 0-4-0VBT | VC | DeW | | 1893 |

Gauge : 1'11½".

| | - | 4wDM | | RH | 327904 | 1951 |

Gauge : 1'10¾".

| | CHARLES | 0-4-0ST | OC | HE | 283 | 1882 |
| | HUGH NAPIER | 0-4-0ST | OC | HE | 855 | 1904 |

NORTH WALES TRAMWAY MUSEUM, TAL-Y-CAFN, GOODS YARD, near LLANDUDNO JUNCTION
Gauge : 1'3".   (SH 787716)

| | - | 4wPM | | L | 20886 | 1941 |

QUARRY TOURS LTD, LLECHWEDD SLATE MINE, BLAENAU FFESTINIOG
Gauge : 2'0".   (SH 699468)

	-	4wBE		BEV	308	1921
	-	4wBE		BEV	323	1921
No.4	THE ECLIPSE	0-4-0WE		Greaves		1927
	Rebuild of	0-4-0ST	OC	WB	1445	1895
	THE COALITION	0-4-0WE		Greaves		1930
	Rebuild of	0-4-0ST	OC	WB	1278	1890
	-	4wPM		Greaves		c1936
MBS 387		4wBE		LMM	1053	1950
MBS 236		4wBE		LMM	1066	1950
	-	4wBE		WR	C6765	1963 +
LE/12/65		4wBE		WR	C6766	1963
LE/12/64		4wBE		WR		
	-	4wBE		WR		

+ Carries plate E6807/1965 in error.

RHEILFFORDD LLYN PADARN, CYFYNGEDIG // LLANBERIS LAKE RAILWAY LTD,
GILFACHDDU, LLANBERIS
Gauge : 1'11½".   (SH 586603)

No.1	ELIDIR	0-4-0ST	OC	HE	493	1889
No.2	THOMAS BACH	0-4-0ST	OC	HE	849	1904
No.3	DOLBADARN	0-4-0ST	OC	HE	1430	1922
No.5	HELEN KATHRYN	0-4-0WT	OC	Hen	28035	1948
	-	4wDM		MR	7927	1941 Dsm
No.9	DOLGARROG	4wDM		MR	22154	1962
No.11	GARRET	4wDM		RH	198286	1940
No.10	BRAICH	4wDM		RH	203031	1941
No.8	YD No.AD689 TWLL COED	4wDM		RH	268878	1952
	-	4wDM		RH	425796	1958 Dsm
No.3	164/137	4wDM		RH	441427	1961
No.19	LLANELLI	4wDM		RH	451901	1961

SNOWDON MOUNTAIN RAILWAY LTD, LLANBERIS
Gauge : 2'7½".   (SH 582597)

2	ENID	0-4-2T	OC R/A SLM	924	1895
3	WYDDFA	0-4-2T	OC R/A SLM	925	1895
4	SNOWDON	0-4-2T	OC R/A SLM	988	1896
5	MOEL SIABOD	0-4-2T	OC R/A SLM	989	1896
6	PADARN	0-4-2T	OC R/A SLM	2838	1922
7	RALPH	0-4-2T	OC R/A SLM	2869	1923
8	ERYRI	0-4-2T	OC R/A SLM	2870	1923
9	NINIAN	0-4-0DH	R/A HE	9249	1986
10	YETI	0-4-0DH	R/A HE	9250	1986

On the steam locos drive is through the rack gear,
with the "driving wheels" free to rotate on their axles.

TALYLLYN RAILWAY CO // Y RHEILFFORDD TALYLLYN, TYWYN
Gauge : 2'3".   (SH 590008)   (Loco shed at Pendre Station)

No.1	TALYLLYN		0-4-2ST	OC	FJ	42	1865
No.2	DOLGOCH		0-4-0WT	OC	FJ	63	1866
No.3	SIR HAYDN		0-4-2ST	OC	HLT	323	1878
No.4	EDWARD THOMAS		0-4-2ST	OC	KS	4047	1921
No.6	DOUGLAS		0-4-0WT	OC	AB	1431	1918
No.7	TOM ROLT		0-4-2T	OC	Pendre		1974
		Rebuild of			AB	2263	1949
No.9A			0-4-0DM		HE	4135	1950 Dsm
No.9	ALF		0-4-0DM		HE	4136	1950
6292			4wDH		HE	6292	1967 Dsm a
DIESEL No.5	MIDLANDER		4wDM		RH	200792	1940
DIESEL No.8	MERSEYSIDER		4wDH		RH	476108	1964
	-		4wDH		RH	476109	1964 Dsm
19			2w-2PMR		Towyn		1952 Dsm +
	TOBY		2w-2PMR		Towyn		1954

+ Converted into a flat wagon.
a Frames utilised in a hydraulic press.

R. WATSON-JONES, 36 Y BERLLAN, PENMAENMAWR
Gauge : 3'0".   (   )

-	4wDM	RH	202987	1941

WELSH HIGHLAND LIGHT RAILWAY (1964) LTD // RHEILFFORDD YSGAFN UCHELDIR CYMRU, GELERT FARM, PORTHMADOG
Gauge : 2'0".   (SH 571393)

No.1	RUSSELL	2-6-2T	OC	HE	901	1906
	NANTMOR	0-6-0WT	OC	OK	9239	1921
	PEDEMOURA	0-6-0WT	OC	OK	10808	1924
No.7	KAREN	0-4-2T	OC	P	2024	1942
No.3	MOEL TRYFAN	0-4-2T	OC	WB	3023	1953
No.5	SNOWDON RANGER	0-4-2T	OC	WB	3050	1953
No.5		4wDM		HE	6285	1968
482	WEIGHTON	4wDH		HE	7535	1977
36	CNICHT	4wDM		MR	8703	1941
6	JONATHAN	4wDM		MR	11102	1959
8	13	4wDM		MR	60S333	1966

9	KATHERINE	4wDM	MR	60S363	1968
No.10		4wDM	RH	191658	1938
	-	4wDM	RH	237914	1946
1	GLASLYN	4wDM	RH	297030	1952
		Rebuilt	W.H.R.		1981
2	KINNERLEY	4wDM	RH	354068	1953
L5		4wDM	RH	370555	1953

MID GLAMORGAN

INDUSTRIAL SITES

BRITISH COAL
        See Section Four for full details.

FORD MOTOR CO LTD, BRIDGEND FACTORY
Gauge : 4'8½".   (SS 935783)

-	0-6-0DH	HC	D1377	1966
	Rebuilt	HE	9039	1979

PRESERVATION SITES

BRECON MOUNTAIN RAILWAY CO LTD // RHEILFFORDD MYNYDD BRYCHEINIOG, PANT
Gauge : 1'11½".

        For details see entry under Powys.

CAERPHILLY RAILWAY SOCIETY LTD, HAROLD WILSON INDUSTRIAL ESTATE,
VAN ROAD, CAERPHILLY
Gauge : 4'8½".   (ST 163865)

41312		2-6-2T	OC	Crewe		1952
	FORESTER	0-4-0ST	OC	AB	1260	1911
	VICTORY	0-4-0ST	OC	AB	2201	1945
	DESMOND	0-4-0ST	OC	AE	1498	1906
28		0-6-2T	IC	Cdf	306	1897
No.2	HAULWEN	0-6-0ST	IC	VF	5272	1945
		Rebuilt		HE	3879	1961
(D2178)	2	0-6-0DM		Sdn		1962
	-	0-6-0DM		HE	5511	1960
	DEIGHTON	0-4-0DE		YE	2731	1959

MICHAEL J STOKES, LITTLE WEST NARROW GAUGE GARDEN RAILWAY,
LITTLE WEST RESIDENTIAL HOME, SOUTHERDOWN near BRIDGEND
Gauge : 2'0".   (SS 877739)

No.4	LITTLE OWL	2w-2DM	MJ.Stokes		1986
	CILGWYN	4wDM	RH	175414	1936
5		4wDM	RH	181820	1936

MID GLAMORGAN COUNTY COUNCIL//CYNGOR SIR CANOL MORGANNWG,
NANTYMOEL NURSERY SCHOOL, WAUNLLWYD TERRACE, NANTYMOEL, OGMORE VALE
Gauge : 2'9½".　(SS 936923)

		4wDM	RH	496040	1963
-					

RHONDDA HERITAGE PARK, LEWIS MERTHYR, COED COE ROAD, TREHAFOD
Gauge : 4'8½".　(ST 040912)

-		0-6-0DH	AB	496	1964
-		4wDM	RH	441936	1960

POWYS

BRECON MOUNTAIN RAILWAY CO LTD // RHEILFFORDD MYNYDD BRYCHEINIOG
PONTSTICILL STATION & PANT STATION (MID GLAMORGAN)
Gauge : 1'11½".　(SO 063120)

2		4-6-2	OC	BLW	61269	1930	
77		2-6-2+2-6-2T	4C	Hano	10629	1928	
4	SAN JUSTO	0-4-2ST	OC	HC	639	1902	
5	SANTA ANA	0-4-2ST	OC	HC	640	1902	
	SYBIL	0-4-0ST	OC	HE	827	1903	
-		0-4-0VBT	VC	H&B			
GRAF SCHWERIN-LöWITZ		0-6-2WT	OC	Jung	1261	1908	
No.9		0-4-0WT	OC	OK	12722	1936	
-		0-4-0VBT	VC	Redstone		1905	
-		0-6-0DH	BMR	001	1987 +		
No.6	BALDWIN	4wDM	MR	7902	1939	
	RHYDYCHEN	4wDM	MR	11177	1961	

    +　Built from parts supplied by BD.

CRAFT CENTRE, ERWOOD RAILWAY STATION, ERWOOD, near BUILTH WELLS
Gauge : 4'8½".　(　　　　)

600		0-4-0DM	JF	22878	1939

WELSHPOOL & LLANFAIR LIGHT RAILWAY PRESERVATION CO LTD, LLANFAIR CAEREINION
Gauge : 2'6".　(SJ 107069)

No.2	(823)	THE COUNTESS	0-6-0T	OC	BP	3497	1902	
No.8		DOUGAL	0-4-0T	OC	AB	2207	1946	
		ORION	2-6-2T	OC	AT	2369	1948	
10	699.01	SIR DREFALDWYN	0-8-0T	OC	FB	2855	1944	
No.14		85	2-6-2T	OC	HE	3815	1954	
No.12		JOAN	0-6-2T	OC	KS	4404	1927	
No.6		MONARCH	0-4-4-0T	4C	WB	3024	1953	
7		CHATTENDEN						
YARD No.690			0-6-0DM		Bg/DC	2263	1949	
-			0-4-0DM		HE	2245	1941	Dsm a
No.11	FERRET	YARD No.86	0-4-0DM		HE	2251	1940	

-	4wDM		RH	191680	1938 Dsm
-	2w-2PMR		Wkm	2904	1940

a In use as a mobile compressor.

J. WOODRUFFE. RHIW VALLEY LIGHT RAILWAY,
LOWER HOUSE, MANAFON, near WELSHPOOL
Gauge : 1'3".   (SJ 143028)

POWYS	0-6-2T	OC	SL	20	1973
MONTY	4wPM		Jaco		1985

SOUTH GLAMORGAN

<u>INDUSTRIAL SITES</u>

A.E. KNILL & CO LTD, No.1 DOCK, BARRY
Gauge : 4'8½".   (ST 118676)

No.3229/25	4wDM		FH	2601	1942 OOU

ALLIED STEEL & WIRE LTD, CASTLE & TREMORFA WORKS AND ROD MILL, CARDIFF
Gauge : 4'8½".   (ST 196757, 209761, 210757)

370	CARLISLE	0-6-0DE	YE	2755	1959 OOU
371	KIN-KENADON	0-6-0DE	YE	2763	1959 Dsm
372	BAMBOROUGH	0-6-0DE	YE	2760	1959
378		0-6-0DE	YE	2633	1957 Dsm
379	CAERLEON	0-6-0DE	YE	2620	1956
380	CAMELOT	0-6-0DE	YE	2640	1957
390	AMESBURY	0-6-0DE	YE	2756	1959 +
391	ASTOLAT	0-6-0DE	YE	2630	1956
392	SARUM	0-6-0DE	YE	2619	1956
394	CUNETIO	0-6-0DE	YE	2770	1959
396	LEMANIS	0-6-0DE	YE	2761	1959
397		0-6-0DE	YE	2757	1959
398	CALLEVA	0-6-0DE	YE	2769	1959

+ Carries plate YE 2758/1959 in error.

BLUE CIRCLE INDUSTRIES LTD, ABERTHAW CEMENT WORKS
Gauge : 4'8½".   (ST 033675)

125	4wDH		TH	213V	1969 OOU

DOW CORNING LTD, CARDIFF ROAD, BARRY
Gauge : 4'8½".   (ST 142685)

14412	4wDM		R/R Unilok	2183	1982

NATIONAL POWER, ABERTHAW POWER STATION
(A Division of C.E.G.B.)
Gauge : 4'8½".   (ST 024658)

	-	0-6-0DM	EES		8199	1963	
	-	0-6-0DM	P		5014	1959	OOU

POWELL DUFFRYN WAGON CO LTD
Maindy Works, Parkfield Place
Gauge : 4'8½".   (ST 172782)

	-	4wDM	FH	3814	1956	
49		4wDM	FH	3959	1961	OOU
	GILLIAN	4wDH	TH	182V	1967	

Radyr Works
Gauge : 4'8½".   (ST 139798)

1		0-4-0DH	RH	512462	1965
	-	D	RH		

PRESERVATION SITES

BUTETOWN HISTORIC RAILWAY SOCIETY, WALES RAILWAY CENTRE, BUTE ROAD, CARDIFF
Gauge : 4'8½".   (ST 192749)

2861		2-8-0	OC	Sdn		1918
4115		2-6-2T	OC	Sdn		1936
5227		2-8-0T	OC	Sdn		1924
5539		2-6-2T	OC	Sdn		1928
6686		0-6-2T	IC	AW	974	1928
7927	WILLINGTON HALL	4-6-0	OC	Sdn		1950
44901		4-6-0	OC	Crewe		1945
48518		2-8-0	OC	Don		1944
80150		2-6-4T	OC	Bton		1956
92245		2-10-0	OC	Crewe		1958
	SIR GOMER	0-6-0ST	OC	P	1859	1932
	-	0-6-0DH		EEV	D1198	1967
107		0-6-0DH		NB	27932	1959

HOLIDAY INN HOTEL, MILL LANE, CARDIFF
Gauge : 4'8½".   (ST 185759)

9629		0-6-0PT	IC	Sdn	1946	Pvd

NATIONAL MUSEUM OF WALES, INDUSTRIAL AND MARITIME MUSEUM, //
AMGUEDDFA CENEDLAETHOL CYMRU, AMGUEDDFA DIWYDIANNOL A MOROL,
BUTE CRESCENT, BUTETOWN, CARDIFF
Gauge : 4'8½".   (ST 192745)

52/001		0-4-0F	OC	AB	1966	1929
	-	0-4-0F	OC	AB	2238	1948

P.D. No.10	0-6-0ST	IC	HC	544	1900
-	0-4-0ST	OC	RSHN	7705	1952

Gauge : 4'4".

-	4wG		NMW	1981

Gauge : 3'0".

-	4wDM		RH	187100	1937 +

+ Currently under renovation elsewhere.

Gauge : 2'0".

4	4wBE	GB	

WEST GLAMORGAN

INDUSTRIAL SITES

B.P. CHEMICALS (U.K.) LTD, BAGLAN BAY WORKS
Gauge : 4'8½".   (SS 744924)

ZZ44	0-6-0DH	EEV	D3989	1970
Z267	0-6-0DH	EEV	D4003	1971

B.P. OIL LLANDARCY REFINERY LTD, LLANDARCY
Gauge : 4'8½".   (SS 718960)

5	0-6-0DH	TH	157V	1965	
6	0-6-0DH	TH	194V	1968	
7	0-6-0DH	TH	230V	1971	OOU
8	0-6-0DH	TH	246V	1973	

BRITISH COAL
See Section Four for full details.

BRITISH STEEL PLC, STRIP PRODUCTS DIVISION
Port Talbot Works, Port Talbot
Gauge : 4'8½". (SS 773885, 775861, 781871)

1	4wWE	(BD	3748	1979		
		(GECT	5476	1979		
2	4wWE	(BD	3749	1979		
		(GECT	5477	1979		
6	0-4-0WE	GB	2802	1958		
6A	0-4-0WE	GB			c	
F	0-4-0WE	GB	2801	1958		
19	4wDE	WB	3003	1951	b	Dsm
20	4wDE	WB	2972	1953	b	Dsm

501		0-4-0DE	BT/WB	3066	1954	a 00U
502		0-4-0DE	BT/WB	3067	1954	+a00U
503		0-4-0DE	BT/WB	3068	1954	00U
504		0-4-0DE	BT/WB	3069	1954	a 00U
505		0-4-0DE	BT/WB	3070	1954	d 00U
508		0-4-0DE	BT/WB	3098	1956	a 00U
509		0-4-0DE	BT	3099	1956	00U
510		0-4-0DE	BT	3100	1956	a 00U
511		0-4-0DE	BT	3101	1956	a 00U
512		0-4-0DE	BT	3102	1956	+a00U
513		0-4-0DE	BT	3103	1957	+a00U
514		0-4-0DE	BT	3120	1957	+a00U
801		Bo-BoDE	AL	77120	1950	d 00U
803		Bo-BoDE	AL	77777	1950	d 00U
805		Bo-BoDE	AL	77779	1950	00U
901		Bo-BoDE	BT/WB	3063	1955	
902		Bo-BoDE	BT/WB	3064	1955	
903		Bo-BoDE	BT/WB	3065	1955	
904		Bo-BoDE	BT/WB	3137	1957	
905		Bo-BoDE	BT/WB	3138	1957	
906		Bo-BoDE	BT/WB	3139	1957	
907		Bo-BoDE	BT/WB	3140	1957	
908		Bo-BoDE	BT/WB	3141	1957	
909		Bo-BoDE	BT	3142	1957	
910		Bo-BoDE	BT	3143	1957	
951		Bo-BoDE	BT	3111	1957	
952		Bo-BoDE	BT	3112	1957	d 00U
953		Bo-BoDE	BT	3113	1957	

+ Converted to slave units for use with locos 501/8/9/10/11.
a Stored in Refractories Engineering Department (SS 777867).
b Converted to a Brake Tender Runner.
c Either GB 2591/1955 or 2737/1956.
d In store at Crane Shed.

Velindre Works, Llangyfelach, Swansea
Gauge : 4'8½".   SS 642997)

712		4wDE	BT/WB	2974	1953	00U
506		0-4-0DE	BT/WB	3071	1954	
507		0-4-0DE	BT/WB	3072	1954	
LD1		0-4-0DE	BT/WB	3097	1956	00U
114		0-4-0DH	NB	27878	1962	00U

FORD MOTOR CO LTD, CRYMLYN BURROWS, SWANSEA
Gauge : 4'8½".   (SS 696932)

P624 TS	4wDH	TH	163V	1966	00U

INCO EUROPE LTD, CLYDACH
Gauge : 4'8½".   (SN 696013)

P 0026	M.N.C. No.1	0-4-0ST	OC	P	1345	1914 Pvd
C 0120	.	0-4-0DH		EEV	D1205	1967

MARCROFT ENGINEERING, PORT TENNANT WORKS
(A Division of CAIB U.K. Ltd)
Gauge : 4'8½". (SS 681933)

	JACK LEE	0-4-0DM	Bg	3590	1962 OOU
2		4wDM	FH	3601	1953

NEATH CARGO TERMINAL, GIANTS GRAVE, BRITON FERRY
Gauge : 4'8½". (SS 735948)

-	0-4-0DH	NB	28038	1961

PARC LEVEL COLLIERY CO LTD, PARC LEVEL COLLIERY, PEN-RHIW-FAWR, YSTALYFERA
Gauge : 2'0". (SN 736103)

WENDY	4wDM	RH	432647	1959
PEARL	4wDM	RH	432648	1959
JANET	4wDM	RH	504546	1963

PRESERVATION SITES

DENNIS DAVIES, CLYDACH
Gauge : 1'11½". (   )

DIANA	0-4-0T	OC	KS	1158	1917

D.J. HITCHMAN, QUAY ROAD, NEATH
Gauge : 3'0". (   )

-	4wDH	HE	8819	1979

Loco in store.

SWANSEA INDUSTRIAL & MARITIME MUSEUM, COAST LINES WAREHOUSE,
SOUTH DOCKS, SWANSEA
Gauge : 4'8½". (SS 659927)

1426	SIR CHARLES	0-4-0F	OC	AB	1473	1916
		0-6-0ST	OC	P	1426	1916
		Incorporates parts of		P	1187	
	-	0-4-0DM		RSHD/WB	7910	1963 +

+ Currently in store at Yorkshire Imperial Metalworks.

SWANSEA VALE RAILWAY PRESERVATION SOCIETY,
SIX PIT JUNCTION, LLANSAMLET, SWANSEA
Gauge : 4'8½".   Locos are kept at :

| | | Six Pit Junction | (SS 682968) | | | |
| | | Upper Bank Works | (| | | |

4270		2-8-0T	OC	Sdn		1919
9642		0-6-0PT	IC	Sdn		1946
	LLANTANAM ABBEY	0-6-0ST	OC	AB	2074	1939
	-	0-6-0ST	IC	HE	3829	1955
2		0-4-0DH		NB	27941	1961
No.1		4wDM		RH	476143	1963

THE VALE OF NEATH RAILWAY SOCIETY, ABERDULAIS and ABERGARWED
Gauge : 4'8½". (SS 762992, 817024)

	PAMELA	0-6-0ST	IC	HE	3840	1956
	SIR JOHN	0-6-0ST	OC	AE	1680	1914
	LORD CAMROSE	0-6-0ST	OC	AE	2008	1930
(12061)	4	0-6-0DE		Derby		1949
	-	0-4-0DM		HC	D1246	1961
2	0A229	4wDM		RH	394014	1956

WEST GLAMORGAN COUNTY COUNCIL // CYNGOR SIR GORLLEWIN MORGANNWG,
CEFN COED COLLIERY MUSEUM // AMGUEDDFA GWAITH GLO CEFN COED,
BLAENANT COLLIERY, CRYNANT
Gauge : 4'8½". (SN 786034)

| | CEFN COED | 0-6-0ST | IC | HE | 3846 | 1956 Pvd |

Gauge : 2'0".

| | - | 4wDH | | HE | 8812 | 1978 |

SECTION 4 BRITISH COAL

The current surface locomotive stock of British Coal is listed here under Group headings.

Locomotives working entirely underground have been omitted as have those temporarily on the surface for repair, transfer, etc.

Those working partly underground and partly on the surface (drift mines, etc) have been included with the surface locos.

SCOTTISH GROUP

(Headquarters :).

Locomotives are kept at the following location :-

LOTHIAN

BG   (NT 274652)   Bilston Glen Colliery, Loanhead.   (Closed)

Gauge : 3'0".

   C/H 1116   0-4-0DM FLP   HE   4634 1954 BG

NORTH EAST GROUP

(Headquarters : Ryhope Road, Sunderland, Tyne & Wear).

Locomotives are kept at the following locations :-

DURHAM

Da   (NZ 436480)   Dawdon Colliery, Seaham Harbour.
Mu   (NZ 400470)   Murton Colliery, Murton. (Part of Hawthorn Colliery)
STC   (NZ 412496)   Area Training Centre, New Seaham.
SH   (NZ 390458)   South Hetton (Hawthorn Colliery), South Hetton.
VT   (NZ 425500)   Vane Tempest Colliery, Seaham.

NORTHUMBERLAND

ANW   (NZ 260882)   National Workshops, Ashington.
BD   (NZ 299925)   Bewick Drift Stockyard, Lynemouth.
Ly   (NZ 301904)   Lynemouth Colliery, Lynemouth.

TYNE & WEAR

Wm   (NZ 393580)   Wearmouth Colliery, Sunderland.

All locos are allocated a 'plant' number, and this is being applied when
the locos are overhauled. For ease of reference this number is shown
whether carried or not by the loco.

Gauge : 4'8½".

No.72	2304/59	0-6-0DH		AB	488	1964	Ly	OOU
	2120/211	0-6-0DH		AB	514	1966	SH	OOU
	2100/520	0-6-0DH		AB	583	1973	SH	OOU
	2100/521	0-6-0DH		AB	584	1973	Ly	
		Rebuilt		AB	6718	1987		
		Rebuilt		HAB	6917	1990		
No.594	20/104/997	0-6-0DH		AB	594	1974	SH	
	20/110/705	0-6-0DH		AB	604	1976	SH	OOU
	20/110/708	0-6-0DH		AB	609	1976	SH	
	20/110/706	0-6-0DH		AB	612	1976	SH	
	20/110/711	0-6-0DH		AB	615	1977	Ly	
		Rebuilt		AB	6719	1987		
	20/110/89	0-6-0DH		AB	647	1979	SH	
	2111/125	0-6-0DH		HE	6612	1965	Wm	
No.502	2668/502	0-6-0DH		HE	6613	1965	Wm	OOU
No.503	2668/503	0-6-0DH		HE	6614	1965	Wm	
	9101/66	0-6-0DH		HE	6662	1966	VT	

Gauge : 3'0".

No.1	9307/110	4wBE		CE	5921	1972	BD
No.3	20/030/22	4wBE		CE	B1561	1977	BD
No.2	20/270/34	4wBE		CE	B3060	1983	BD
	20/110/738	4wDH	FLP	CE	B2293	1983	BD
		Rebuild of		CE	B0190	1974	

Gauge : 2'6".

	20/109/088	4wDH		CE	B1819D	1978	Wm
	2109/41	4wBE	FLP	(EE	1809	1952	Wm
				(Bg	3363	1952	
2214/398	20/122/734	4wBE	FLP	(EE	3155	1961	Wm
				(RSHD	8295	1961	

Gauge : 2'2".

-		4wDM		RH	392107	1955	ANW

Gauge : 2'0".

	20/400/5	4wBE		CE	B0451	1975	Mu	
	20/400/23	4wBE		CE	B1539	1977	Mu	
L5 2476	2207/456	4wBE	FLP	(EE	2476	1958	STC	
				(RSHN	7980	1958		
2519	2103/35	4wBE	FLP	(EE	2519	1958	STC	OOU
				(Bg	3500	1958		
DM804	2216/282	0-6-0DM	FLP	HC	DM804	1951	Da	
VICTORIA 842	2216/286	0-6-0DM	FLP	HC	DM842	1954	STC	
	2202/345	0-6-0DM	FLP	HC	DM1065	1959	STC	OOU
TYNESIDE GEORGE								
L1 DM1119	2305/54	0-6-0DM	FLP	HC	DM1119	1958	STC	
L2 HELEN	2203/348	0-6-0DM	FLP	HC	DM1247	1961	STC	
DM1366	2304/67	0-6-0DM	FLP	HC	DM1366	1965	STC	
L4 LAURA	20/180/6	4wDM		HE	6348	1975	STC	
	20/120/16	4wDM	FLP	HE	7080	1971	ANW	

NORTH YORKSHIRE GROUP

(Headquarters : P.O. Box 13, Allerton Bywater, Castleford, West Yorkshire).

Locomotives are kept at the following locations :-

NORTH YORKSHIRE

GW	(SE 525318)	Gasgoine Wood Colliery, South Milford.
Kl	(SE 527233)	Kellingley Colliery, Knottingley.
KTC	(SE 529231)	Kellingley Training Centre, Knottingley.

SOUTH YORKSHIRE

BM	(SE 365061)	Barnsley Main Colliery, Barnsley.
HM	(SE 419060)	Houghton Main Colliery, Little Houghton.

WEST YORKSHIRE

AB	(SE 421279)	Allerton Bywater Colliery, Allerton Bywater.
Pm	(SE 260115)	Parkmill Colliery, Clayton West. (Closed)
PW	(SE 451226)	Prince of Wales Colliery, Pontefract.
Sh	(SE 383202)	Sharlston Colliery, New Sharlston.

Gauge : 4'8½".

108		0-4-0DH		S	10089	1962	Kl
109		0-4-0DH		S	10118	1962	Kl
110		0-4-0DH		S	10119	1962	Kl
111		0-4-0DH		S	10120	1963	Kl

Gauge : 3'0".

-		0-6-0DM	FLP	HB	DM1421	1972	KTC Dsm

Gauge : 2'6".

-		4wBE	FLP	CE	B3340	1987	GW
-		4w-4wBE	FLP	CE	B3467	1988	KTC
18	DM 890	0-4-0DM	FLP	HC	DM890	1955	PW
-		0-6-0DM	FLP	HC	DM1024	1959	ANW
-		0-6-0DM	FLP	(HC	DM1435	1979	KTC
				(HE	8583	1979	
-		0-6-0DM	FLP	(HC	DM1436	1979	KTC
				(HE	8584	1979	
6638		4wDM		HE	5694	1960	
		Rebuilt		HE	6638	1967	Kl
6639		4wDM		HE	5695	1960	
		Rebuilt		HE	6639	1967	Kl
7		4wDH	FLP	HE	7491	1979	KTC
1-44-251	8566	4wDH	FLP	(HE	8566	1981	KTC
				(AB	618	1981	
1-44-84A	FOGGWELL FLYER	4wDH	FLP	(HE	8567	1981	PW
				(AB	619	1981	

1-44-87A	BULLFROG'S BULLET	4wDH	FLP	_(HE	8568 1981	PW	
				(AB	620 1981		
3	-	4wDH	FLP	HE	8802 1987	KTC	
		4wBE	FLP	HE	9152 1986	KTC	
	THOMAS	4wDH	FLP	HE	9270 1987	GW	
	-	4wDM		SMH	60SD752 1979	AB	OOU
2-11-41B	754	4wDM		SMH	60SD754 1980	AB	
	-	4wDM		SMH	60SD755 1980	AB	

Gauge : 2'2".

TL 42	4wDM		HE	6273 1965	HM	
HM4723E	4wDM		HE	7274 1973	HM	
-	4wDH		HE	7530 1977	HM	

Gauge : 2'0".

-	4wBE	FLP	CE	B3086 1984	BM	
-	4wDM		HE	6631 1965	Pm	Dsm
-	4wDM	FLP	HE	7386 1976	S1	

SOUTH YORKSHIRE GROUP

(Headquarters : St. Georges, Thorne Road, Doncaster, South Yorkshire).

Locomotives are kept at the following locoations :-

NOTTINGHAMSHIRE

Mn	(SK 609783)	Manton Colliery, Manton.
So	(SK 558809)	Shireoaks Colliery, Shireoaks.

SOUTH YORKSHIRE

AM	(SE 558138)	Askern Colliery, Askern.
Bn	(SE 571076)	Bentley Colliery, Bentley.
BTC	(SE 569074)	Bentley Training Centre, Bentley.
Dn	(SE 517867)	Dinnington Colliery, Dinnington.
Mb	(SK 550925)	Maltby Colliery, Maltby.

Gauge : 4'8½".

No.20	MANTON	0-6-0DM	HC	D1121 1958	Mn	OOU
No.4	DL 4	0-6-0DM	HC	D1152 1959	Mn	OOU

Gauge : 3'0".

2		0-6-0DM	FLP	HC	DM1120 1958	BTC
1		0-4-0DM	FLP	HE	3614 1948	BTC
3		0-6-0DM	FLP	HE	4816 1955	BTC
8505		4wDH	FLP	HE	8505 1981	BTC

Gauge : 2'6".

-	4wBE		CE	B3434A 1987	Mb

Gauge : 2'3".

-	0-4-0DM	FLP	HE	3573 1947	Bn

Gauge : 2'0½".

524/64	4wBE	FLP	CE	B2970A 1982	So

Gauge : 2'0".

No.2	524/12	4wBE	FLP	CE	B1574F 1978	Dn
	-	0-4-0DM	FLP	HC	DM749 1949	AM

WESTERN GROUP

(Headquarters :).

Locomotives are kept at the following locations :-

GREATER MANCHESTER

Ac	(SD 798012)	Agecroft Colliery, Pendlebury.
Ps	(SD 651006)	Parsonage Colliery, Leigh.
PTC	(SD 653007)	Parsonage Training Centre, Leigh.

MERSEYSIDE

Pk	(SJ 600947)	Parkside Colliery, Newton-le-Willows.

STAFFORDSHIRE

Fl	(SJ 914420)	Florence Colliery, Longton.	
HH	(SJ 884423)	Hem Heath Colliery, Trentham.	
KTC	(SJ 887433)	Kemble Training Centre, Trentham.	
TMS	(SJ)	Trentham Machinery Stores.	(Closed)

Gauge : 4'8½".

63.000.402	L0402	0-6-0DE	YE	2660	1957	Ac	00U

Gauge : 2'6".

8	-	4wBE	FLP	Bg	3557	1961	KTC	
8		4wBE	FLP	Bg	3608	1965	KTC	
2416		4wBE	FLP	_(EE	2416	1957	PTC	
				(RSH	7935	1957		
	-	0-6-0DM	FLP	HC	DM1238	1960	PTC	
	NEWTON	4wDH		HE	8975	1979	Pk	
63.000.323		4wDM		RH	441945	1959	Fl	
63.000.315		4wDM		RH	441946	1959	Pk	+ 00U
63/000/339	LP 61/47	4wDH		RH	512994	1965	Fl	
63/000/324		4wDM		RH 7002/0867/3	1967		Fl	
No.1		4wDH		SMH	101T022	1982	HH	
No.2		4wDH		SMH	101T023	1985	HH	

+ Carries plate 441944 in error.

Gauge : 2'0".

63.000.347	4wDM	FLP	RH	433390	1959	TMS	00U
63.000.369	4wDM	FLP	RH	497760	1963	Ps	

NOTTINGHAMSHIRE GROUP

(Headquarters : Edwinstowe, Mansfield, Nottinghamshire).

Locomotives are kept at the following locations :-

NOTTINGHAMSHIRE

BH	(SK 552594)	Berry Hill Park, Mansfield.
BTC	(SK 696743)	Bevercotes Training Centre, Bevercotes, and
	(SK 700728)	Lound Hall, Bevercotes.
Cp	(SK 594632)	Clipstone Colliery, Clipstone.
Cv	(SK 603505)	Calverton Colliery, Calverton.
Gl	(SK 613438)	Gedling Colliery, Gedling.
Mf	(SK 570616)	Mansfield Coal Preperation Plant, Mansfield.
Ot	(SK 659675)	Ollerton Colliery, Ollerton.

Gauge : 4'8½".

D21	47453	0-6-0DH	RR	10270	1967	Ot	
	-	0-6-0DH	S	10072	1961	Cv	
109189	BRIERLEY	0-6-0DH	S	10157	1963	Cv	
	-	0-6-0DH	S	10160	1963	Gl	
	JULIAN	0-6-0DH	S	10161	1964	Gl	
	-	4wDH	TH	172V	1966	Mf	
D13	14814	4wDH	TH	175V	1966	Cp	00U
28533		4wDH	TH	178V	1967	Cp	
	CHARLES	0-6-0DH	YE	2940	1965	Cv	

Gauge : 2'4".

27		4wBE	FLP	CE	B3044 1983	BTC	
-		4wBE	FLP	CE	B3501 1990	BTC	
-		0-4-0DM	FLP	HB	DM1420 1972	BTC	Dsm
6		0-6-0DM	FLP	HC	DM970 1956	BTC	
-		0-6-0DM	FLP	HC	DM1286 1962	BTC	

Gauge : 2'0".

-		4wDM	RH	222068 1943	BH	Pvd

Trapped Rail System.

Becorit 'Roadrailer' (400mm)

-		1adBE	FLP	BGB 25/400/002 1987	Cp
-		1adBE	FLP	BGB 25/400/003 1987	Cp

CENTRAL GROUP

(Headquarters : Cole Orton Hall, Cole Orton, Leicestershire).

Locomotives are kept at the following locations :-

DERBYSHIRE

Sk (SK 530670) Shirebrook Colliery, Shirebrook.
STC (SK 285193) Swadlincote Training Centre, Swadlincote.

LEICESTERSHIRE

Rw (SK 312162) Rawdon Colliery, Moira.

STAFFORDSHIRE

LH (SK 060168) Lea Hall Colliery, Rugeley.
Lt (SJ 973126) Littleton Colliery, Huntingdon.

WARWICKSHIRE

DM (SP 258899) Daw Mill Colliery, Over Whitacre.

WEST MIDLANDS

Cv (SP 322843) Coventry Colliery, Keresley.

Gauge : 4'8½".

63/000/305	No.2	0-6-0DH	EEV	D1120 1966	LH	
63.000.441	WESTERN ENTERPRISE 6wDE		GECT	5421 1977	Lt	
63.000.442	WESTERN PIONEER	6wDE	GECT	5422 1977	Cv	
63.000.443	WESTERN PROGRESS	6wDE	GECT	5468 1978	Lt	
63.000.444	WESTERN JUBILEE	6wDE	GECT	5478 1979	Cv	
63.000.445	WESTERN QUEEN	6wDE	GECT	5479 1979	Lt	
63.000.446	WESTERN KING	6wDE	GECT	5480 1979	Lt	
DL15 10		0-4-0DH	HC	D1344 1965	Sk	
NCB 13 1964		0-6-0DH	HE	6294 1964	Rw	
NCB 12 1964		0-6-0DH	HE	6295 1964	Rw	
NCB SM AREA No.24 1969		0-6-0DH	HE	6690 1969	Cv	0OU +
63/000/316 No.3D		0-6-0DH	HE	7181 1970	LH	
COVENTRY MINE No.1		0-6-0DH	HE	7396 1974	Cv	0OU +
COVENTRY COLLIERY No.2		0-6-0DH	HE	7494 1976	Cv	0OU +
-		0-6-0DH	RR	10187 1964	DM	
No.64/2		0-6-0DH	RR	10188 1964	DM	
-		0-6-0DH	RR	10240 1965	LH	
-		0-6-0DH	RR	10255 1966	LH	
DL11		0-6-0DH	YE	2910 1963	Cv	0OU +

+   Property of Wilmott Bros. Ltd, Ilkeston.

Gauge : 2'6".

5074		4wBE	FLP	CE	5074 1965	LH	
15/19		4wBE	FLP	CE	5097 1966	LH	
-		4wBE	FLP	CE	B3135A 1986	STC	
-		4wBE	FLP	GB	6090 1963	LH	
-		4wDH		HE	8825 1978	LH	
-		4wDH		HE	8826 1978	LH	Dsm
63/000/447 7		4wDH		HE	8971 1979	Lt	
-		4wDH		HE	8973 1979	LH	
-		4wDM		HE	9041 1982	Lt	
63.000.364		4wDM		RH	441948 1959	LH	0OU a
63.000.310		4wDM		RH	476112 1962	LH	0OU
15		4wDM		RH	506491 1964	Lt	0OU

a   Known to British Coal as 436862 (in error) and carries
    plate with that number.

Gauge : 2'0".

5 6 2986		4wBE	FLP	GB	2986 1960	STC	
		Rebuilt		GB	420460 1979		

SOUTH WALES GROUP

(Headquarters : Tredomen, Ystrad Mynach, Mid Glamorgan.)

Locomotives are kept at the following locations :-

DYFED

Cn   (SN 494077)   Cynheidre Colliery, Five Roads.   (Closed)

GWENT

BTC   (ST 158982)   Britannia Training Centre, Pengam.

MID GLAMORGAN

T1   (ST 102995)   Trelewis Colliery, Treharris.   (Closed)

WEST GLAMORGAN

B1   (SN 869054)   Blaengwrach Colliery, Glyn Neath.   (Closed)
Bt   (SN 785032)   Blaenant Colliery, Crynant.   (Closed)

Gauge : 3'0".

-		0-4-0DM	FLP	HC	DM1314	1963	BTC
-		0-4-0DM	FLP	HE	6696	1966	BTC

Gauge : 2'0".

7			0-4-0DM	FLP	HE	6623	1973	B1	Dsm
390/11	145/10	521/3	4wDM	FLP	RH	398064	1956	Cn	OOU
-			4wDM		RH	487966	1963	Cn	Dsm

Trapped Rail Systems :-

   Becorit 'Roadrailer' (400mm).

-	1adCEH	FLP	BGB	30/400/001 1981	T1

   UMM 'Mineranger' System (500mm).

-	2adDH	FLP	UMM	0024 1983	Bt

COAL PRODUCTS DIVISION

(Headquarters : Coal House, Harrow-on-the-Hill, Greater London).

COAL PRODUCTS LTD (Wholly owned subsidiary company).

Locomotives are kept at the following locations :-

DURHAM MANAGEMENT UNIT.

TYNE & WEAR
MC   (NZ 316627)   Monkton Coking Plant, Hebburn.

MIDLANDS MANAGEMENT UNIT.

DERBYSHIRE
AvC   (SK 394678)   Avenue Coking & Chemical Works, Wingerworth.

WEST MIDLANDS
CvH   (SP 318845)   Coventry Home Fire Plant, Keresley.

WALES MANAGEMENT UNIT.

MID GLAMORGAN
CwC   (ST 066865)   Cwm Coking Plant, Llantwit Fardre.

Gauge : 4'8½".

Shunting Locos.

(D2099	03099)	0-6-0DM	Don		1960	MC	OOU
(D2139)	No.1	0-6-0DM	Sdn		1960	MC	OOU
	-	0-4-0DH	HC	D1345	1970	AvC	OOU
	-	0-4-0DH	HC	D1388	1970	AvC	OOU
No.7		0-6-0DH	HE	6973	1969	AvC	
	-	0-6-0DH	HE	7305	1973	MC	
HOMEFIRE No.4		0-6-0DH	RR	10239	1965	CvH	
No.2	38/400/081	0-6-0DH	S	10148	1963	CvH	
No.3	38/400/084	0-6-0DH	S	10150	1963	CvH	
No.6		4wDH	TH	199V	1969	AvC	
	-	4wDH	TH	219V	1970	AvC	OOU

Coke Oven Locos.

	-	0-4-0WE	GB	2180	1948	CwC	OOU
1		4wWE	GB	2508	1955	AvC	
2		4wWE	GB	2509	1955	AvC	
1		0-4-0WE	GB	2690	1957	CwC	
2		0-4-0WE	GB	2691	1957	CwC	
	-	4wWE	TH	313V	1985	MC	

OPENCAST EXECUTIVE

Locomotives listed may be the property of the site operating contractor
(listed alongside each location), of the opencast executive, or be hired by
the contractor from a third party.

Locomotives are kept at the following locations :-

SCOTTISH REGION

FIFE

Wf   (NT 196983)   Westfield Disposal Point, Kinglassie.
                           (Crouch Mining Ltd).

NORTHERN REGION

NORTHUMBERLAND

Wd   (NZ 237957)   Widdrington Disposal Point, Widdrington.
                           (Crouch Mining Ltd).

CENTRAL (NORTH) REGION

DERBYSHIRE

Ox   (SK 469741)   Oxcroft Disposal Point, Stanfree, near Clowne.
                           (Hargreaves Industrial Services Ltd).

SOUTH YORKSHIRE

WS   (SK 368980)   Wentworth Stores, Hartley, near Wentworth.

WEST YORKSHIRE

BO   (SE 300164)   British Oak Disposal Point, Crigglestone.
                           (Hargreaves Industrial Services Ltd).

SOUTH WALES REGION

DYFED

CB   (SN 424059)   Coed Bach Disposal Point, Kidwelly.
                           (Powell Duffryn Coal Preparation Ltd).

WEST GLAMORGAN

GCG   (SN 713120)   Gwaun-cae-Gurwen Disposal Point, Gwaun-cae-Gurwen.
                           (Powell Duffryn Coal Preparation Ltd).
On   (SN 843105)   Onllwyn Disposal Point, Onllwyn.
                           (Crouch Mining Ltd).

Gauge : 4'8½".

(D2037)	03037	0-6-0DM	Sdn		1959	0x	
(D3023)	08016	0-6-0DE	Derby		1953	BO	
(D3179)	08113	0-6-0DE	Derby		1955	GCG	
(D3765)	08598	0-6-0DE	Derby		1959	GCG	
(D2996)	07012	0-6-0DE	RH	480697	1962	CB	
MP 342		0-6-0DH	EEV	D924	1966	On	
	DEREK CROUCH	0-6-0DH	EEV	D1201	1967	Wd	
MP 201		0-6-0DH	EEV	D1202	1967	On	
MP 202		0-6-0DH	EEV	D1230	1969	On	
MP 224		0-6-0DH	EEV	3994	1970	Wd	
4/33	6678	0-4-0DH	HE	6678	1968	WS	OOU
1/13		0-6-0DH	HE	7410	1976	WS	OOU
	-	0-6-0DH	HE	8979	1979	0x	
468048		0-6-0DE	RH	468048	1963	CB	
11348/C	WL No.1	4wDH	RR	10268	1967	Wf	
11348/L	WL No.2	4wDH	RR	10269	1967	Wf	

MINING RESEARCH AND DEVELOPMENT ESTABLISHMENT

Locomotives are kept at the following location :-

DERBYSHIRE

STS   (SK 285193)   Swadlincote Test Site, Swadlincote.

Gauge : 2'6".

-		DE	GMT		1976	STS +

+ Experimental 3-car articulated linear motor manriding train
  - converted to DH in 1981.

Other locomotives and personnel carriers are here from time to time for tests and evaluation.

SECTION 5

MINISTRY OF DEFENCE
Army Railway Organisation

Locomotives are used at the following locations; those depots marked +
normally work their traffic with permanent local stock but also use Army
Railway Organisation locomotives as the occasion demands (see entry under
appropriate County). Depot types are :-

C.A.D.	=	Central Ammunition Depot.
C.O.D.	=	Central Ordnance Depot.
C.V.D.	=	Central Vehicle Depot.
D.O.E.	=	Department of the Environment.
P.E.E.	=	Proof & Experimental Establishment.
R.A.F.	=	Royal Air Force.
R.A.O.C.	=	Regional Depot Royal Army Ordnance Corps.
R.N.A.D.	=	Royal Naval Armament Depot.
R.O.F.	=	Royal Ordnance Depot.

ASH		(SO 932338)	Ashchurch, Gloucestershire.
BIS		(SP 581203, 583199)	C.O.D. Graven Hill, Bicester.
			Also n.g. locomotives occasionally present.
BMR		(SP 581203)	C.O.D. Bicester Military Railway, Oxfordshire.
CMK	+	(ST 982302)	R.A.F. No.11 Maintenance Unit, Chilmark & Dinton, Wilts. (Integrated System)
DHL	+	(SU 276266)	R.N.A.D. Dean Hill, Hampshire.
DON		(SJ 696133)	C.O.D. Donnington, Shropshire.
ER		(NY 246656)	R.O.F. East Riggs, Dumfries & Galloway.
ESK		(SD 083927)	P.E.E. Eskmeals, Cumbria.
HES		(SE 523540)	322 Engineer Park, Hessay, North Yorkshire.
INC		(NS 677758)	P.E.E. Inchterf, Milton of Campsie, near Kirkintilloch, Strathclyde.
KIN		(SP 373523, 374524)	C.A.D. Kineton, Warwickshire.
LEU	+	(NO 453212)	R.A.F. Leuchars, Fife.
LM		(SP 153473, 155476)	Engineer Resources, Long Marston, near Stratford-Upon-Avon, Warwickshire.
LON		(NY 355676)	C.A.D. Longtown, Cumbria.
			(One railcar outstationed at Smalmstown, NY 367687)
LUD		(SU 261507)	C.V.D. Ludgershall, Wiltshire.
LUL		( )	Lulworth Ranges, East Lulworth, Dorset.
LYD	+	(TR 033198)	D.O.E. Lydd Gun Ranges, Lydd, Romney Marsh, Kent. (n.g. only)
MCH		(SU 395103)	Marchwood Military Port, Hampshire.
MOL		(SO 503467)	R.A.O.C. Moreton-on-Lugg, Hereford & Worcester.
RAD	+	(SJ 784545)	Royal Ordnance Plc, Radway Green, Alsager, Cheshire.
SHO		(TM 946856)	P.E.E. New Ranges, Shoeburyness, Southend-on-Sea, Essex.
TRE	+	(SM 970324)	R.N.A.D. Trecwm, Dyfed.

Gauge : 4'8½".

198	ROYAL ENGINEER	0-6-0ST	IC	HE	3798	1953	LM
2144	(03144)						
	WESTERN WAGGONER	0-6-0DM		Sdn		1961	LM
(D3937)	08769	0-6-0DE		Derby		1960	
(D6552)	33034	Bo-BoDE		BRCW	DEL154	1961	LUD +
123	A7 15548	0-4-0DM		Bg/DC	2157	1941	RAD
200		0-4-0DM		AB	358	1941	DHL
201	FROG 61/30289 9903/3396	0-4-0DM		AB	362	1942	RAD
202		0-4-0DM		AB	357	1941	LON
226		0-4-0DM		_(VF	5261	1945	INC
				⁻(DC	2180	1945	
230		0-4-0DM		_(VF	5265	1945	LM
				⁻(DC	2184	1945	
234		0-4-0DM		AB	370	1945	DHL
235		0-4-0DM		AB	371	1945	LM
236		0-4-0DM		AB	372	1945	TRE
242		4wDH		RR	10242	1966	CMK
		Rebuilt		AB		1986	
244		4wDH		RR	10244	1966	CMK
		Rebuilt		AB	6528	1987	
249		0-4-0DM		_(VF	5258	1945	HES
				⁻(DC	2177	1945	
252	RIVER SARK	4wDH		TH	270V	1977	LON
253		4wDH		TH	271V	1977	DON
254		4wDH		TH	272V	1977	MOL
255		4wDH		TH	273V	1977	LUD
256	MARLBOROUGH	4wDH		TH	274V	1977	DON
257	TELA	4wDH		TH	275V	1978	LM
258		4wDH		TH	298V	1981	LUD
259		4wDH		TH	299V	1981	LM
260		4wDH		TH	300V	1982	LUD
261		4wDH		TH	301V	1982	LM
263		4wDH		TH	303V	1982	LUD
264	RIVER ESK	4wDH		TH	306V	1983	LON
265		4wDH		TH	307V	1983	KIN
266		4wDH		TH	308V	1983	MCH
267		4wDH		TH	309V	1984	MOL
268		4wDH		TH	310V	1984	MCH
269		4wDH		TH	311V	1984	MCH
270	GREENSLEEVES	4wDH		TH	V318	1987	BIS
271	STOREMAN	4wDH		TH	V324	1987	BIS
272	ROYAL PIONEER	4wDH		TH	V320	1987	BIS
273	EDGEHILL	4wDH		TH	321V	1987	KIN
274	WAGGONER	4wDH		TH	322V	1987	KIN
275	SAPPER	4wDH		TH	V323	1988	BIS
276	CONDUCTOR	4wDH		TH	V319	1988	BIS
277	KINETON	4wDH		TH	V332	1988	KIN
278	COPPICE	4wDH		TH	V333	1988	KIN
420		0-6-0DH		RH	459515	1961	DON
421		0-6-0DH		RH	459516	1961	ASH
422		0-6-0DH		RH	459517	1961	ASH
423		0-6-0DH		RH	459518	1961	SHO
424		0-6-0DH		RH	459520	1961	SHO
425	RIVER TAY	0-6-0DH		RH	459519	1961	LEU
426	B	0-6-0DH		RH	459521	1961	ER
427		0-6-0DH		RH	466616	1961	DON
428	C	0-6-0DH		RH	466617	1961	LON
429	A RIVER ANNAN	0-6-0DH		RH	466618	1961	LON
430		0-6-0DH		RH	466621	1961	SHO

431	MUNCASTER CASTLE	0-6-0DH	RH	466622 1962	ESK	
432		0-6-0DH	RH	466623 1962	DON	
433		0-6-0DH	RH	468043 1963	ER	
434	MILLOM CASTLE	0-6-0DH	RH	468044 1963	ESK	
435		0-6-0DH	RH	468045 1963	LUD OOU	
436 D		0-6-0DH	RH	468046 1963	ER	
440	RIVER EDEN	0-6-0DH	RH	468041 1962	LEU	
(9116)		2w-2DMR	CE	5427 1968	BIS DsmT	
9117		4wDMR	BD	3706 1975	SHO	
9118		4wDMR	BD	3707 1975	BIS OOU	
9119		4wDMR	BD	3708 1975	KIN	
9120		4wDMR	BD	3709 1975		
9121		4wDMR	BD	3710 1975	KIN	
9122		4wDMR	BD	3711 1975		
(9123)		4wDMR	BD	3712 1975	KIN DsmT	
9124		4wDMR	BD	3713 1975	BIS OOU	
9126	1	4wDMR	BD	3742 1976		
9127		4wDMR	BD	3743 1976	KIN	
9128	2	4wDMR	BD	3744 1976	LON	
9129		4wDMR	BD	3745 1976	SHO	
9150		4wDMR	BD	3746 1976	KIN	
9248		2w-2DMR	DC	1895	SHO DsmT	
	ANNA	4wRE	Wkm	11547 1987	LUL	
	FIONA	4wRE	Wkm	11548 1987	LUL	
	BELLA	4wRE	Wkm	11549 1987	LUL	
	DEBBIE	4wRE	Wkm	11550 1987	LUL	
	ENID	4wRE	Wkm	11551 1987	LUL	
	CLAIRE	4wRE	Wkm	11552 1987	LUL	
	-	4wDMR	Wkm	11621 1986	LUL	

+ Privately preserved - stored on M.O.D. premises.

Gauge : 60cm.

NG 47		4wDH	BD	3699 1973	ER	
NG 48		4wDH	BD	3700 1973	ER	
	DOLLY PARTON	4wDM	MR	8641 1941	ER	
758220	BLACK BARRON	4wDM	MR	8745 1942	ER	
LOD/758022	PENELOPE	4wDM	MR	8826 1943	LYD	
	FRED	4wDM	MR	8857 1944	LYD	
LOD/758221		4wDM	MR	8886 1944	LYD Dsm	
AD39	L.R. No.3	4wDM	RH	223696 1944	SHO Dsm	
34		4wDM	HE	7009 1971	ER	
AD 35		4wDM	HE	7010 1971	LYD	
		Rebuilt	AB	6941 1988		
AD 36	GALLOWAYS	4wDM	HE	7011 1971	LYD	
AD 37		4wDM	HE	7012 1971	LYD	
		Rebuilt	AB	6014 1988		
38		4wDM	HE	7013 1971	ER	
NG 46		4wDH	BD	3698 1973	ER	
NG 49		4wDH	BD	3701 1973	ER	

+ Currently under renovation at TH works.

SECTION 6 IRELAND

ULSTER

```
ANTRIM        270
ARMAGH        270
DOWN          271-272
FERMANAGH     +
LONDONDERRY   272
TYRONE        273
```

+ No known locomotives exist.

REPUBLIC OF IRELAND

```
CARLOW        +
CAVAN         +
CLARE         273
CORK          273-274
DONEGAL       274
DUBLIN        274-275
GALWAY        275-276
KERRY         276
KILDARE       276-277
KILKENNY      +
LEITRIM       +
LEIX/LAOIS    277-278
LIMERICK      278
LONGFORD      278
LOUTH         +
MAYO          278
MEATH         279
MONAGHAN      +
OFFALY        279-280
ROSCOMMON     280
SLIGO         280
TIPPERARY     280-281
WATERFORD     281
WESTMEATH     282
WEXFORD       282
WICKLOW       +
```

+ No known locomotives exist.

ANTRIM

INDUSTRIAL SITES

BULRUSH PEAT CO LTD, RANDALSTOWN
Gauge : 75cm.

2		4wDM	MR	40S307	1967

NORTHERN IRELAND RAILWAYS LTD, LISBURN GOODS YARD, LISBURN
Gauge : 5'3".

1		0-6-0DE	EEV	D1266	1969 Pvd

PRESERVATION SITES

SHANES CASTLE RAILWAY, SHANES CASTLE, RANDALSTOWN
Gauge : 3'0".

No.3	SHANE	0-4-0WT	OC	AB	2265	1949
	NANCY	0-6-0T	OC	AE	1547	1908
No.1	TYRONE	0-4-0T	OC	P	1026	1904
	-	4wDM		MR	11039	1956
No.6	RORY	4wDH		MR	102T007	1974
No.4		4wDH		SMH	102T016	1976
	-	2w-2PMR		Wkm	7441	1956

RAILWAY PRESERVATION SOCIETY OF IRELAND, WHITEHEAD DEPOT
Gauge : 5'3".

171	SLIEVE GULLION	4-4-0	IC	BP	5629	1913
		Rebuilt		Dundalk	42	1938
461		2-6-0	IC	BP	6112	1922
No.85	MERLIN	4-4-0	3C	BP	6733	1932
No.27	LOUGH ERNE	0-6-4T	IC	BP	7242	1949
4		2-6-4T	OC	Derby		1947
186		0-6-0	IC	SS	2838	1879
No.3	R.H. SMYTH	0-6-0ST	OC	AE	2021	1928
23		4wDM		FH	3509	1951
4		4wDM		RH	382827	1955
1		4w-4wDMR		NCC		1933

ARMAGH

PRESERVATION SITES

PEATLANDS COUNTRY PARK, DERRYHUBBERT
Gauge : 3'0".

	-	4wDM	FH	3719	1954
	HENRY NEIL	4wDM	Schöma	1727	1955

DOWN

CONTAINER REFURBISHING CO LTD, NEWRY
Gauge : 3'0".

| | | 4wDM | L | 36745 | 1951 |

HARLAND & WOLFF LTD, QUEENS ISLAND SHIPYARD, BELFAST
Gauge : 5'3".

| No.79 | | 4wBE/RE | EE | 517 | 1921 OOU |

NORTHERN IRELAND RAILWAYS LTD, ADELAIDE YARD, BELFAST
Gauge : 5'3".

| | | 4wDM | R/R Unilok | A114 c1965 OOU |

PRESERVATION SITES

DOWNPATRICK AND ARDGLASS RAILWAY PROJECT, DOWNPATRICK
Gauge : 5'3".

No.3	GUINNESS	0-4-0ST	OC	HC	1152	1919
	-	0-4-0T	OC	OK	12475	1934
	-	0-4-0T	OC	OK	12662	1935
(G613)		4wDH		Dtz	57226	1962
421	W.F. GILLESPIE O.B.E.	6wDH		Inchicore		1961
E432		6wDH		Inchicore		1961

ULSTER FOLK & TRANSPORT MUSEUM, BELFAST TRANSPORT MUSEUM,
WITHAM STREET, BELFAST 4
Gauge : 5'3".

30		4-4-2T	IC	BP	4231	1901
93		2-4-2T	IC	Dundalk	16	1895
800	MAEVE	4-6-0	3C	Inchicore		1939
74	DUNLUCE CASTLE	4-4-0	IC	NB	23096	1924
No.1		0-6-0ST	OC	RS	2738	1891
8178		2w-2DMR		Dundalk		1932

Gauge : 3'0".

2		0-4-0Tram	OC	K	T84	1883
2		0-4-0T	OC	P	1097	1906
2	KATHLEEN	4-4-0T	OC	RS	2613	1887
11	PHOENIX	4wDM		AtW	114	1928
		Rebuilt		Dundalk		1932
10		0-4-0+4wDMR		WkB		1932

Gauge : 1'10".

| 20 | | 0-4-0T | IC | Spence | | 1905 |

ULSTER FOLK & TRANSPORT MUSEUM, TRANSPORT MUSEUM, CULTRA
Gauge : 2'0".

		4wDM	FH			
		4wDM	HE	3127	1943	
		4wPM	MR	246	1916	+
		4wDM	MR	9202	1946	+

+ Not currently on public display.

LONDONDERRY

INDUSTRIAL SITES

BULRUSH PEAT CO LTD, NEWFERRY ROAD, BELLAGHY, MAGHERAFELT
Gauge : 75cm.

3		4wDM	MR	22220	1964
1		4wDM	MR	40S309	1968
		4wDH	Schöma	4978	1988
		4wDH	Schöma	4979	1988
		4wDH	Schöma	4992	1988

J. McGILL, EGLINTON
Gauge : 1'10".

		4wDM	FH

C. TENNANT (N.I.) LTD, GLENCONWAY, DUNGIVEN
Gauge : 3'0".

		4wDM	RH	218030	1942

PRESERVATION SITES

FOYLE VALLEY RAILWAY COMPANY, LONDONDERRY
Gauge : 3'0".

No.4	MEENGLASS	2-6-4T	OC	NW	828	1907	
No.5	DRUMBOE	2-6-4T	OC	NW	829	1907	a
6	COLUMBKILLE	2-6-4T	OC	NW	830	1907	
2	BLANCHE	2-6-4T	OC	NW	956	1912	
1	"PUP"	2-2-0PMR		A&O		1906	
3		2-4w-2PMR		DC		1926	+
12		0-4-0+4wDMR		WkB/Dundalk		1934	
18		0-4-0+4wDMR		WkB/Dundalk		1940	

+ Now unmotorised.
a Currently under renovation at Enterprise Ulster Pennyburn
   site.

TYRONE

ULSTER MINERALS LTD, CURRAGHINALT, near GORTIN
Gauge : 2'0".

-	4wBE	WR		525801	1988 OOU
	Rebuild of WR			(G7125	1967?)

Site on care and maintenance basis from 3/1989.

CLARE

INDUSTRIAL SITES

IARNROD EIREANN
Ennis Depot
Gauge : 5'3".

416A	2w-2PMR	Wkm	8919	1962

Ennis Station
Gauge : 3'0".

5	0-6-2T	OC	D	2890	1892 Pvd

+ Currently under restoration at Shannon, Co Clare.

CORK

INDUSTRIAL SITES

IRISH STEEL HOLDINGS LTD, HAULBOWLINE ISLAND, COBH
Gauge : 4'8½".

1A01		4wDM	R/R	Unilok	1996 1980
1A02		4wDM	R/R	Unilok	1997 1980
-		4wDM	R/R	Unilok	2218 1984

PRESERVATION SITES

GREAT SOUTHERN RAILWAY PRESERVATION SOCIETY, MALLOW
Gauge : 5'3".

131		4-4-0	IC	NR	5757	1901
G616		4wDH		Dtz	57227	1961
G617		4wDH		Dtz	57229	1961
	-	4wDM		RH	252843	1948
2		4wDM		RH	312425	1951
	JZA 979	4wDM		Scammell		1960
	-	2w-2PMR		DC	1495	1927
2509		0-4-0+4wDMR		WkB		1947

Gauge : 3'0".

C42	2w-2PMR	Wkm		7129	1955

IARNROD EIREANN, CORK (KENT) STATION
Gauge : 5'3".

36	2-2-2	IC	Bury		1848

ROSMINIAN FATHERS, UPTON, INNISHANNON
Gauge : 2'0".

-	4wDM	S/O	RH	264244	1949

DONEGAL

INDUSTRIAL SITES

DONEGAL PEAT DEVELOPMENT CO, BELLANAMORE, near FINTOWN (Closed)
Gauge : 2'0".

-	4wPM	MR	7944	1943	Dsm

GLENTIES TURF CO-OP SOCIETY LTD, ADARA ROAD, GLENTIES
Gauge : 2'0".

LM20A	4wDM	RH	243387	1946	
LM26E	4wDM	RH	248458	1946	
LM264	4wDM	RH	371535	1954	
LM198	4wDM	RH	398076	1956	
(C16)	2w-2PMR	Wkm	4806	1948	Dsm

DUBLIN

INDUSTRIAL SITES

ARTHUR GUINNESS, SON & CO (DUBLIN) LTD, ST. JAMES GATE BREWERY, DUBLIN
Gauge : 1'10".   RTC.

17	0-4-0T	IC	Spence	1902	Pvd
-	4wDM		FH		Dsm
-	4wDM		FH		OOU
-	4wDM		FH		OOU
-	4wDM		FH		OOU

FH's are four of 3068/1947; 3255/1948; 3444, 3446, 3447, 3449/1950.

IARNROD EIREANN
Connolly Depot, Dublin
Gauge : 5'3".

	415A	2w-2PMR	Wkm	8918	1962
2	414A	2w-2DMR	Wkm	8920	1962

<u>Inchicore Works, Dublin</u>
Gauge : 5'3".

B103	Co-CoDE	BRCW		1956
B113	Bo-BoDE	Inchicore		1950
B114	Bo-BoDE	Inchicore		1951
425	6wDH	Inchicore		1962
429	6wDH	Inchicore		1962
601	4wDH	Dtz	56118	1956

Locos privately preserved; stored at this location.

PRESERVATION SITES

<u>TRINITY COLLEGE, SCHOOL OF ENGINEERING, DUBLIN</u>
Gauge : 1'9".

| - | 0-6-0 | IC | T.Kennan | 1855 |

GALWAY

INDUSTRIAL SITES

BORD NA MONA.
<u>Attymon // Ath Tiomain</u>
        Site is 6 miles east of Athenry on the L99 road 1½ miles south
        of Attymon Junction Station.
        For loco details see separate Section.

<u>Clonkeen // Cluain Chaoin</u>
        A sub-shed of Attymon.
        Site is 1½ miles west of Attymon Junction Station on the minor
        road to Athenry.
        For loco details see separate Section.

<u>Derryfadda // Doire Fhada</u>
        Site is 2 miles south west of Ballyforan on the road from
        Ballyforan to Kilglass.
        For loco details see separate Section.

<u>GALWAY METAL CO, ORANMORE</u>
Gauge : 5'3".   RTC.

| K 801 | 0-8-0DH | MAK | 800028 | 1954 Dsm |

<u>UNILOKOMOTIVE LTD, INTERNATIONAL DIVISION, MERVUE INDUSTRIAL ESTATE, GALWAY</u>
        New locomotives under construction occasionally present.

PRESERVATION SITES

GALWAY MINIATURE RAILWAY, (LEISURELAND EXPRESS), SALTHILL, GALWAY
Gauge : 1'3".

382		0-8-0PH	S/O	SL	73 35	1973

WESTRAIL (TUAM) LTD, TUAM STATION, TUAM
Gauge : 5'3".

90	0-6-0T	IC	Inchicore		1875
E428	6wDH		Inchicore		1961
No.3	0-4-0DM		RH	395302	1956

KERRY

PRESERVATION SITES

GREAT SOUTHERN & WESTERN RAILWAY PRESERVATION SOCIETY
Blennerville, near Tralee
Gauge : 3'0".

No.5T	2-6-2T	OC	HE	555	1892

Fenit
Gauge : 5'3".

1	4wDM	RH	305322	1951

KILDARE

INDUSTRIAL SITES

BORD NA MONA
Ballydermot // Baile Dhiarmada
       Site is 1½ miles north of Rathnagan on the road from Kildare to
       Edenderry.
       Serves B.S.L. Allenwood and Irish C.E.C.A.
       For loco details see separate Section.

Kilberry // Cill Bheara
       Site is 3 miles north west of Athy on Monastervin road.
       Serves Kilberry Fertiliser Factory.
       For loco details see separate Section.

Lullymore // Loilgheach Mor
       Site is 3 miles from Allenwood on Edenderry road.
       Serves Lullymore Briquetting Factory.
       For loco details see separate Section.

Timahoe // Tigh Mochua
        Site is 3 miles west of Prosperous on the Clane to Edenderry
        road, at the village of Corduff.
        Serves B.S.L. Allenwood.
        For loco details see separate Section.

Ummeras //
        Site is 4 miles north of Monasterevin on the road from
        Monasterevin to Bracknagh.
        For loco details see separate Section.

BORD SOLATHAIR AN LEACTREACHAIS,
ALLENWOOD SOD-PEAT BURNING POWER STATION, NAAS
Gauge : 3'0".   (Closed)

4		4wDM	RH	300518	1950
3		4wDM	RH	314222	1951
2		4wDM	RH	314223	1951
1		4wDM	RH	326051	1952

D. KARNEY, KILLINA CARBURY
Gauge : 3'0".

-	4wDM	HE	6075	1961 OOU

No.1 MAINTENANCE COMPANY, IRISH ARMY, CURRAGH
Gauge : 2'0".

-	4wDM	MR	8970 c1945

LEIX/LAOIS

INDUSTRIAL SITES

BORD NA MONA
Coolnamona // Cúil Na Moná
        Site is 3½ miles south of Portlaoise on the road to Abbeyleix.
        Serves the Coolnamona Fertiliser Factory.
        For loco details see separate Section.

PRESERVATION SITES

IRISH STEAM PRESERVATION SOCIETY, STRADBALLY HALL, STRADBALLY
Gauge : 3'0".

No.2		0-4-0WT	OC	AB	2264	1949
No.4	NIPPY	4wDM		FH	2014	1936
	-	4wDM		HE	2280	1941
4		4wDM		RH	326052	1952
C39		2w-2PMR		Wkm	6861 c1954

STRADBALLY STEAM MUSEUM, STRADBALLY
Gauge : 1'10".

    15   0-4-0T   IC Spence   1912 +

            + Carries plate dated 1895, in error.

LIMERICK

INDUSTRIAL SITES

LIMERICK WAGON WORKS, LIMERICK
Gauge : 5'3".

    G611   4wDH   Dtz   57225 1962 Pvd

LONGFORD

INDUSTRIAL SITES

BORD NA MONA,
Mountdillon // Cnoc Dioluin
            Site is 2 miles east of Lanesborough on the N63 road to
            Longford, and consists of the Mountdillon, Derryaroge, Begnagh,
            Derryad, Derryaghan & Corlea Bogs.
            Locos are also outstationed at Lanesborough Power Station.
            For loco details see separate Section.

MAYO

INDUSTRIAL SITES

BORD NA MONA,
Bangor Erris // Beannchar Iorrais
            Site is 2 miles west of Bangor off the road to Belmullet, on the
            minor road to Srahmore.
            For loco details see separate Section.

Tionnsca Abhainn Einne
            Site is 9 miles west of Crossmolina on the road to Belmullet.
            This, the Oweniny River Project, is always spoken of as
            "T.A.E.".
            Serves B.S.L. Bellacorick.
            For loco details see separate Section.

PRESERVATION SITES

WESTPORT HOUSE COUNTRY ESTATE, WESTPORT
Gauge : 1'3".

            -               2-6-0DH   S/O SL   80.10.89   1989

MEATH

TARA MINES, NAVAN
Gauge : 5'3".

       -             4wDM   R/R Unilok   1944 1977

OFFALY

BORD NA MONA
Bellair // Baile Ard
>Site is 3 miles north of Ballycumber on the road between
>Ballycumber and Moate.
>For loco details see separate Section.

Blackwater // Uisce Dubh
>Site is 1½ miles east of Shannonbridge on the road to Cloghan,
>and consists of the Blackwater, Kilmacshane, Garryduff, Lismanny
>& Culliaghmore Bogs.
>Serves B.S.L. Shannonbridge.
>For loco details see separate Section.

Boora // Buarach
>Site is 6 miles east of Cloghan on the Tullamore road, and
>consists of the Noggusboy, Derries, Turraun, Pollagh, Oughter,
>Boora, Derrybrat, Drinagh & Clongawney More Bogs.
>Serves B.S.L. Ferbane and Derrinlough Briquetting Plant.
>For loco details see separate Section.

Clonsast // Cluain Sosta
>Site is 3 miles north of Portarlington on the road to
>Rochfortbridge, and consists of the Clonsast, Derrylea,
>Derryounce & Garryhinch Bogs.
>For loco details see separate Section.

Derrygreenagh // Doire Dhraigneach
>Site is 2 miles south of Rochfortbridge on the Rhode road,
>and consists of the Derryhinch, Drumman, Derryarkin, Ballybeg,
>Cavemount, Esker, Mount Lucas, Ballycon, Derrycricket &
>Cloncreen Bogs.
>Serves B.S.L. Rhode and Croghan Briquetting Plant.
>For loco details see separate Section.

Lemanaghan // Liath Manchain
>Site is at Ferbane, off the Birr to Athlone road.
>For loco details see separate Section.

```
BORD SOLATHAIR AN LEACTREACHAIS,
PORTARLINGTON SOD-PEAT BURNING POWER STATION    (Closed)
Gauge : 3'0".

     2                        4wDM        RH        249525  1947 OOU
     1                        4wDM        RH        249526  1947
     3                        4wDM        RH        279604  1949
          -                   4wDM        RH        422566  1958 OOU
          -                   4wDM        RH        422567  1958


ERIN PEAT PRODUCTS LTD
Site near Birr, on L 113 road south of Taylors Cross      R.T.C.
Gauge : 2'0".

          -                   4wDM        MR          8749  1942 OOU


Derrinlough Works
Gauge : 2'0".

     E.I.B.  39              4wDM        MR          7361  1939 Dsm
```

ROSCOMMON

<u>INDUSTRIAL SITES</u>

WESTERN INDUSTRIES (BOYLE) LTD, KEELOGUES LIMESTONE QUARRY, BOYLE
Gauge : 2'0".   RTC.

```
          -                   4wDM        FH                         OOU
      .
```

SLIGO

<u>INDUSTRIAL SITES</u>

McTIERNAN BROS, GLEN BALLINSHEE COLLIERY, GEEVAGH
Gauge : 2'0".

```
          -                   0-4-0BE     WR
```

TIPPERARY

<u>INDUSTRIAL SITES</u>

BORD NA MONA
Littleton (Ballydeath) // Baile Dhaith
          Site is 1½ miles south of Littleton, near Thurles.
          Serves B.S.L. Littleton.
          For loco details see separate Section.

Templetouhy // Teampall Tuaithe
        Site is 3 miles south east of Templetouhy on the Templemore to
        Urlingford road.
        Serves B.S.L. Littleton.
        For loco details see separate Section.

COMHLUCHT SIUICRE EIREANN TEO, THURLES SUGAR FACTORY
Gauge : 5'3".

-	4wDM	RH	312424	1951	Dsm

MOGUL OF IRELAND LTD, SHALEE SILVER MINES, near NENAGH (Closed)
Gauge : 2'6".

11	4wDH	CE	5879/2	1971 OOU
12	4wDH	CE	5879/1	1971 OOU
13	4wDH	CE	5952A	1972 OOU
14	4wDH	CE	5952B	1972 OOU
15	4wDH	CE	5960A	1973 OOU
16	4wDH	CE	5960B	1973 OOU
17	4wDH	CE	5960C	1973 OOU
18	4wDH	CE	5960D	1973 OOU

POPE BROS LTD, PEAT WORKS, near URLINGFORD
Gauge : 2'0".   RTC.

D3	4wDM	HE	2659	1942 OOU

TIPPERARY ANTHRACITE, LICKFINN, BALLINUNTY, THURLES
Gauge : 2'0".   (Closed)

-	4wBE	CE	B0145B	1973 OOU

WATERFORD

INDUSTRIAL SITES

IARNROD EIREANN, WATERFORD
Gauge : 5'3".

5	417A	2w-2DMR	Wkm	8917	1962

PRESERVATION SITES

TRAMORE MINIATURE RAILWAY, TRAMORE
Gauge : 1'3".

278	2-8-0PH	S/O	SL	22	1973

WESTMEATH

BORD NA MONA
Ballivor // Baile Iomhair
          Site is 6 miles west of Ballivor on road from Trim to Mullingar.
          For loco details see separate Section.

Coolnagan // Cúil Na Gcon
          Site is 5 miles west of Castlepollard on the road to Granard,
          2 miles south of the village of Coole.
          For loco details see separate Section.

MIDLAND IRISH PEAT MOSS LTD, MULLINGAR
Gauge : 2'0".

	-	4wDH	AK	No.9	1983
	-	4wDM	FH	2306	
	-	4wDM	FH		
	-	4wDM	LB	54183	1964
	-	4wDM	MR	7304	1938
No.4		4wDM	MR	9239	1947
	-	4wDM	MR	9543	1950
	-	4wDM	R&R	84	1938
	-	4wDM	RH	193974	1938
	-	4wDM	Schöma	1676	1955

PRESERVATION SITES

RAILWAY PRESERVATION SOCIETY OF IRELAND, MULLINGAR DEPOT
Gauge : 5'3".

No.184	0-6-0	IC	Inchicore	1880

WEXFORD

INDUSTRIAL SITES

TEDCASTLE McCORMACK LTD, COURTOWN BRICK & TILE WORKS, COURTOWN   (Closed)
Gauge : 1'8".

-	4wDM	RH	264237	1948 Dsm

Bord na Mona   Irish Turf Board

The Bord operates rail systems on peat bogs throughout the country, and locomotives are kept at the locations listed below.

Gauge : 3'0" (except those marked + which are 2'0").

Locations : (For more details see the relevant County Section).

A	Attymon, Co. Galway.
Bd	Ballydermot, Co. Kildare.
Be	Bellair, Co. Offaly.
Bi	Ballivor, Co. Westmeath.
Bl	Blackwater, Co. Offaly.
Bn	Bangor Erris, Co. Mayo.
Bo	Boora, Co. Offaly.
Cg	Coolnagan, Co. Westmeath.
Ck	Clonkeen, Co. Galway.
Cm	Coolnamona, Co. Laois.
Cs	Clonsast, Co. Offaly.
De	Derryfadda, Co. Galway.
Dg	Derrygreenagh, Co. Offaly.
K +	Kilberry, Co. Kildare.
Le	Lemanaghan, Co. Offaly.
Li	Littleton, Co. Tipperary.
Lu	Lullymore, Co. Kildare.
M	Mountdillon, Co. Longford.
TAE	Tionnsca Abhainn Einne, Co. Mayo.
Te	Templetouhy, Co. Tipperary.
Ti	Timahoe, Co. Kildare.
U	Ummeras, Co. Kildare.

18		4wDM	RH	211687	1941	Lu	OOU
LM 12		4wDM	Wcb	40331	1945	K	
LM 13 D		4wDM	RH	198251	1939	Be	
LM 14 D		4wDM	RH	198290	1940	Cs	
LM 15		4wDM	RH	198326	1940	Li	
LM 16 B		4wDM	RH	200075	1940	K	+
LM 17 G		4wDM	RH	242901	1946	A	
LM 18 G		4wDM	RH	242902	1946	TAE	
LM 21 A		4wDM	RH	243392	1946	K	+
LM 23 E		4wDM	RH	244788	1946	Cs	
LM 24 E		4wDM	RH	244870	1946	Cs	
LM 25 E	7	4wDM	RH	244871	1946	Cs	
LM 27 G	8	4wDM	RH	249524	1947	Ti	
LM 28 G	12	4wDM	RH	249543	1947	Ti	
LM 29 G		4wDM	RH	249544	1947	M	
LM 30 G	11	4wDM	RH	249545	1947	Ti	
LM 31 E		4wDM	RH	252232	1947	Cg	
LM 32 E	4	4wDM	RH	252233	1947	Ti	
LM 33 E	2	4wDM	RH	252234	1947	Ti	Dsm
LM 34 E	1	4wDM	RH	252239	1947	A	
LM 35 E		4wDM	RH	252240	1947	Cg	
LM 36 E	2	4wDM	RH	252241	1947	M	
LM 37 E	3	4wDM	RH	252245	1947	M	

LM 38	E		4wDM	RH	252246	1947	Ck	
LM 39	E		4wDM	RH	252247	1947	M	
LM 40	E		4wDM	RH	252251	1947	M	
LM 41	E		4wDM	RH	252252	1947	A	
LM 42	C		4wDM	RH	252849	1947	K	+
LM 46			4wDM	RH	259184	1948	Bd	
LM 47	F		4wDM	RH	259185	1948	K	
LM 48	F		4wDM	RH	259186	1948	Bd	
LM 49	F		4wDM	RH	259189	1948	TAE	
LM 50			4wDM	RH	259190	1948	Bd	
LM 51	F		4wDM	RH	259191	1948	Cg	
LM 52	F	15	4wDM	RH	259196	1948	K	
LM 53	F		4wDM	RH	259197	1948	Bd	
LM 54	F		4wDM	RH	259198	1948	Cm	
LM 55	F		4wDM	RH	259205	1948	Bd	
LM 56	F	7	4wDM	RH	259204	1948	Ti	
LM 57	F		4wDM	RH	259203	1948	Bd	
LM 58	F		4wDM	RH	259206	1948	Li	
LM 59			4wDM	RH	259737	1948	Cs	
LM 60	F		4wDM	RH	259738	1948	Cs	
LM 61	F		4wDM	RH	259739	1948	Bi	
LM 62	F		4wDM	RH	259743	1948	Bi	
LM 63	F		4wDM	RH	259744	1948	Bi	
LM 64	F		4wDM	RH	259745	1948	Cs	
LM 65			4wDM	RH	259749	1948	Li	
LM 66		5	4wDM	RH	259750	1948	M	
LM 67	F	17	4wDM	RH	259751	1948	Ti	
LM 68			4wDM	RH	259752	1948	Li	
LM 69	F		4wDM	RH	259755	1948	Bi	
LM 70	F		4wDM	RH	259756	1948	Cg	
LM 71	F		4wDM	RH	259757	1948	M	
LM 72	F		4wDM	RH	259758	1948	M	
LM 73		8	4wDM	RH	259759	1948	M	
LM 74	F	6	4wDM	RH	259760	1948	Ti	
LM 75	R		4wDM	RH	326047	1952	Lu	OOU
LM 76			4wDM	RH	326048	1952	Be	
LM 77	H		4wDM	RH	329680	1952	Bd	
LM 78	H		4wDM	RH	329682	1952	Cs	
LM 79	H	10	4wDM	RH	329683	1952	Ti	
LM 80	H	13	4wDM	RH	329685	1952	Ti	
LM 81	H		4wDM	RH	329686	1952	Cs	
LM 82			4wDM	RH	329688	1952	Bd	
LM 83		1	4wDM	RH	329690	1952	K	
LM 84	J	14	4wDM	RH	329691	1952	Ti	
LM 85	J		4wDM	RH	329693	1952	M	
LM 86	J		4wDM	RH	329695	1952	M	
LM 87	J		4wDM	RH	329696	1952	Li	
LM 88	J	16	4wDM	RH	329698	1952	Ti	
LM 89	J		4wDM	RH	329700	1952	Cg	
LM 90	J		4wDM	RH	329701	1952	Bi	
LM 91			4wDM	RH	371962	1954	Bo	
LM 92	L		4wDM	RH	371967	1954	Le	
LM 93	T		4wDM	RH	373376	1954	Bd	
LM 94	T		4wDM	RH	373377	1954	Bd	
LM 95	T		4wDM	RH	373379	1954	Lu	
LM 96	L		4wDM	RH	375314	1954	TAE	
LM 97	T		4wDM	RH	375332	1954	Lu	OOU
LM 98	T		4wDM	RH	375335	1954	Cs	
LM 99	U		4wDM	RH	375336	1954	Cm	
LM 100			4wDM	RH	375341	1954	Cs	
LM 101	U		4wDM	RH	379059	1954	Bo	

LM 102		4wDM	RH	379076	1954	Be	
LM 103		4wDM	RH	375318	1954	Bo	
LM 104	8	4wDM	RH	375322	1954	TAE	
LM 105 U		4wDM	RH	375344	1954	Bl	
LM 106 U		4wDM	RH	375345	1954	M	
LM 107 U		4wDM	RH	379055	1954	Bo	
LM 108 U		4wDM	RH	379061	1954	TAE	
LM 109 U		4wDM	RH	379064	1954	Bo	
LM 110 U		4wDM	RH	379066	1954	Bo	
LM 111		4wDM	RH	379079	1954	Bl	
LM 112		4wDM	RH	375699	1954	K	+
LM 113 U		4wDM	RH	379068	1954	Dg	
LM 114 U		4wDM	RH	379070	1954	Bo	
LM 115 U		4wDM	RH	379073	1954	Be	
LM 116 M	15	4wDM	RH	379077	1954	TAE	
LM 117 U		4wDM	RH	379910	1954	Bo	
LM 118 V		4wDM	RH	379913	1954	Bo	
LM 119 V		4wDM	RH	379916	1954	TAE	
LM 120 V		4wDM	RH	379917	1954	Cm	OOU
LM 121 V		4wDM	RH	379922	1954	Dg	
LM 122 V		4wDM	RH	379923	1954	Bo	
LM 123		4wDM	RH	379925	1954	Bn	
LM 124 N		4wDM	RH	379081	1954	Cs	
LM 125 O		4wDM	RH	379084	1954	Cs	
LM 126 W		4wDM	RH	379927	1954	Bo	
LM 127 W		4wDM	RH	379928	1954	Bo	
LM 128 X		4wDM	RH	383260	1955	Bl	
LM 129 X		4wDM	RH	383264	1955	M	
LM 130 P		4wDM	RH	382812	1955	TAE	
LM 131		4wDM	RH	379086	1955	Bo	
LM 132 P		4wDM	RH	382809	1955	Cs	
LM 133 P		4wDM	RH	379090	1955	Dg	
LM 134 Q		4wDM	RH	382811	1955	Dg	
LM 135 Q		4wDM	RH	382814	1955	Dg	a
LM 136 Q		4wDM	RH	382815	1955	Dg	
LM 137 Q	6	4wDM	RH	382817	1955	TAE	
LM 138 Q		4wDM	RH	382819	1955	Bn	
LM 139 Q		4wDM	RH	392137	1955	Te	
LM 140 Q		4wDM	RH	392139	1955	Bi	
LM 141 Q		4wDM	RH	392142	1955	Bo	
LM 142 Q		4wDM	RH	392145	1955	Bo	
LM 143		4wDM	RH	392148	1956	Bl	
LM 144 Q		4wDM	RH	392149	1956	Dg	
LM 145 X		4wDM	RH	394023	1956	Dg	
LM 146		4wDM	RH	394024	1956	Lu	
LM 147 X		4wDM	RH	394025	1956	Lu	
LM 148 X		4wDM	RH	394026	1956	Dg	
LM 149 X		4wDM	RH	394028	1956	Dg	
LM 150 X		4wDM	RH	394027	1956	Dg	
LM 151 Q		4wDM	RH	392150	1956	Bl	
LM 152 Q		4wDM	RH	392151	1956	Bi	
LM 153 Q		4wDM	RH	394029	1956	Dg	
LM 154 X		4wDM	RH	394030	1956	Bl	
LM 155 X		4wDM	RH	394031	1956	M	
LM 156 X		4wDM	RH	394032	1956	Dg	
LM 157 X		4wDM	RH	394033	1956	Dg	
LM 158 X		4wDM	RH	394034	1956	Dg	
LM 159 X		4wDM	RH	402174	1956	Dg	
LM 160 X		4wDM	RH	402176	1956	Dg	
LM 161 X		4wDM	RH	402175	1956	Bo	
LM 162 X		4wDM	RH	402177	1956	TAE	

LM 163	X		4wDM	RH	402178	1956	Bo
LM 164	Q		4wDM	RH	392152	1956	Ck
LM 165			4wDM	RH	402179	1956	Dg
LM 166	Q		4wDM	RH	402977	1956	Bi
LM 167	Q		4wDM	RH	402978	1956	Cm
LM 168	Q		4wDM	RH	402980	1956	Dg
LM 169	Q		4wDM	RH	402981	1956	Dg
LM 170	Q		4wDM	RH	402982	1956	Te
LM 171	Q		4wDM	RH	402983	1956	Le
LM 172	Q		4wDM	RH	402984	1956	Cg
LM 173	Q		4wDM	RH	402985	1957	Le
LM 174	Q		4wDM	RH	402986	1957	TAE
LM 175			0-4-0DM	RH	420042	1958	Bo
LM 176			0-4-0DM	BnM		1961	Dg
LM 177		5	4wDM	RH	218037	1943	Li
LM 178			0-4-0DM	Dtz	57120	1960	Bo
LM 179			0-4-0DM	Dtz	57121	1960	Bo
LM 180			0-4-0DM	Dtz	57122	1960	Bo
LM 181			0-4-0DM	Dtz	57123	1960	Bo
LM 182			0-4-0DM	Dtz	57126	1960	TAE Dsm
LM 183			0-4-0DM	Dtz	57127	1960	Bo
LM 184			0-4-0DM	Dtz	57130	1960	Bl
LM 185			0-4-0DM	Dtz	57131	1960	Bo
LM 186			0-4-0DM	Dtz	57132	1960	Bo
LM 187			0-4-0DM	Dtz	57133	1960	Bo
LM 188			0-4-0DM	Dtz	57124	1960	Dg
LM 189		2	0-4-0DM	Dtz	57125	1960	Dg
LM 190			0-4-0DM	Dtz	57128	1960	Dg
LM 191			0-4-0DM	Dtz	57129	1960	Dg
LM 192			0-4-0DM	Dtz	57134	1960	Dg
LM 193			0-4-0DM	Dtz	57135	1960	Dg
LM 194		7	0-4-0DM	Dtz	57136	1960	Bl
LM 195			0-4-0DM	Dtz	57137	1960	TAE Dsm
LM 196			0-4-0DM	Dtz	57138	1960	TAE
LM 197			0-4-0DM	Dtz	57139	1960	Dg
LM 199			0-4-0DM	HE	6232	1962	Dg
LM 200			0-4-0DM	HE	6233	1962	Bo
LM 201			0-4-0DM	HE	6234	1962	TAE
LM 202			0-4-0DM	HE	6235	1962	TAE
LM 203			0-4-0DM	HE	6236	1962	Bo
LM 204			0-4-0DM	HE	6237	1963	Dg
LM 205			0-4-0DM	HE	6238	1963	TAE
LM 206			0-4-0DM	HE	6239	1963	Bo
LM 207			0-4-0DM	HE	6240	1963	TAE
LM 208			0-4-0DM	HE	6241	1963	Dg
LM 209			0-4-0DM	HE	6242	1963	TAE
LM 210			0-4-0DM	HE	6243	1963	Dg
LM 211			0-4-0DM	HE	6244	1963	Dg
LM 212			0-4-0DM	HE	6245	1963	Bn
LM 213			0-4-0DM	HE	6246	1963	Dg
LM 214			0-4-0DM	HE	6247	1963	TAE
LM 215			0-4-0DM	HE	6248	1963	Dg
LM 216			0-4-0DM	HE	6249	1963	Bo
LM 217			0-4-0DM	HE	6250	1963	Bl
LM 218			0-4-0DM	HE	6251	1963	Dg
LM 219			0-4-0DM	HE	6252	1963	Bo
LM 220			0-4-0DM	HE	6253	1963	Bo
LM 221			0-4-0DM	HE	6254	1963	Dg
LM 222			0-4-0DM	HE	6255	1963	Bl
LM 223			0-4-0DM	HE	6256	1963	Dg
LM 224			4wDM	RH	375317	1954	K +

LM 225	0-4-0DM	HE	6304	1964	Dg
LM 226	0-4-0DM	HE	6305	1964	M
LM 227	0-4-0DM	HE	6306	1964	Bo
LM 228	0-4-0DM	HE	6307	1964	Bo
LM 229	0-4-0DM	HE	6308	1964	Bl
LM 230	0-4-0DM	HE	6309	1964	Bl
LM 231	0-4-0DM	HE	6310	1964	Bl
LM 232	0-4-0DM	HE	6311	1964	Dg
LM 233	0-4-0DM	HE	6312	1965	Bo
LM 234	0-4-0DM	HE	6313	1965	Bl
LM 235	0-4-0DM	HE	6314	1965	Bo
LM 236	0-4-0DM	HE	6315	1965	Dg
LM 237	0-4-0DM	HE	6316	1965	Cm
LM 238	0-4-0DM	HE	6318	1965	Bd
LM 239	0-4-0DM	HE	6317	1965	Dg
LM 240	0-4-0DM	HE	6319	1965	Cs
LM 241	0-4-0DM	HE	6320	1965	Bd
LM 242	0-4-0DM	HE	6321	1965	M
LM 243	0-4-0DM	HE	6322	1965	Bo
LM 244	0-4-0DM	HE	6323	1965	Bo
LM 245	0-4-0DM	HE	6324	1965	Bo
LM 246	0-4-0DM	HE	6325	1965	Bo
LM 247	0-4-0DM	HE	6326	1965	M
LM 248	0-4-0DM	HE	6328	1965	M
LM 249	0-4-0DM	HE	6327	1965	M
LM 250	0-4-0DM	HE	6329	1965	M
LM 251	0-4-0DM	HE	6330	1965	M
LM 252	0-4-0DM	HE	6331	1965	M
LM 253	0-4-0DM	Dtz	57834	1965	Bl
LM 254	0-4-0DM	Dtz	57835	1965	Bl
LM 255	0-4-0DM	Dtz	57838	1965	Cm
LM 256	0-4-0DM	Dtz	57837	1965	Bo
LM 257	0-4-0DM	Dtz	57836	1965	Dg
LM 258	0-4-0DM	Dtz	57839	1965	M
LM 259	0-4-0DM	Dtz	57840	1965	Bo
LM 260	0-4-0DM	Dtz	57841	1965	Bo
LM 261	0-4-0DM	Dtz	57842	1965	Bl
LM 262	0-4-0DM	Dtz	57843	1965	M
LM 263	4wDM	RH	7002/0600-1	1968	+
LM 265	4wDM	RH	375696	1954	K +
LM 266	0-4-0DM	HE	7232	1971	Bo
LM 267	0-4-0DM	HE	7233	1971	M
LM 268	0-4-0DM	HE	7234	1971	Li
LM 269	0-4-0DM	HE	7235	1971	Bo
LM 270	0-4-0DM	HE	7237	1971	M
LM 271	0-4-0DM	HE	7236	1971	Ti
LM 272	0-4-0DM	HE	7239	1971	TAE
LM 273	0-4-0DM	HE	7246	1972	Bo
LM 274	0-4-0DM	HE	7238	1971	M
LM 275	0-4-0DM	HE	7240	1972	Cm
LM 276	0-4-0DM	HE	7241	1972	Lu
LM 277	0-4-0DM	HE	7242	1972	Ti
LM 278	0-4-0DM	HE	7243	1972	Bl
LM 279	0-4-0DM	HE	7244	1972	Bd
LM 280	0-4-0DM	HE	7245	1972	Bl
LM 281	0-4-0DM	HE	7247	1972	Bl
LM 282	0-4-0DM	HE	7248	1972	Bl
LM 283	0-4-0DM	HE	7250	1972	Ti
LM 284	0-4-0DM	HE	7249	1972	Bd
LM 285	0-4-0DM	HE	7253	1972	Ti
LM 286	0-4-0DM	HE	7254	1972	Bl

LM 287		0-4-0DM	HE	7255	1972	Bl	
LM 288		0-4-0DM	HE	7256	1972	Bl	
LM 289	LM 1	0-4-0DM	HE	7252	1972	Dg	
LM 290	2	0-4-0DM	HE	7251	1972	Dg	
LM 291		4wDM	RH	421428	1958	Ti	OOU
LM 292		0-4-0DM	HE	8529	1977	Bl	
LM 293		0-4-0DM	HE	8530	1977	Bl	
LM 294		0-4-0DM	HE	8531	1977	Bl	
LM 295		0-4-0DM	HE	8532	1977	Bl	
LM 296		0-4-0DM	HE	8534	1977	Bl	
LM 297		0-4-0DM	HE	8533	1977	Bo	
LM 298		0-4-0DM	HE	8538	1977	Bo	
LM 299		0-4-0DM	HE	8537	1977	Lu	
LM 300		0-4-0DM	HE	8535	1977	M	
LM 301		0-4-0DM	HE	8536	1977	Bd	
LM 302		0-4-0DM	HE	8539	1977	M	
LM 303		0-4-0DM	HE	8540	1977	Dg	
LM 304		0-4-0DM	HE	8543	1977	Bn	
LM 305		0-4-0DM	HE	8544	1977	Cs	
LM 306		0-4-0DM	HE	8541	1977	Bo	
LM 307		0-4-0DM	HE	8542	1977	Bo	
LM 308		0-4-0DM	HE	8546	1977	Cs	
LM 309		0-4-0DM	HE	8545	1977	Te	
LM 310		0-4-0DM	HE	8547	1977	Dg	
LM 311		0-4-0DM	HE	8930	1980	Bl	
LM 312		0-4-0DM	HE	8550	1977	Bl	
LM 313		0-4-0DM	HE	8549	1977	Bl	
LM 314		0-4-0DM	HE	8548	1977	Li	
LM 315		0-4-0DM	HE	8922	1979	Bl	
LM 316		4wDM	SMH	60SL741	1980	U	
LM 317		4wDM	SMH	60SL742	1980	U	
LM 318		0-4-0DM	HE	8925	1979	M	
LM 319		0-4-0DM	HE	8926	1979	Bo	
LM 320		0-4-0DM	HE	8939	1980	Bl	
LM 321		0-4-0DM	HE	8927	1980	Li	
LM 322		0-4-0DM	HE	8923	1979	Bl	
LM 323		0-4-0DM	HE	8924	1979	Bl	
LM 324		0-4-0DM	HE	8931	1980	Dg	
LM 325		0-4-0DM	HE	8942	1981	Li	
LM 326		0-4-0DM	HE	8932	1980	De	
LM 327		0-4-0DM	HE	8933	1980	Cm	
LM 328		0-4-0DM	HE	8935	1980	Li	
LM 329		0-4-0DM	HE	8936	1980	Te	
LM 330		0-4-0DM	HE	8937	1980	M	
LM 331		0-4-0DM	HE	8934	1980	De	
LM 332		0-4-0DM	HE	8551	1977	Li	
LM 333		0-4-0DM	HE	8940	1981	Li	
LM 334		0-4-0DM	HE	8941	1981	Bl	
LM 335		0-4-0DM	HE	8938	1980	Li	
LM 336		0-4-0DM	HE	8943	1981	Bl	
LM 337		0-4-0DM	HE	8944	1981	Lu	
LM 338		0-4-0DM	HE	8945	1981	Li	
LM 339		0-4-0DM	HE	8946	1981	Dg	
LM 340		0-4-0DM	HE	8928	1980	Bl	
LM 341		4wDM	SMH	60SL740	1980	Cm	
LM 342		0-4-0DM	HE	8929	1980	Bl	
LM 343		4wDM	SMH	60SL746	1980	De	
LM 344		4wDM	SMH	60SL751	1980	M	
LM 345		4wDM	SMH	60SL749	1980	TAE	
LM 346		4wDM	SMH	60SL747	1980	A	
LM 347		4wDM	SMH	60SL750	1980	Cg	

No.		Type	Builder	Works No	Year	Loc	Notes
LM 348		4wDM	SMH	60SL744	1980	Bo	
LM 349		4wDM	SMH	60SL743	1980	Be	
LM 350		4wDM	SMH	60SL748	1980	De	
LM 351		4wDM	SMH	60SL745	1980	K	
LM 352		4wDM	FH	3989	1962	K	+ OOU
LM 353		4wDM	RH	497771	1963	K	+
LM 354		4wDH	Gleismac			Bl	
LM 355		4wDH	Gleismac			M	
LM 356		4wDH	?Gleismac			Li	
LM 357		4wDH	Gleismac			Bl	
LM 358		4wDH	?Gleismac				
LM 359		4wDH	Gleismac			M	
LM 360		4wDH	?Gleismac				
LM 361		4wDH	DundalkE			TAE	
LM 362		4wDH	DundalkE	LM362	c1984	Bn	
LM 363		4wDH	DundalkE	LM363	c1984	Bo	
LM 364		4wDH	?DundalkE			Bl	
LM 365		4wDH	DundalkE	LM365	c1984	De	
LM 366		4wDH	?DundalkE			Bl	
LM 367		4wDH	DundalkE	LM367	c1984	De	
LM 368		4wDH	Gleismac	LM368	c1984	Bl	
LM 369		4wDH	DundalkE			Bo	
LM 370		4wDH	?Gleismac			Bo	
LM 371		4wDH	?Gleismac			Bl	
LM 372		4wDH	Gleismac			Bl	
LM 373		4wDH	?Gleismac			Bl	
LM 374		4wDH	HE	9239	1984	M	
LM 375		4wDH	HE	9240	1984	M	
LM 376		4wDH	HE	9241	1984	M	
LM 377		4wDH	HE	9243	1984	M	
LM 378		4wDH	HE	9242	1984	Bl	
LM 379		4wDH	HE	9251	1985	Lu	
LM 380		4wDH	HE	9252	1985	Cm	
LM 381		4wDH	HE	9253	1985	Bl	
LM 382		4wDH	HE	9254	1985	Dg	
LM 383		4wDH	HE	9255	1985	Bn	
LM 384		4wDH	HE	9256	1985		
LM 385		4wDH	HE	9257	1986	TAE	
LM 386		4wDH	HE	9258	1986	Bl	
LM 387		4wDH	HE	9259	1986	Bo	
		4wDH	HE	9272	1986	Lu	
C 11		4wPMR	BnM			Cs	
(C 28)		2w-2PMR	Wkm	4818	1948	Lu	Dsm
C 35		4wPMR	BnM	1		Ti	
(C 36)	36	4wPMR	BnM	2		Bo	
(C 43)		2w-2PMR	Wkm	7130	1955	Bl	
C 45		2w-2PMR	Wkm	7132	1955	Dg	
C 47		4wPMR	BnM	3		Cs	
C 48		4wPMR	BnM	4		Cg	
C 49		4wPMR	BnM	5		M	
C 50		4wPMR	BnM	6		Bd	Dsm
C 51		4wPMR	BnM	7	1958	Bi	
C 52		4wPMR	BnM	8		TAE	OOU
C 53		4wPMR	BnM	9		Dg	
C 54		4wPMR	BnM	10		Bl	
C 55		2w-2PMR	Wkm	7680	1957	Ck	OOU
C 56		2w-2PMR	Wkm	7681	1957	Te	OOU
C 58		2w-2PMR	Wkm	8730	1960	Bl	OOU
(C 61)		4wPMR	BnM			Li	
(C 62)		4wDMR	BnM		1960	M	
C 63		4wPMR	BnM		1972	A	

C 64		4wPMR	BnM	1972	Li
C 65		4wPMR	BnM	1972	M
C 66		4wPMR	BnM	1972	Cm
C 67		4wPMR	BnM	1972	Dg
C 68		4wPMR	BnM	1972	Bd
C 69		4wPMR	BnM	1972	Ti
C 70		4wPMR	BnM	1972	TAE
C 71		4wPMR	BnM	1972	M
C 72		4wPMR	BnM	1972	Bi
C 73		4wPMR	BnM	1972	Cg
C 74	74	4wPMR	BnM	1972	Bo
C 75		4wPMR	BnM	1972	Bl
C 76		4wPMR	BnM	1972	Dg
C 77		2w-2PMR	BnM	1972	Cm
C 78		4wPMR	BnM	1972	Bo
C 79		4wPMR	BnM	1972	Bd
C 80		4wPMR	BnM	1972	Bl
F 210		4wDM	BnM		Li
F 222	BOG FOREMAN	4wDM	BnM	1983	Bl
F 230	CIR 862	4wDM	BnM		Bl
F 308		4wDM	Massey Ferguson		Bo
F 321		4wDM	Massey Ferguson		Bo
F 348		4wDM	BnM		Bl
F 349		4wDM	BnM		Bl
F 353		4wDM	BnM	1983	Bl
F 360		4wDM	BnM	1983	Li
F 630	341 NRI	4wDM	BnM		Bl
F 635		4wDM	BnM		Bo
F 842	CIR 861	4wDM	BnM		Bl
F 851	726 XZ	4wDM	BnM		Bl
F 878		4wDM	BnM	c1975	Bl
-		4wDM	BnM	c1975	Bl

+ 2' gauge.
a Carries plate 382841.

SECTION 7

BRITISH RAILWAYS DEPARTMENTAL STOCK

PART ONE

BRITISH RAIL (MAINTENANCE) LTD.
BRITISH RAIL RESEARCH AND DEVELOPMENT DIVISION.

PART TWO

BATTERY LOCOMOTIVES.
DE-ICING AND SANDITE UNITS.
DIESEL SHUNTING LOCOMOTIVES.
GENERAL PURPOSE MAINTENANCE VEHICLES AND PERSONNEL CARRIERS.
MISCELLANEOUS POWERED DEPARTMENTAL VEHICLES.
ROUTE LEARNING UNITS AND INSPECTION SALOONS.
TRAINING LOCOMOTIVES.
TRAIN HEATING UNITS.

PART ONE

BRITISH RAIL (MAINTENANCE) LTD
Derby Locomotive Works, Derbyshire
Gauge : 4'8½". (SK 364353)

	-	4wBE	BREL		c1960
602		4wBE	BREL		c1975
	-	4wBE	BREL		c1980
1700	(DRT 81110)	4w-2DM	Taylor-Hubbard	1490	1940

Eastleigh Works, Hampshire
Gauge : 4'8½". (SU 457185)

D2991	0-6-0DE	RH	480692	1962 +

    + In use as a stationary generator.

Glasgow Works, Strathclyde
Gauge : 4'8½". (NS 605665)

2777	4wDM	R/R	Unilok	2091	1982

Wolverton Works, Buckinghamshire
Gauge : 4'8½". (SP 812413)

-	4wDM	SMH	103GA078	1978 OOU

BRITISH RAIL RESEARCH AND DEVELOPMENT DIVISION

Although most of the vehicles detailed in this section can be seen at the location listed, they go where they are required and so will of course be found at other locations as well.

Derby Technical Centre, Derbyshire
Gauge : 4'8½".   (SK 365350)

31790	(D5861 31326)	A1A-A1ADE	BT	397	1962
47971	(D1616 47480)				
	ROBIN HOOD	Co-CoDE	Crewe		1964
ADB 968021	(E3044 84009)	Bo-BoWE	NB	27801	1960
97403	(D172 46035)				
	IXION	1Co-Co1DE	Derby		1962
97404	(D182 46045)	1Co-Co1DE	Derby		1962 OOU
97409	(D60 45022)				
	LYTHAM ST ANNES	1Co-Co1DE	Crewe		1962 OOU a
97410	(D30 45029)	1Co-Co1DE	Derby		1961 OOU
97411	(D42 45034)				
	TOPAZ	1Co-Co1DE	Derby		1961 OOU
97412	(D50 45040)	1Co-Co1DE	Crewe		1962 OOU
97413	(D114 45066)				
	AMETHYST	1Co-Co1DE	Crewe		1961 OOU a
47972	(D1646 47545)	Co-CoDE	Crewe		1965
47973	(D1614 47561)				
	DERBY EVENING TELEGRAPH	Co-CoDE	Crewe		1964
RDB 965344	DRC 730J	4wDM	R/R Unimog		1970
RDB 975010	(M79900) IRIS	2-2w-2w-2DMR	Derby		1956
RDB 975089	(M 50396)	2-2w-2w-2DMR	PR		1957 OOU
ADB 975813	(W 43001)	4w-4wDE	Crewe		1972 OOU
ADB 975844	(S 64305) 4002	4w-4wRER	York		1971 OOU +
ADB 975849	(S 62426)	4w-4wRER	York		1971 OOU +
ADB 975850	(S 62429)	4w-4wRER	York		1971 OOU +
ADB 975851	(S 64304) 4002	4w-4wRER	York		1971 OOU +
(ADB 977650)	210 001 60201	4w-4DER	Derby C&W		1981 OOU
(ADB 977649)	210 002 60200	4w-4DER	Derby C&W		1981 OOU
(ADB 977669)	61001	4w-4wWE	Elh		1956
(ADB 977673)	61006	4w-4wWE	Elh		1956
(ADB 977677)	61021	4w-4wWE	Elh		1956
DB 999600		4w-4wDHR	York		1987
DB 999601		4w-4wDHR	York		1987

+   Stored at Clapham Junction (TQ 269754).
a   Stored at Tinsley Yard, Sheffield (SK 402897).

Mickleover, Derbyshire
Gauge : 4'8½".   (SK 307359)

999507	LABORATORY 20	4wDMR	Wkm	8025	1958

Old Dalby, Leicestershire
Gauge : 4'8½".   (SK 679240)

DB 998900		2w-2DMR	Bg/DC	2267	1950 OOU
RDB 998901		2w-2DMR	Bg/DC	2268	1950 OOU

PART TWO

BATTERY LOCOMOTIVES

Former Watford Line motor vehicles converted to battery locomotives.

Gauge : 4'8½".

EASTERN REGION

	HE	(TQ 313887)	Hornsey T.M.D., London.				
97703	(M 61182)		4w-4wRE/BER	Afd/Elh	1958	HE	
			Rebuilt	Don	1980		
97704	(M 61185)		4w-4wRE/BER	Afd/Elh	1958	HE	
			Rebuilt	Don	1980		
97705	(M 61184)		4w-4wRE/BER	Afd/Elh	1958	HE	
			Rebuilt	Don	1980		
97706	(M 61189)		4w-4wRE/BER	Afd/Elh	1958	HE	
			Rebuilt	Don	1980		
97707	(M 61166)		4w-4wRE/BER	Afd/Elh	1958	HE	
			Rebuilt	Don	1975		
97708	(M 61173)		4w-4wRE/BER	Afd/Elh	1958	HE	
			Rebuilt	Don	1975		
97709	(M 61172)		4w-4wRE/BER	Afd/Elh	1958	HE	
			Rebuilt	Don	1975		OOU
97710	(M 61175)		4w-4wRE/BER	Afd/Elh	1958	HE	
			Rebuilt	Don	1975		OOU

LONDON MIDLAND REGION

	BD (SJ 296904)	Birkenhead North, Merseyside.			
DB 97701	(M 61136)	4w-4wRE/BER	Afd/Elh	1958	BD
		Rebuilt	Wolverton	1974	
DB 97702	(M 61139)	4w-4wRE/BER	Afd/Elh	1958	BD
		Rebuilt	Wolverton	1974	

DE-ICING UNITS

Gauge : 4'8½".

LONDON MIDLAND REGION

	BD (SJ 296904)	Birkenhead North, Merseyside.			
	WD (SU 222828)	Willesden, Greater London.			
ADB 977345	(M 61178)	4w-4wRER	Afd/Elh	1957	BD
ADB 977347	(M 61180)	4w-4wRER	Afd/Elh	1957	BD
ADB 977349	(M 61183)	4w-4wRER	Afd/Elh	1957	BD
(ADB 977385)	M 61148	4w-4wRER	Afd/Elh	1957	WD

Each autumn some units are temporarily reallocated for de-leafing and
Sandite duties.

BI (TQ 305058)	Brighton, East Sussex.
BM (SZ 104921)	Bournemouth, Dorset.
CL (TQ 995425)	Chart Leacon, Kent.
EH (SU 461182)	Eastleigh, Hants.
FR (SU 658001)	Fratton, Hampshire.
GI (TR 782684)	Gillingham, Kent.
RE (TR 373657)	Ramsgate, Kent.
SU (TQ 333678)	Selhurst, London.
WP (TQ 255723)	Wimbledon Park, London.

001	ADB 977365	(S 10726 S)	4w-4RER	Elh	1938	BM
			Rebuilt	Stewarts Lane	1968	
	ADB 977368	(S 10500 S)	4w-4RER	Elh	1940	
			Rebuilt	Stewarts Lane	1969	
002	ADB 977367	(S 10499 S)	4w-4RER	Elh	1940	EH
			Rebuilt	Stewarts Lane	1968	OOU
	ADB 977366	(S 10497 S)	4w-4RER	Elh	1940	
			Rebuilt	Stewarts Lane	1968	
003	ADB 975594	(S 12658 S)	4w-4RER	Lancing/Elh	1950	SU
			Rebuilt	Selhurst	1978	
	ADB 975595	(S 10994 S)	4w-4RER	Lancing/Elh	1941	
			Rebuilt	Selhurst	1978	
004	ADB 975586	(S 10907 S)	4w-4RER	Lancing/Elh	1947	EH
			Rebuilt	Selhurst	1977	
	ADB 975587	(S 10908 S)	4w-4RER	Lancing/Elh	1947	
			Rebuilt	Selhurst	1977	
005	ADB 975588	(S 10981 S)	4w-4RER	Lancing/Elh	1941	WP
			Rebuilt	Selhurst	1977	
	ADB 975589	(S 10982 S)	4w-4RER	Lancing/Elh	1941	
			Rebuilt	Selhurst	1977	
006	ADB 975590	(S 10833 S)	4w-4RER	Lancing/Elh	1948	WP
			Rebuilt	Selhurst	1978	
	ADB 975591	(S 10834 S)	4w-4RER	Lancing/Elh	1948	
			Rebuilt	Selhurst	1978	
007	ADB 975592	(S 10993 S)	4w-4RER	Lancing/Elh	1941	GI
			Rebuilt	Selhurst	1978	
	ADB 975593	(S 12659 S)	4w-4RER	Lancing/Elh	1950	
			Rebuilt	Selhurst	1978	
008	ADB 975596	(S 10844 S)	4w-4RER	Lancing/Elh	1948	CL
			Rebuilt	Selhurst	1979	
	ADB 975597	(S 10987 S)	4w-4RER	Lancing/Elh	1941	
			Rebuilt	Selhurst	1979	
009	ADB 975598	(S 10989 S)	4w-4RER	Lancing/Elh	1941	BI
			Rebuilt	Selhurst	1980	
	ADB 975599	(S 10990 S)	4w-4RER	Lancing/Elh	1941	
			Rebuilt	Selhurst	1980	
010	ADB 975600	(S 10988 S)	4w-4RER	Lancing/Elh	1941	BI
			Rebuilt	Selhurst	1980	
	ADB 975601	(S 10843 S)	4w-4RER	Lancing/Elh	1948	
			Rebuilt	Selhurst	1980	
011	ADB 975602	(S 10991 S)	4w-4RER	Lancing/Elh	1941	RE
			Rebuilt	Selhurst	1980	
	ADB 975603	(S 10992 S)	4w-4RER	Lancing/Elh	1941	
			Rebuilt	Selhurst	1980	

```
012   ADB 975604  (S 10939 S)   4w-4RER    Lancing/Elh    1947 FR
                                 Rebuilt    Selhurst       1981
      ADB 975605  (S 10940 S)   4w-4RER    Lancing/Elh    1947
                                 Rebuilt    Selhurst       1981
013   ADB 975896  (S 11387 S)   4w-4RER    Lancing/Elh    1950 RE
                                 Rebuilt    Selhurst       1982
      ADB 975897  (S 11388 S)   4w-4RER    Lancing/Elh    1950
                                 Rebuilt    Selhurst       1982
015   ADB 977531  (S 14047)    4w-4RER    Lancing/Elh    1951 WP
                                 Rebuilt    Selhurst       1987
      ADB 977532  (S 14048)    4w-4RER    Lancing/Elh    1951
                                 Rebuilt    Selhurst       1987
016   ADB 977533  (S 14273)    4w-4RER    Lancing/Elh    1951 WP
                                 Rebuilt    Selhurst       1987
      ADB 977534  (S 14384)    4w-4RER    Lancing/Elh    1951
                                 Rebuilt    Selhurst       1987
017   ADB 977566  (S 65312)    4w-4RER    Elh            1955 EH
      ADB 977567  (S 65314)    4w-4RER    Elh            1955
1066  ADB 977376  (S 60002)    4-4wDER    Afd/Elh        1957 SU
      ADB 977377  (S 60003)    4-4wDER    Afd/Elh        1957
```

SANDITE UNITS

Gauge : 4'8½".

ANGLIA REGION

```
            CC (TM 175154)    Clacton, Essex.
            EM (TQ 432848)    East Ham, London.
            IL (TQ 444868)    Ilford, London.

ADB 977599 61073 996    4w-4wWE    York    1958 CC
ADB 977602 61228 997    4w-4wWE    York    1958 IL
ADB 977605 61062 998    4w-4wWE    York    1958 EM
```

EASTERN REGION

```
            HT (NZ 278657)    Heaton, Newcastle, Tyne & Wear.
            NL (SE 328330)    Neville Hill, Leeds, West Yorkshire.

TDB 977535 (53259)    2-2w-2w-2DMR    MC    1957 HT
TDB 977536 (53295)    2-2w-2w-2DMR    MC    1957 HT
```

DIESEL SHUNTING LOCOMOTIVES

SCOTTISH REGION
Gauge : 4'8½".

K (NS 423385)   Kilmarnock Works, Strathclyde.

97701	DR 76105	DB 966017	4wDE	Matisa	2655	1975	K	+
			Rebuilt	Kilmarnock		1986		
97702	DR 76212	(DB 965281)	4w-4wDH	Plasser	13	1966	K	+
			Rebuilt	Kilmarnock		1986		OOU
97703	DR 76104	DB 966016	4wDE	Matisa	2654	1975	K	+
			Rebuilt	Kilmarnock		1988		

+ Carry 977xx in error, should read 976xx.

WESTERN REGION
Gauge : 4'8½".

GL (SO 841185)   Gloucester M.P.D., Gloucester.
LA (SX 554563)   Laira M.P.D. Plymouth, Devon.
RG (SU 706740)   Reading C.C.E. Yard, Berkshire.
RR (ST 138798)   Radyr C.C.E. Yard, South Glamorgan.

97650		0-6-0DE	RH	312990	1952	RG OOU
97651	C.C.E.PLANT No.83651	0-6-0DE	RH	431758	1959	GL OOU
97652	C.C.E.PLANT No.83652	0-6-0DE	RH	431759	1959	LA OOU
97653	C.C.E.PLANT No.83653	0-6-0DE	RH	431760	1959	RR
97654	C.C.E.PLANT No.83654	0-6-0DE	RH	431761	1959	RG

Severn Tunnel Emergency Train Loco, Sudbrook Sidings, Gwent.
Gauge : 4'8½".   (ST 503874)

97806	(D4105 09017)	0-6-0DE	Hor		1961

GENERAL PURPOSE MAINTENANCE VEHICLES AND PERSONNEL CARRIERS

The location shown for each trolley is its "base location" but trolleys go where they are required and so can be found working elsewhere.

Gauge : 4'8½".

ANGLIA REGION

N (TG 247078)   Norwich, Norfolk.

DX 98504	4wDHR	Plasser	52792	1985	N

EASTERN REGION

B	(NZ 588074)	Battersby, North Yorkshire.
BD	(SE 268883)	Bedale, North Yorkshire.
D	(NZ 294138)	Darlington, Durham.
DM	(SE 573038)	Doncaster Marshgate, South Yorkshire.
G	(NZ 784057)	Glaisdale, North Yorkshire.
GD	(NZ 251634)	Gateshead, Tyne & Wear.
HO	(SE 244392)	Horsforth, West Yorkshire.
M	(SE 048118)	Marsden, West Yorkshire.
NFB	(NZ 243663)	Newcastle Forth Bank, Tyne & Wear.
R	(SE 369128)	Royston, West Yorkshire.
S	(SK 386878)	Sheffield, South Yorkshire.
T	(NZ 458185)	Thornaby, Cleveland.
W	(NZ 898107)	Whitby, North Yorkshire.

Ref 1	Ref 2	Ref 3	Type	Builder	No.	Year	Loc	Notes
-			2w-2PMR	Wkm	730	1932	GD	DsmT
(DE 900856)	DB 965045	(DX 68011)	2w-2PMR	Wkm	7073	1955	D	
(DE 320489)	DB 965071	(DX 68013)	2w-2PMR	Wkm	7586	1957	T	
(DE 320491)	DB 965073	(DX 68014)	2w-2PMR	Wkm	7588	1957	D	DsmT
(DE 320493)	(DB 965075)		2w-2PMR	Wkm	7590	1957		Dsm
(DE 320496)	(DB 965078)	(DX 68015)	2w-2PMR	Wkm	7593	1957	B	DsmT
(DE 320497)	DB 965079	(DX 68016)	2w-2PMR	Wkm	7594	1957	G	
(DE 320498)	DB 965080	(DX 68017)	2w-2PMR	Wkm	7595	1957	D	
DE 320501	DB 965083	(DX 68019)	2w-2PMR	Wkm	7598	1957	D	OOU
(DE 320503)	(DB 965085)	(DX 68020)	2w-2PMR	Wkm	7600	1957	D	DsmT
(DE 320514)	DB 965096	(DX 68022)	2w-2PMR	Wkm	7611	1957		
(DE 320515)	DB 965097	(DX 68044)	2w-2PMR	Wkm	7612	1957	D	
(DE 320517)	DB 965099	(DX 68023)	2w-2PMR	Wkm	7614	1957	D	DsmT
(DE 320520)	DB 965102	(DX 68024)	2w-2PMR	Wkm	7617	1957	BD	DsmT
DB 965949	PLANT No. 68/005 DX		2w-2PMR	Wkm	10645	1972	HO	
DB 965950	PLANT No. 68/006		2w-2PMR	Wkm	10646	1972	NFB	
DB 965951	68/007	(DX 68003)	2w-2PMR	Wkm	10647	1972	M	
DB 965952	PLANT No.68/008 SUE		2w-2PMR	Wkm	10648	1972	W	
DB 965987	PLANT No.68/010		2w-2PMR	Wkm	10731	1974	D	
	68/038		2w-2DMR	Wkm	1947	1935	DM	+
			Rebuilt	DonM	1982			
	68/043		2w-2DMR	Wkm			DM	+
			Rebuilt	DonM	1983			
-			2w-2PMR	Wkm			R	Dsm
	DX 98505		4wDHR	Plasser	52793	1985	S	
	DR 98506		4wDHR	Plasser	52794	1985	NFB	

+ Converted to Tunnel Inspection Units.

LONDON MIDLAND REGION

CK	(NY 373602)	Carlisle Kingmoor, Cumbria.
CN	(SD 496711)	Carnforth, Lancashire.
GB	(SJ 928976)	Guide Bridge, Manchester.
M	(SH 745014)	Machynlleth, Powys.
N	(SO 112913)	Newtown, Powys.
NOT	(SK 563388)	Nottingham, Nottinghamshire.
NR	(SP 759595)	Northampton, Northants.
W	(SP 008982)	Walsall, West Midlands.
WA	(TQ 111975)	Watford, Hertfordshire.

72107	DX 68025 DB 965392	2w-2DMR	Matisa	D8 006	1971	
		Rebuilt	Crewe		1980	CK	OOU
	DX 98203A	4wDMR	Plasser	52531A	1983	GB
	(DX)98205A	4wDMR	Plasser	52760A	1985	WA
	DX 98300A	4wDMR	Geismar G.780.001		1985	NOT
	DX 98301A	4wDMR	Geismar G.780.002		1985	W
	DX 68302A	4wDMR	Geismar G.780.003		1985	MAN
	DX 98303A	4wDMR	Geismar G.780.004		1985	W
	DX 68304A	4wDMR	Geismar G.780.005		1985	CN
	(DX)68700	4wDHR	Permaquip	001	1985	W
	DX 68701	4wDHR	Permaquip	003	1986	W
	DX 68702	4wDHR	Permaquip	007	1986	W
	DX 68703	4wDHR	Permaquip	008	1986	NR
	(DX)68800	4wDHR	Permaquip	001	1985	N
	(DX)68801	4wDHR	Permaquip	002	1985	M
	(DX)68802	4wDHR	Permaquip	003	1985	M
	(DX)68803	4wDHR	Permaquip	004	1985	M

C	(NN 385252)	Crianlarich, Central.
CA	(NS 953455)	Carstairs, Strathclyde.
D	(NH 553585)	Dingwall, Highland.
DA	(NT 139778)	Dalmeny, Lothian.
DE	(NO 393294,	
	NO 394260)	Dundee, Tayside.
DU	(NT 682785)	Dunbar, Lothian.
GJ	(ND 153592)	Georgemas Junction, Highland.
GL	(NM 898809)	Glenfinnan, Highland.
GR	(NS 816786)	Greenhill Upper Junction, Central.
HBT	(ND 126581)	Halkirk Ballast Tip, Highland.
I	(NJ 775219)	Inverurie, Grampian.
P	(NO 113231)	Perth, Tayside.
PO	(NS 598626)	Polmadie, Strathclyde.
S	(NS 654644)	Shettleston, Strathclyde.
SB	(NN 222814)	Spean Bridge, Highland.
SL	(NT 228714)	Slateford, Lothian.
ST	(NX 073603)	Stranraer, Dumfries.

-		2w-2PMR	Wkm			HBT Dsm
(DX)68030		2w-2DMR	Matisa D8 005	1971	SL	
		Rebuilt	Kilmarnock	1979		
(DX)68031		2w-2DMR	Matisa D8 014	1971		
		Rebuilt	Kilmarnock	1981	DE	
(DX)68035		2w-2DMR	Permaquip FBT 1	1982	DA	
DX 68036		2w-2DMR	Permaquip TB 001	1982	DA	
(DX)98100	(DB 966025)	2w-2DMR	Schöma 4016	1974	C	
DX 98101	DB 966027	2w-2DMR	Schöma 4017	1974	GJ	
DX 98201A		4wDMR	Plasser 52465A	1982	ST	
DX 98204A		4wDMR	Plasser 52759A	1984	SL	
DX 98212A		4wDMR	Plasser 52985A	1986	I	
DR 98213		4wDMR	Plasser 53187A	1988	CA	
DR 98214		4wDMR	Plasser 53188A	1988	DU	
DX 98402		4w-4wDHR	Plasser 8948	1979	GR	
		Rebuilt	Kilmarnock	1987		
DX 98500		4wDHR	Plasser 52788	1985	GJ	
DX 98501		4wDHR	Plasser 52789	1985	D	
DX 98502		4wDHR	Plasser 52790	1985	SB	
DX 98503		4wDHR	Plasser 52791	1985	C	
DX 98600		4wDMR	Permaquip 001	1985	C	
DX 68704		4wDHR	Permaquip 002	1986		
DX 68705		4wDHR	Permaquip 004	1986	S	
DX 98706		4wDHR	Permaquip 010	1986	S	
(DX)98707		4wDHR	Permaquip 012	1986	DE	
DX 68811		4wDHR	Permaquip 001	1987	D	
DX 68812	KYH 862X	4wDMR R/R	Minilok	c1984	P	
(DX 68813)	A855 MUA	4wDMR R/R	Zweiweg 1019	1983	S	
(DX)68901	E179 HUY	4wDM R/R	Bruff	1987		
DX 68904	E710 GUY	4wDMR R/R	Bruff	1987	PO	
DX 68905	F653 RAB	4wDMR R/R	Bruff	1988	GL	

a The body of one of these trolleys is in Strathcarron Station Yard.

SOUTHERN REGION

AD	(TR 021415)	Ashford Plant Depot, Kent.		
B	(SU 642527)	Basingstoke, Hampshire.		
BH	(SU 301019)	Brockenhurst, Hampshire.		
CE	(TR 146573)	Canterbury East, Kent.		
EC	(TQ 327659)	East Croydon, London.		
EH	(SU 461182)	Eastleigh, Hampshire.		
FE	(TR 234368)	Folkestone East, Kent.		
HA	(TQ 303154)	Hassocks, East Sussex.		
HG	(TQ 393743)	Hither Green, London.		
HS	(TQ 815096)	Hastings, East Sussex.		
PU	(TQ 315615)	Purley, London.		
RE	(TR 371658)	Ramsgate, Kent.		
SS	(TQ 654676)	Sole Street, Kent.		
SV	(TQ 522555)	Sevenoaks, Kent.		
SW	(TQ 515684)	Swanley, Kent.		
T	(TQ 592459)	Tonbridge, Kent.		
TB	(TQ 287364)	Three Bridges, West Sussex.		
W	(TQ 309799)	Waterloo North Sidings, London.		
WP	(TQ 254720)	Wimbledon Park, London.		

DS 52	DX 68081	2w-2PMR	Wkm	7031 1954	HS	DsmT
DS 3317	DX 68084	2w-2PMR	Wkm	(7824 1957?)	WP	DsmT
DB 965143	DX 68093	2w-2PMR	Wkm	7974 1958		DsmT
DS 965336	DX 68071	2w-2DMR	Wkm	10343 1969	AD	OOU
DB 965990	DX 68073	2w-2DMR	Wkm	10705 1974	AD	OOU
DB 965991	DX 68075	2w-2DMR	Wkm	10707 1974	PU	
DB 965992	DX 68078	2w-2DMR	Wkm	10708 1974	AD	OOU
DB 965993	DX 68080	2w-2DMR	Wkm	10706 1974	HS	
DB 966031	(DX)98082	2w-2DMR	Wkm	10839 1975	WP	
DB 966033	DX 68086	2w-2DMR	Wkm	10841 1975	CE	
DB 966034	DX 68088	2w-2DMR	Wkm	10842 1975	FE	
DB 966035	DX 68090	2w-2DMR	Wkm	10843 1975	SW	OOU
	DX 98207A	4wDMR	Plasser 52762A 1985		HG	
	DR 98209A	4wDMR	Plasser 52764A 1985		TB	
	DX 98211A	4wDMR	Plasser 52766A 1985		EH	
	DR 98215	4wDMR	Plasser 53192A 1988		EC	
	DR 98216	4wDMR	Plasser 53193A 1988		EH	+
	DR 98217	4wDMR	Plasser 53194A 1988		HG	
	DR 98218	4wDMR	Plasser 53195A 1988		HG	
	DR 98219	4wDMR	Plasser 53196A 1988		TB	
	DR 98220	4wDMR	Plasser 53197A 1988		EH	
	(DX)98401	4wBE	Permaquip 001 1987		W	
PM 002		4wDHR	Permaquip 004 1989		EH	
No.3	DX 68708	4wDHR	Permaquip 006 1986		BH	
	DX 68710	4wDHR	Permaquip 011 1986		B	
	DX 68805	4wDMR	Permaquip 006 1986		SV	
	DX 68806	4wDMR	Permaquip 007 1986		SS	
	DX 68807	4wDMR	Permaquip 008 1986		T	
	DX 68808	4wDMR	Permaquip 009 1986		RE	
	DX 68809	4wDMR	Permaquip 010 1986		HA	
	DX 68810	4wDMR	Permaquip 011 1986		PU	
	(DX)68902 E192 JAB	4wDMR R/R	Bruff 1987		EH	
	(DX)68903 E193 JAB	4wDMR R/R	Bruff 1987		TB	

+ Converted to Drain Flushing Unit.

WESTERN REGION

B	(ST 566718, ST 609722)	Bristol, Avon.	
EJ	(SX 937937)	Exmouth Junction, Devon.	
N	(SN 752974)	Neath, West Glamorgan.	
NE	(ST 306882)	Newport, Gwent.	
PR	(SU 799027)	Princes Risborough, Buckinghamshire.	
RG	(SU 703739)	Reading, Berkshire.	
SW	(SS 658593)	Swansea, West Glamorgan.	

PWM 2831	DX 68004	2w-2PMR	Wkm	5009	1949	N	OOU
(PWM 3960)		2w-2PMR	Wkm	6945	1955	EJ	Dsm
PWM 4303	(DX)68007	2w-2PMR	Wkm	7506	1956	EJ	
PWM 4305	(DX)68009	2w-2PMR	Wkm	7508	1956	EJ	
T003	CEPS 68097	4wDHR	Permaquip T003		1988	PR	
	(DX)98202	4wDMR	Plasser 52530A		1983	SW	
	DX 98206A	4wDMR	Plasser 52761A		1985	B	
	DX 98208A	4wDMR	Plasser 52763A		1985	EJ	
	DX 98210A	4wDMR	Plasser 52765A		1985	NE	
	DR 98221	4wDMR	Plasser 53198A		1988	RG	
	DX 68709	4wDHR	Permaquip	005	1986	B	
No.1	DX 68711	4wDHR	Permaquip	009	1986	RG	
F591 RWP	CEPS 68906	4wDMR R/R	Bruff		1988	B	

MISCELLANEOUS POWERED DEPARTMENTAL VEHICLES

083569	(S19S)	4w-4RER	UC	1930	a	
DB 975007	(E 79018)	2-2w-2w-2DMR	Derby	1954	b	OOU
DB 977391	(51433)	2-2w-2w-2DMR	MC	1959	b	
DB 977392	(53167)	2-2w-2w-2DMR	MC	1959	b	
DB 977393	(53246)	2-2w-2w-2DMR	MC	1959	b	OOU
018(ADB 977290	(S 65318)	4w-4RER	Elh	1954	c	
(ADB 977291	(S 65324)	4w-4RER	Elh	1954		
019 ADB 977068	(S 14549)	4w-4RER	Lancing/Elh	1954	c	OOU
021(ADB 977304	(S 65317)	4w-4RER	Elh	1954	d	
(ADB 977305	(S 65322)	4w-4RER	Elh	1954		
(025) (ADB 977294	(S 11305 S)	4w-4RER	Lancing/Elh	c1951	e	OOU
4623 (ADB 977295	(S 11306 S)	4w-4RER	Lancing/Elh	c1951		OOU
050 ADB 977296	(S 65319)	4w-4RER	Elh	1954	e	
053 ADB 977505	(S 65321)	4w-4RER	Elh	1954	e	
054 ADB 977506	(S 65323)	4w-4RER	Elh	1954	e	
062(ADB 977559	(S 65313)	4w-4RER	Elh	1954	c	
(ADB 977560	(S 65320)	4w-4RER	Elh	1954		
080(	S 61035	4w-4RER	Elh	1957	e	
(	S 61342	4w-4RER	Elh	1957	e	
935 ADB 977640	61463	4w-4wWE	York	1960	f	
1054 ADB 977207	(S 61658)	4w-4RER	Afd/Elh	1958	e	
DB 977394	(S 65316)	4w-4RER	Elh	1954	e	
2015 (DB 977602	(S 62483)	4w-4wRER	York	1974	b	
(DB 977603	(S 62482)	4w-4wRER	York	1974	b	OOU
6022(ADB 977609)	S 65414	4w-4RER	Elh	1958	e	

a Static Stores Coach based at Sandown, Isle of Wight.
b Ultrasonic Test Units.
c Stores Units for use between various S.R. depots.
d R.M. & E.E. Lathe Shunter based at East Wimbledon Depot.
e R.M. & E.E. Tractor Unit.
f East Coast Main Line Instruction Unit.

ROUTE LEARNING UNITS AND INSPECTION SALOONS

These units work wherever required on the region to which they are allocated.

The E.R. units are maintained at Cambridge.
The L.M.R. units are maintained at Tyseley.
The Sc.R. units are maintained at Haymarket.
The S.R. uses units from the W.R. and L.M.R. when required.
The W.R. units are maintained at Reading.

TDB 975023	(W 55001)	O1	2-2w-2w-2DMR	GRC	1958	WR
975042	(M 55019)	L119	2-2w-2w-2DMR	GRC	1958	LMR a
DB 975349	(E 51116)		2-2w-2w-2DMR	GRC	1957	ER +
975540	(W 55016)	T02	2-2w-2w-2DMR	GRC	1958	LMR a
TDB 975659	(W 55035)	L135	2-2w-2w-2DMR	Pressed Steel	1960	WR
TDB 977223	(Sc55007)		2-2w-2w-2DMR	GRC	1958	ScR OOU
TDB 977607	(51464)	336	2-2w-2w-2DMR	MC	1959	ScR a
TDB 977608	(51525)	336	2-2w-2w-2DMR	MC	1959	ScR a
TDB 977651	(53290)	302	2-2w-2w-2DMR	MC	1957	ScR a
TDB 977652	(53197)	302	2-2w-2w-2DMR	MC	1957	ScR a

+ These vehicles are the power cars of twin-car units.
a These vehicles are also used for Sandite duties when required.

TRAINING LOCOMOTIVES

ED	(NS 603677)	Eastfield, Glasgow, Strathclyde.
HO	(SE 407896)	Holbeck, Leeds, West Yorkshire.
IL	(TQ 445869)	Ilford, London.
MM	(SP 22x32x)	Home Office, Fire Service College, Moreton-in-Marsh, Gloucestershire.
TI	(SK 407896)	Tinsley Yard, Sheffield, South Yorkshire.
TO	(SK 485352)	Toton, Nottinghamshire.

ADB 968024	45017	23	1Co-Co1DE	Derby		1961	TI OOU
(ADB 968027	25912)	D7672					
	TAMWORTH CASTLE		Bo-BoDE	Derby		1967	HO
ADB 968028	(27024	D5370)	Bo-BoDE	BRCW	DEL 213	1962	ED OOU
(ADB 968029)	20001	(D8001)	Bo-BoDE	(EE	2348	1957	TO
				(VF	D375	1957	
(ADB 968030)	33018	(D6530)	Bo-BoDE	BRCW		1960	MM
	20188	(D8188)	Bo-BoDE	(EE	3669	1967	IL
				(VF	D1064	1967	

TRAIN HEATING UNITS

Former Capital Stock locomotives converted to E.T.H. pre-heating units.

CK	(NY 411546)	Carlisle Upperby.
OOC	(TQ 218823)	Old Oak Common, London.

97250	(25310 D7660)	ETHEL 1	Bo-BoDE	Derby		1966	OOC OOU
97251	(25305 D7655)	ETHEL 2	Bo-BoDE	BP	8065	1966	CK
97252	(25314 D7664)	ETHEL 3	Bo-BoDE	Derby		1966	CK

INDUSTRIAL RAILWAY SOCIETY

This book has been produced by the INDUSTRIAL RAILWAY SOCIETY (founded in 1949 as the Birmingham Locomotive Society, Industrial Locomotive Information Section). The Society caters for those interested in privately owned locomotives and railways. Members receive the INDUSTRIAL RAILWAY RECORD in addition to a bi-monthly bulletin containing topical news and amendments to the Society produced Pocket Books. Access is available to a well stocked library and visits are arranged to industrial railways, etc.

If you are interested in industrial railways why not join the Society now. Further details can be obtained by sending a stamped addressed envelope to

B. Mettam, 27 Glenfield Crescent, Newbold, Chesterfield S41 8SF.

INDUSTRIAL RAILWAY RECORD

This profusely illustrated magazine contains articles of lasting interest concerning a wide variety of industrial locomotives and railways of all gauges, at home and abroad. Accurate line drawings of locomotives and rolling stock, together with carefully prepared maps, are a regular feature.

The INDUSTRIAL RAILWAY RECORD is available to non-members by direct subscription. Send to the following address for full details -

R.V. Mulligan, Owls Barn, The Chestnuts, Aylesbeare, Exeter EX5 2BY